Lowly
Origin

JONATHAN KINGDON

Lowly Origin

Where, When,
and Why Our
Ancestors First
Stood Up

PRINCETON UNIVERSITY PRESS
Princeton and Oxford

Copyright © 2003 by Princeton University Press
Published by Princeton University Press, 41 William Street,
Princeton, New Jersey 08540

In the United Kingdom: Princeton University Press, 3 Market Place,
Woodstock, Oxfordshire OX20 1SY

Second printing, and first paperback printing, 2004
Paperback ISBN 0-691-12028-5

The Library of Congress has cataloged the cloth edition of this book as follows

Kingdon, Jonathan.
Lowly origin : where, when, and why our ancestors first stood up /
Jonathan Kingdon.
p. cm.
Includes bibliographical references and index.
ISBN 0-691-05086-4 (cl : alk. paper)
1. Fossil hominids. 2. Bidpedalism—Origin. 3. Human beings—Ori-
gin. 4. Human evolution. I. Title.
GN282 .K54 2003
599.93'8—dc21   2002072852

British Library Cataloging-in-Publication Data is available

This book has been composed in Stone

Printed on acid-free paper. ∞

pup.princeton.edu

Printed in the United States of America

10 9 8 7 6 5 4 3 2

To Minnie, Dorothy, Elena, and Afra.

Much loved mothers, and

All their Sisters.

Contents

List of Figures

List of Tables

Acknowledgments

When my parents took me, as a very small boy, to Olduvai Gorge and to the nearby giant crater of Ngorongoro, none of us could have foreseen what an enduring influence these eroding plains and volcanoes would have on my later life and work. Nor could we have anticipated how a few months in Natal and frequent family holidays on the East African coast and its hinterland (from Lamu to Mtwara) would later help shape my thinking about human evolution. Readers will not have to cover many pages to appreciate how many of the ideas in this book are informed by a life lived in places that are haunted by the bones of extinct beings yet also alive with living fossils.

Nonetheless, it has to be people, their discoveries, their enthusiasms, their visions, and their published ideas that have inspired me and driven this book to completion. Over several decades of living in Africa, Asia, Europe, Australia, and America, I have been extraordinarily privileged to have met an inspiring diversity of scientists and naturalists. In the depth and range of their minds and in the generosity of their intercourse I have found the means and the courage to attempt a new account of bipedal origins while simultaneously exploring the saga of my own origins as a representative but singular human animal.

Few of those who first inspired me are still alive. However, impressions are often at their most intense and fresh in youth, and this is the place to remember and acknowledge where and whence primary enthusiasms and influences began. Of those directly relevant to the theme of this book, I fondly remember Dorothy and Teddy Kingdon, Bill Bishop, Reg Moreau, Julian Huxley, Desmond Vesey Fitzgerald, Hugh Elliot, Francois Bourliere,

L.S.B. and Mary Leakey, Peter Miller, Max Costa, Philip Leedal, Phyl Ginner, Charles Jackson, and Sonia Cole.

I owe debts of gratitude to those that have inspired me with their own work and insights but have also examined, taken an interest in, or (sometimes without knowing it) helped shape some of the ideas in this book. Among them are Matt Cartmill, Yves Coppens, Richard Dawkins, Tim Flannery, Colin Groves, Mike Hammer, John Harris, Alison Jolly, Cliff Jolly, Adriaan Kortlandt, Bob Martin, Pat Shipman, Elwyn and Friederun Simons, Chris Stringer, and Caroline Tutin. I am especially grateful to Steve Wainwright, Alan Walker, Laura Snook, and Ryk Ward for their friendship, support, and critical interest. One Kampala evening, quite some years ago, my colleague Tag El Sir Ahmed urged me to write and illustrate a book about the human family. "Self-made Man" was my earlier response, but Tag will be surprised by this extension of that memory and by the lasting viability of the seed he planted. My family continue to be a source of support and stimulus.

It has been my privilege to participate in, share or simply observe the work of learned institutions in many countries. The Institute of Biological Anthropology and Department of Zoology at Oxford University have been my academic homes for many years, and I treasure my colleagues and my association with them. Earlier, Makerere University and the National Museum in Kampala, Uganda, provided home, inspiration, and dynamo for a particularly active and stimulating period of my life and career. At Chiromo, Nairobi University, my colleagues, especially Reino Hofmann and Malcolm Coe, always made me welcome and provided laboratory space and facilities. I have kept in touch with many of my Makerere and Nairobi colleagues. More recently, Duke University in North Carolina, Skidmore College in New York State, and Kyoto University in Japan invited me as Visiting Professor, all institutions where the work and ideas of new colleagues were both stimulus and influence in helping shape some of the themes in this book. In particular, I remember Shiro Kondo, Akira Suzuki, and Masao Kawai for their many kindnesses at the Primate Research Institute at Inuyama. Likewise, Knut Schmidt-Nielsen, Elwyn and Friederun Simons, Peter Klopfer, Dan Livingstone, Richard Kay, Lesley Digby, Carel van Schaik, and many others gave me the warmest of welcomes and much collegiate stimulus at Duke. While I struggled with the first draft of this book at Skidmore, Tad, Phyl, John, Kris, Kathy, Megan—indeed, most of the faculty—made my spell in Saratoga Springs a delight and also provided much practical assistance. I am grateful to Richard Leakey for a fellowship with the Kenya National Museums and, much earlier on, for less formal welcomes from Louis and

Mary to view the research work of the then Coryndon Museum. Meave and Louise Leakey maintain, with equal distinction and outstanding grace, the splendid traditions begun by their family. All who work on human origins owe the Leakeys many thanks for their pioneering initiatives.

I have been a regular visitor to the British Museum of Natural History since boyhood and am grateful to Captain C.H.B. Grant and Robert Hayman for identifying my specimens and, among many others, Peter Andrews, Alan Gentry, and Chris Stringer for the opportunity to work and visit. For much shorter visits to the South African Museum in Pretoria, the Musee d'Histoire Naturelle and Musee de l'Homme in Paris, I remember Bob Brain, Elizabeth Vrba, Yves Coppens, Drs. G. and F. Petter, and Francois Bourliere with gratitude. Tim Flannery was my host at the Australian Museum in Sidney, and he and Mike Archer introduced me to the dynamic community of evolutionary biologists in Australia, an association facilitated by a research fellowship with C.S.I.R.O. in Atherton and Canberra. I have been the beneficiary of brief visits to the collections of the Museums of Natural History in New York, Washington, and Los Angeles, and I am grateful to many scientists there.

As a Scientific Fellow of the Zoological Society of London over many years, I have come to greatly value the Society's meetings, library, and collections. Likewise, the Royal Geographic Society has been the venue for many important projects, and I have been fortunate enough to have participated in several of their expeditions and other activities, both in London and abroad. Nigel and Shane Winser have been leading lights in all these initiatives and have also taken a personal interest in my own biogeographic and evolutionary enterprises. Over the years the Welcome Trust has been the principal supporter of my work on African Mammal and Primate evolution and the "Atlas of Evolution in Africa." I remember my association with the Trust and its former director, Peter Williams, with respect, gratitude, and affection.

One catalyst for embarking on this book was an invitation to speak at a Festschrift at Barcelona University organized by Daniel Turbon, Jordi Serralonga, and Victoria Medina to honor the Catalan polymath, Jordi Sabater Pi. It was at this conference, in 1997, that I presented a preliminary outline of the theme of this book.

The references serve to identify authors that have been consulted, but among the names that I must thank are many who have given me practical assistance or have played a part in the development of my ideas through their work or personal contact: L. Aiello, K. Albrooke, J. Altman, S. Bahuchet, C. Bangham, S. Bearder, A. Behrensmeyer, M. Black, C. Boesch, N. Bolwig, G. Brauer, T. Bromage, S. Bunney, T. Butynski, A. Caiger-Smith,

J. Chapell, D. Cheyney, G. Clarke, P. Clarke, S. Clithero, M. Coe, P. Collet, M. Colyn, H. Cronin, J. de Boer, T. Dissotel, C. Erhart, D. Falk, R. Foley, J.-P. Gautier, A. Gautier-Hion, J. Goodall, M. Gosling, J. Gowlett, P. Grubb, A. Hamilton, W. Hamilton, G. Harrington, G. Harrison, T. Harrison, B. Hedges, P. Henderson, A. Hill, F. C. Howell, R. Hughes, R. Inskeep, H. Ishida, C. Janis, D. Johanson, J. Kamminga, T. Kano, A. Kingdon, E. Kingdon, P. A. Kingdon, R. Kingdon, Z. E. Kingdon, Z. F. Kingdon, C. Koenig, M. Kohler, S. Kuroda, L. Leland, M. Lock, J. Lovett, Q. Luke, J. Maynard Smith, J. Maley, E. Meehan, P. Morris, S. Moya-Sola, T. Nishida, E. O'Brien, R. O'Hanlon, C. A. Oppenheim, M. Pagel, I. Parker, C. Pell, C. Peters, M. Pickford, D. Pilbeam, J. Reader, V. Reynolds, M. Ridley, A. Root, C. Ross, T. Rowell, M. Ruvolo, J. Ryle, R. Savage, B. Senut, R. Seyfarth, E. Siefert, R. Southwood, C.-B. Stewart, T. Struhsaker, J. Terborgh, C. Tickell, S. Tomkins, P. Trezise, T. White, A. Whiten, B. Wood, R. Wrangham, J. Yamagiwa, A. Zahavi, A. Zihlman, S. Zuckerman.

In the production of this book I am grateful to Sam Elworthy of Princeton University Press for his editorial work and helpful advice and for steering this book through a crowded publishing schedule; thanks also to Sarah Harrington, Jack Repchek, Linny Schenck, Allison Aydelotte, Carmina Alvarez-Gaffin, and Dimitri Karetnikov. Torstein Olsen transformed my sketches into splendid professional maps. I thank Steve Tomkins for providing a clean copy of the "Punch" cartoon on page 7 together with a commentary that has been incorporated into the caption. I have had much help with the practicalities of finding references, coping with computers, and preparing manuscripts; I thank Eric Siefert, Megan Moran, Laura Tobias, Laura Snook, Michael Hoffmann, and the librarians of several libraries, notably in Oxford, Skidmore, Duke, and at the London Zoological Society.

Some years ago, writing of my field guide, Tim Flannery said he sensed that it was some sort of love letter to a place that had nurtured and instructed me while giving me space to grow. That solitary privilege may not be far below the surface of this work too, but I am conscious of a swelling multitude joining in the search for a truer history set in a world that stretches far beyond Africa. I am thankful for a sense of deep fellowship as we search for traces of our ancestors' passage through the landscapes of our home: not a forbidding mineral Globe but the living Oikos of Ecology. I am grateful to all the scientists and naturalists that are helping us learn to be less alien from the world into which we have been born; steps along the way to a homecoming.

CHAPTER 1

Preface to a Self-portrait from the Center of the World

*W*hy Lowly Origin? Peculiarity of bipedalism and role of geography and ecology in explaining it. Evolution by increments. Hypotheses and definitions. The beginnings of bipedalism dated to about 6 million years ago (mya), originating in East African coastal forests. "Evolution by river basin." Separate fore-/hindlimb origins. Bipedalism as the criterion for all hominins. Bipedalism and brain develop separately.

*C*harles Darwin, in the final words of his "Descent of Man" (1871), put it this way: "[I]t seems to me, that man with all his noble qualities—with his god-like intellect which has penetrated into the movements and constitution of the solar system—with all these exalted powers—Man still bears in his bodily frame the indelible stamp of his lowly origin" (1).

Darwin was referring to many more than one or two stages of human evolutionary history. In the preceding pages, he had invoked wormlike, fishlike, and reptilian ancestries, and it was as much to these as to four-footed primates that he contrasted a soaring intellect, exalted powers, and noble, upright qualities.

For that most eminent of Victorians—no less than for any member of another culture, past or present, historic or prehistoric—uprightness (or, more prosaically, bipedalism) was a primary and definitive difference between humans and other animals. How that stance evolved is still a great mystery, and although fragmentary fossils of very early bipeds are, at last, being uncovered, there are still many more questions than answers when it comes to giving life to these broken bones and teeth. Some new ideas about bipedalism, its precursor conditions, as well as some of its consequences are central themes in this book. Although there are many scientific papers and single chapters of books that discuss bipedalism, this is probably the first to be devoted to it as a single dominant theme—the central condition on which human evolution is predicated.

In borrowing Darwin's two concluding words as my title, I invite reflection on a moment or "stage" in human evolution that was both metaphorically and literally "lowly." I attempt to reconstruct, in the light of much new evidence and inference, the appearance, ecology, and geography of those ancestral apes that were not yet bipedal yet must already have been predominantly terrestrial. Ancestors whose nonerect gait put them on the other side of that great conceptual divide between the category "Apes" and what we call "Hominids." I also reflect, but in a much more summary fashion, on the very earliest and even more "lowly" attributes of primitive aquatic vertebrates, because I find some relevance there for hand-brain connections.

The many undeniably apelike features of human gross anatomy were sufficient for Darwin's argument, but modern genetics has greatly extended the depth and reach of his insights. From this very contemporary perspective, his words "still bears in his bodily frame the indelible stamp" reads like a prophecy. You and I now know that almost every step of our evolutionary history is written into every cell of our bodies. My genome includes sequences that date back more than 700 million years, when my ancestor consisted of no more than one cell. Locked into the genetic mosaic that adds up to a living being are huge numbers of indelible or "undeleted" genetic particles that demonstrate a patrimony that goes back not just to apes but to the start of life on Earth. In common with every other organism, each one of us is the sum of genetic additions and subtractions on an unbroken thread of life that ties us, step by step, back to that fecund moment of origin, the first and lowliest of all our "beginnings" (figure 1.1).

It can justly be argued that because evolution is the sum of so many tiny genetic increments, any focus on just one event has to be distorting and arbitrary, even for so apparently momentous an event as rising up on

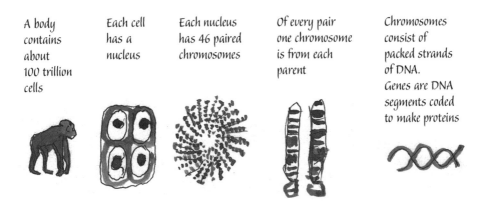

A body contains about 100 trillion cells

Each cell has a nucleus

Each nucleus has 46 paired chromosomes

Of every pair one chromosome is from each parent

Chromosomes consist of packed strands of DNA. Genes are DNA segments coded to make proteins

FIGURE 1.1 Building blocks of life, from DNA and proteins up to organisms.

two legs. To offset such conceptual isolation and to put bipedalism in a broader perspective, I have devised more than one framework to present my ideas. Only multiple frameworks can hint at the scale and difficulty of the enterprise. Our gait may be as plainly factual as our unquestionable existence as primates, mammals, animals; yet the puzzle of why an ape should get up on two legs is inseparable from the larger mystery of our emergence from nature as a culture-bearing species. For all the new fossils, newly mapped genome, and new awareness of the biological roots of human health, reproduction, and material culture, it is our profound and continuing ignorance of nature itself that remains the primary obstacle to self-knowledge. It is not difficult to report new discoveries from the frontiers of science; but it is less easy, as a scientist, to acknowledge that lacking the intellectual tools necessary to understand nature, we lack the means to understand ourselves. In the meantime, my multistranded narrative may hint at some of the many dimensions of human evolution while also expressing a personal confidence that the gap between nature and culture will one day be bridged by one of our greatest cultural achievements: science.

My first, largely symbolic presentation derives from an attempt, in the late 1980s to put together what I envisaged as a "Family Album," a sort of pasted-up scrapbook of my far-flung family, the diaspora of modern humans (2). After publishing it under the title "Self-made Man," I was challenged by a friend, who knew that I was also a painter, to attempt a "self-portrait" painted in both words and images. Not an autoimage of the artist as a young man compared with his middle-aged and elderly self, but rather a self-portrait informed by modern genetics and ecology as well as some less modern palaeontology. A portrait in which the younger self is the minimal vertebrate, an appetite-driven, wriggling backbone attracted hither, repelled thither; the youth an alert mammal-like reptile; the person in his prime a vivacious ape; and the elderly, worldly-wise wizard a contemporary, wholly modern human.

To try and retrace any part of that ancestry can be portrayed as a very personal quest, and there can be few that would deny the self-centeredness of our interest. It is in that spirit that I have adopted the metaphor of self-portrait as a medium to tell the story. But it is a self-portrait that reveals itself by increments. Each is drawn at a different stage of life, and each is set within a different landscape. Lifted out of this succession for special attention is the pivotal event on which human evolution hangs. This is not the arrival of consciousness, the ability to talk, or the evolution of a big brain. (All of these properties seem to have had very protracted histories.) Rather, it is the much more sudden event of walking on two legs, not four. What follows is not only new as an explanation, using new data, but also invokes new ways of approaching the problem of bipedal origins.

By including rudimentary vertebrates, reptiles, and monkeys in my autoportrait, I am expressing my self-awareness of *belonging* to nature, not being inexplicably different. In acknowledging the many qualities that seem more or less unique to me and my kind, I do not forget to remind myself that they must, in every case, be derived from earlier conditions that are typical for primates or other animals. Most of the characteristics that we envisage as uniquely human are actually species-specific amalgams, truly unique recombinations or composites of much more modest, preexistent increments. Some of the many unknowns in our evolutionary history will eventually become more understandable through some such incremental approach.

In such a fragmented biography, the acquisition of bipedal stance can so easily be presented as some sort of portentous coming of age: the moment in which all that followed would change irrevocably. The term hominin (or hominid) that we use to separate all bipeds from their ape cousins certainly reinforces that expectation. Yet, as many newly discovered fossils demonstrate, our monopoly of bipedalism must be seen in the context of numerous extinct bipeds. Since I first began to assemble the material for "Self-made Man," the number of new fossil hominin species has doubled, and what was envisaged as a pagoda tree of human evolution has become a bush that looks more and more like a thicket with numerous pruned branches and a succession of dead ends. While the biogeographic model presented in the following pages contributes new ideas to explain such bewildering diversity, only more fossils from more localities can tell us the true story.

The supposed bell of destiny must be muted by the awareness that not all the apes that became bipedal found themselves on a human trajectory. Getting up on two legs may have rung in a human future for our direct ancestors, but at least some bipeds, including some of the ones best known as fossils, remained "cranial apes." That much is borne out by the fossil record. So, assuming that the distinction is a real one, what was it about our specific lineage that emancipated the earliest members of our branch from being just one more type of bipedal ape?

For clues to that puzzle, I turn to my second, less symbolic framework of ideas, locating my players in a succession of geographic and ecological contexts (without doing violence to fossil facts or the logic of known paleoecology and paleogeography). I seek answers in known anatomical changes that anticipate typically human attributes by diminishing the differences between juveniles and adults, males and females. I suggest corresponding changes in behavior that might have enhanced versatile all-group responses to various unpredictable challenges. Such social and mental versatility would have undermined the more genetically fixed responses of a species in possession of an ecological niche that existed within relatively predictable limits. Step by step, the predetermined behavior of a species with a single niche must have given way to the new competences of a species that could acquire multiple niches through an ever-expanding armory of technology, techniques, and eventually systems of communication to back them up.

For the most part, I have used the often random and accidental provenances of fossils as mere guides to the larger ecological and geographic contexts for human evolution, seeking clues in those details of African biogeography and ecology that we can still retrieve and reconstruct today. I have also sought to put the likely anatomical and behavioral responses of early hominins to a succession of environmental challenges into a sequential and spatial order that is consistent with the fossil record. A full time chart and checklist of fossil hominins has been kept for the last chapter, together with a summary of my conclusions, leaving the rest of the chapters to stress my biogeographic perspectives. Thus the first tie-up between time, place, ecology, and behavior is located on the east African coast, the second and third involve movement into the interior (each involving subtly different but highly significant divergences). The hominin trail leads on into Highvelt and other interior uplands and thence, very much later, to the Atlas Mountains (or Arabia). Each such translocation involved further refinements of bipedalism, from merely functional standing and walking to much later skills in fast running and jumping (3). In addition, there must have been a succession of mental and behavioral adjustments as the habitats and climates of particular populations changed over time. These are some of the disparate strands of analysis within which I have presented my ideas.

Finally, as a specialist in the evolution of mammals, the perspective that I have sustained longest (and reinforced most decisively in this book) is that of the emergence of humans as the evolution of yet another mammal—a very peculiar and special one, true, but in essence just one more African mammal. I have, as long as I can remember, always seen myself in that light and seek here to share that self-image. If the reflection you see is distorted by the mirror I have constructed or by my own deficiencies of vision and knowledge, that is my responsibility. But I take heart from the certainty that I share, with you and with others before us, the impulse to try and make sense of the deeply puzzling animal that stares back at us from the mirror.

I like to think that Charles Darwin, who must have been amused by contemporary cartoons of himself as an ape or the final morph of an egg-larva-pupa transformation (figure 1.2), would have enjoyed the conceit of a hagfish (a primitive, eel-like fish) rendered as a self-portrait. After all, he concluded that the "early ancestors of man, thus seen in the dim recesses of time, must have been as simply, or even still more simply organised than the lancelet or the amphioxus." As if in anticipation of the Human Genome Project, he also invited the idea of reconstructing the past from the realities of the present: "look to man as he exists; and we shall, I think, be able partially to restore the structure of our early progenitors, during successive periods" (1).

Self-portraits require mirrors, but reflections can stare back at surprising moments and from unexpected experiences. For example, among the diversions of my backwoods childhood in Africa were hypnotic audiences over the cadavers of various wild animals while they were being butchered or skinned. Commonest were antelopes, ostriches, or wild pigs being prepared for the pot. Then there was a leopard being carefully skinned for its coat; and a zebra. Least commonplace were species such as an aardvark, a striped hyena, or a monkey, victims of some accident and dismembered or dissected out of pure curiosity.

I especially remember the brutal rending away of a baboon's pungent pelt and the revelation of its stretched-out, pink, pathetic nakedness—like a jarring rip in the invisible curtain that had kept me separate from all other animals. Through the torn skin, its flesh was difficult to dissociate from my own. As a very small child I had once spent some months playing with an equally juvenile baboon, but for all its noisy, toothy de-

FIGURE 1.2 An obituary cartoon from "Punch" of December 6, 1881. The cartoon is a plaudit, with the "evolved gentleman" taking his hat off as a mark of respect to Darwin. Darwin is posed in the dress and attitude of a classical philosopher. The circle, labeled "Time's Meter," provides the frame for a spiral of "evolving forms" with the worm theme probably referring to Darwin's late work on the earthworm.

termination to subordinate me to its ferocious, infantile will, I had somehow kept vestiges of my species-specific distance. Yet here was the racked body of a dead adult that mirrored me. As my own warm, living hands sampled the springy resilience of cool gray fingers I imagined myself suffering the helpless indignities of being played with because I, too, for an instant, was dead. This must remain one of my earliest experiences of see-

FIGURE 1.3 Acting on hunches. A 1958 sketch in which I pondered the posture of a foraging ape.

ing my self-portrait in another animal. Years later, the element of self-
portraiture must have remained when I made anatomical studies and
drawings, not only of a baboon but also of humans.

This book tries to extend that moment of perception; but instead of a
dead baboon, the principal objects in which I seek my own ancestral reflections are fossilized apes and hominins (figure 1.3). Although evolutionary
science takes over from childish intuition to guide my brush and pencil, a
central preoccupation is to try and bridge the gap between my long-lost genetic self as a baboonlike quadruped and the bizarre biped I am today.

Yet another incentive to write this book has been my discomfort with
the terms in which human evolution is often presented. Too often I have
been unable to match stories of one mammal's evolution, that of humans, with what I know of the biology of other African mammals and
their occupation of African landscapes (4). In reaction, I began to ponder
those respects in which the biogeography and ecology of other living
mammals might help illuminate the course of human evolution.

One of the end-products of evolutionary theorizing is a genealogical
tree that places every fossil species in a temporal and relational position
to other known fossil species. Because there are a limited number of fossils and a large number of theorists, the choice of trees embraces almost
every permutation of postulated relationships. Just how different these
trees can be is illustrated in figure 1.4, where some of the more plausible

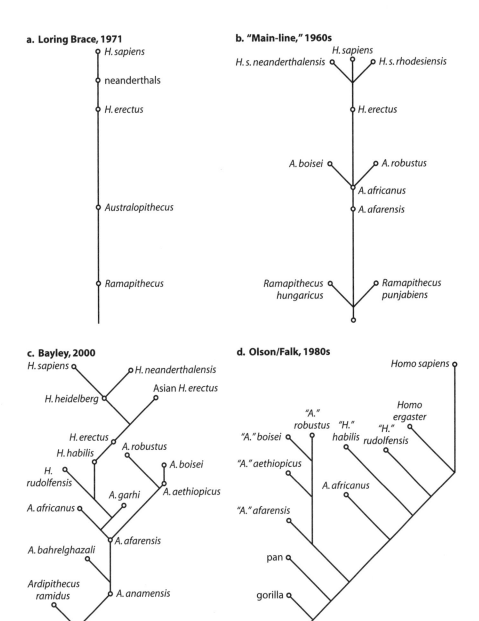

FIGURE 1.4 Four trees representing different assumptions about human evolution. A. Loring Brace in 1971 envisaged a single line extending over 15 million years. B. A typical 1960s "main-line" stem with hominids distinct for 15 million years. C. A 2000 tree with more branches but still a single tree on a 5-million-year time scale. D. The Olson/Falk tree (1981–1988), including apes on an 8- or 9-million-year time scale. (See text page 232 [figure 7.1] for my own conclusions about the human genealogical "bush.")

genealogies are displayed and their authors listed. My own conclusions about the relationship between fossil hominins correspond more closely with those of the authors Olson and Falk (5, 6) than to any others.

In a climate of conspicuous neglect of geography, these anatomically oriented authors have postulated a set of relationships that I have found broadly consistent with both the geographic and ecological patterns that have emerged from my own studies. The doubling in numbers of hominin species discovered since I began my last book on human evolution has itself been a direct stimulus to writing this book. This doubling in numbers has reinforced my discomfort with earlier explanations. Speciation, especially multiple speciation, has to take place in geographic or ecological compartments, and the evolution of discrete animals, plants, and endemic communities has been one of my long-term interests and the subject of many publications. The pages that follow seek patterns of isolation and dispersal that are at least consistent with the broad patterns that I have learned to recognize.

All animals have finite distributions that are loaded with many detailed implications for their ecological adaptations, evolutionary origins, and ability to spread or disperse. Both contemporary and ancient Africa can be understood as a pattern of ecological islands (7). Islands and isolation of any sort are intrinsic aspects of speciation, so the chapters that follow set out to contest the view that "it may be as futile to seek a specific and localized place of origin for hominids as it is for any other group" (8). Discussion of the geography of human evolution has often been so threadbare, abstract, and generalized that our many and different ancestors have no perceptible existence in time and space. There needs to be a fuller acknowledgement and awareness that our forebears were embedded in the same ecological matrices that other mammals are and have been, all with specific and finite distributions.

One of the most striking and surprising peculiarities of equatorial African fauna and flora is the frequency with which forest and nonforest species form pairs.* Among plants, amphibians, birds, and mammals, there are forest species whose closest relative is not another forest-adapted species but a nonforest sibling. These animals and plants apparently owe their primary success to adaptations that are not overwhelmingly governed by the weather. Free of such confining constraints, they would seem

---

*I use the term nonforest because a stereotype has arisen of always contrasting forest with "savanna" or grassland when there are all manner of arid-adapted thicket formations that are emphatically *not* forest and *not* savannas. Furthermore, research on carbon isotopes in plio-pleistocene soils has suggested that open grassy "savannas" became extensive only between 1 and 2 million years ago (9).

to have responded to past oscillations of climate by evolving sibling species so that one or other form can take advantage of whatever climatic phase is currently dominant (7). It stretches the definition of "sibling species" to pair chimpanzees with humans, but it is appropriate to point out that the processes that have generated such sibling pairs may also have played a role in the evolution of both humans and chimpanzees.

Behind the evolution of such pairs are processes that are much more complex than mere two-way traffic between forest and nonforest. The habitats of today's species may differ in many important ways from those of ancestral species, but the fact that modern chimpanzees are forest-dwelling fruit eaters while omnivorous humans live in more open habitats has led to a widespread assumption about the course of human evolution. The favorite image is of forests drying out and the four-legged, forest-living ancestors of humans adapting to more open conditions by becoming erect. This, in my view, must be wrong; chimpanzee ancestors were not always tied to rain forests, and human ancestors could not have moved out into open environments until they were already bipedal. For a less simplistic scenario of bipedal origins, the abstractions of adaptation need to be broken down into increments and the dynamics of speciation related to those displayed by numerous nonhuman organisms. These comparisons suggest that in addition to taking account of climate change, African distribution patterns need to be examined in terms of the continent's surface pattern of ancient swells and basins, rivers and uplands (figure 1.5)

One challenge for species adapting to new or different habitats has been the repetitive drying out and retreat of extensive forests to a network of narrow galleries and riverine strips. During the Plio-Pleistocene, this tended to coincide with each global glaciation and gave a special importance to rivers as focal areas or refuges. With the return of humid climates, forests could expand from their riverine cores and swallow up the intervening country. In the pages that follow, both minor and major rivers and their basins play a central role in my understanding of the relationship between forest and nonforest biota.

My conviction that the human being is intrinsically one more African mammal found expression in the early 1960s, when, building on my Tanganyika childhood, I began an inventory of the mammals of eastern Africa that was, in effect, a series of essays on the evolutionary process and the diversity of its expressions. The multivolume "Atlas of Evolution in Africa" that emerged from my studies included a brief profile of *Homo sapiens* whose "peculiarities have been evolved by fundamentally the same processes that have determined the peculiarity of other mammals"

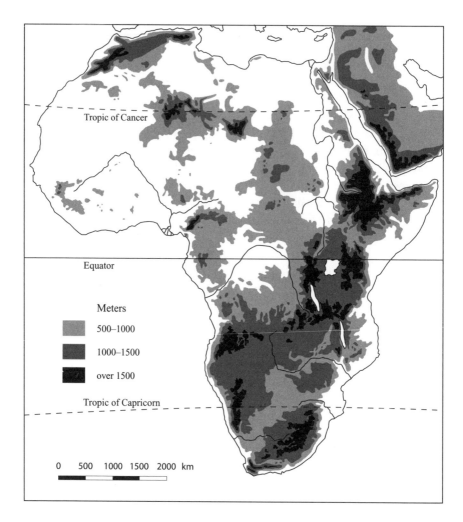

FIGURE 1.5 Africa: surface relief. Note separation into distinct northwest, northeast, east, south, southwest, Cameroon, and Saharan uplands. These areas represent discrete centers of endemism or refuges for numerous organisms adapted to cool or montane conditions. The separation of river basins and river banks has also influenced distributions and evolution. For schematic map of ancient tectonic basins, see figure 4.13.

(10). At the time, I presented this fauna as a microcosm of the mammals of an entire continent. As a sampler of Africa's diverse environments, its position as a corridor reinforced the pivotal status of this region. The fauna is, indeed, central to our understanding of evolution in Africa. In terms of my own evolution, both as a generic human and as an individual, entitling eastern Africa as "Center of the World" is a pardonable exaggeration.

Although my atlas identified some eastern endemics, I noted that the majority of mammal species belonged to a much wider area, some evolved far away perhaps, but drawn in to a region that serves as a corridor between north and south. It is the place where forest communities (stereotypically "western") broadly overlap with those of savannas (likewise, commonly envisaged as "eastern"). This overlapping ensures that eastern Africa has the most diverse mammalian fauna in the whole continent and therefore in the world. The recurrent finding of new and ever more puzzling hominin fossils in east African soils confirms that their biogeographic history must have paralleled that of many other mammal groups. We can safely suppose that many species of prehistoric animals with distant origins were prone to finding their way to this most strategic and attractive of regions, but which species? Discriminating between local endemics and successful immigrants remains as much a challenge for studies of fossil as of living species.

From the time I was preparing my volumes up to the present, the greater part of our knowledge of fossil hominins and their environments has derived from a handful of sites on the eastern side of the continent, including the far southeast. As a result, eastern Africa has come to be portrayed as the archetypal Garden of Eden, the Center of the World for human evolution. Within limits, this updated Biblicism has some truth but, as I have already pointed out, east Africa has been an ancient theater for the excursions of habitats and fauna from very distant parts of Africa. If there is drama in human evolution, the local origins and subsequent travels of "provincials" that then made it big must be a large part of the story. One novelty of this account will be my efforts to identify the possible provenances of those provincials, however far-flung or improbable their places of origin might seem.

Consciousness of geography has been somewhat ambivalent in the literature on human origins. On the one hand, hominin sites are located and described in meticulous detail and, in the minds of many students, easily become equated with place and ecosystem of origin. On the other hand, in many theoretical models geography plays no role at all: everything hangs on the cusp of a tooth or the diameter of a fossa.

This combination is dangerous because fossils usually sample animals only once they are common and widespread. Occasionally, they may actually be in terminal decline. It is much rarer for them to be plausibly close to their place and time of origin. Mentally lifting hominins out of the communities in which they lived is equally distorting.

I see a need to relate what little we know about the evolution of prehistoric humans to more general patterns of mammal distribution. For example, it is important to know if a contemporary specimen comes from the edge or center of its range. At any given moment in time, species may give the illusion of being stable entities. In fact, they may actually be in a state of active dispersal, isolation, or even contraction and decline (11). It is therefore significant to know whether a species, living or extinct, is newly evolved and actively expanding its range; or if it represents a stable, longer established form. Some extinct species such as mollusks, pigs, and rhinos have long been viewed in this perspective, but there are now sufficient hints coming from a rapidly expanding scatter of hominin fossils to merit bolder timings and mappings of their supposed status in space and time. The maps that are offered here take full account of the fossils, but I often attempt to relate their peculiarities, localities, and dates to patterns of distribution and speciation that are inferred from living species. Some of my scene-setting may eventually prove to be misplaced or displaced, but even if some of the suggestions prove to be plain wrong, that will be of less significance than my self-reminding insistence that human evolution was never virtual but, like that of any other organism, had to have taken place in real space and real time.

A lifetime of walking through landscapes with the conscious awareness that my ancestors preceded me there has served to reinforce my determination to share this consciousness and assert its relevance for the way in which we reconstruct and model prehistory in Africa. Superimposed on this ancestor-inhabited world is a contemporary worldview in which new perspectives in genetics have taught us to envisage our ancestry as unbroken threads of DNA winding their way back, past innumerable catastrophes, to their earliest beginnings as single celled organisms. Along the way are billions of mutations, some of which find expression in incremental changes that can be picked up in the fossil record but which also require some knowledge to be studied and interpreted, some imagination to be visualized and portrayed.

Thus a central subject of this book is the rising up of what I call a "Ground Ape" onto two hindlegs, suggesting reasons and a location, in time and place, for that evolutionary moment. Partly because the event is incomprehensible without considering the events that preceded it, I have backtracked the trail that led up to that moment. Then, from the first, unsteady steps of a misshapen ape in an obscure province of Africa, I deduce, on the basis of other species and the inferred peculiarities of these bipeds, *where* the trails of their descendants might have led.

One difference between this account and many that have preceded it is that I try not to amalgamate adaptations. Thus, rearing up onto two of four limbs is seen as but a single adaptation within a long series of preceding and succeeding events, each of which was a discrete, perhaps modest, but essential prerequisite to becoming human. This multifaceted, piecemeal approach has shaped the "dissected" style of my analysis.

A second peculiarity of my own understanding of human evolution can be contrasted with the pictures that are painted in innumerable books, articles, and dioramas representing hominins and other extinct mammals in picturesque "National Park–like" fire-climax savannas. These depictions contradict the evidence that such landscapes became common only 1 to 2 mya (10). My own understanding of early human habitats, as expressed in this book, may derive in part from the years of my childhood and youth spent walking, hunting in, or traversing the "Itigi Thicket" of central Tanganyika. On one foray into this dry but dense and shady habitat, my companion, an entomologist, remarked that this must have been what large parts of Africa must have been like before human-set fires became widespread. His casual remark stuck in my mind, and its likely truth has been borne out by subsequent research.

Another primary difference between my approach and those of my predecessors is that I envisage standing as a relatively inefficient response to an exceptionally benign but very localized environment. This is the exact converse of previous explanations, which attempt to understand bipedalism in terms of improved efficiency under very widespread "savanna" conditions that were more difficult and trying than those in the forests or woodlands that preceded this supposed "ordeal" (12, 13).

Each chapter seeks to identify and portray some outstanding features of a particular ancestral condition. I try to locate innovations in some sort of framework of time and place and correlate changes in behavior with their anatomical and ecological contexts. The settings may be continents, ecoregions, a locality, or the provenance of a single significant fossil; the choice of which depends on sources and quality of evidence to locate particular evolutionary developments in space and time.

Fossil or molecular sources of information and supposed time frames are listed in a conspectus at the beginning of each chapter. The chapters succeed one another and develop, step by step, as an unfolding saga of hominid biogeographic history. The central focus is my analysis of the origins of bipedalism, but I have "topped and tailed" this pivotal event with my larger vision of how straightening the back, standing, walking (slowly, fast), running, and the slow elaboration of hand-eye-mind coordination must all have developed in a long, drawn-out sequence. Thus, both the beginning and final chapters of my story are a continuum of increments, always built on what went before. The evolutionary future is always constrained by its evolutionary past.

My reconstruction of the "East Coast Ground Ape," which owes an important debt to my colleague, Clifford Jolly (14), is essentially an artifact of both analysis and imagination. It represents the assertion that an intermediate form must have existed between quadrupeds and the first bipedal hominins. I contend that traditional attempts to make a single mental leap from four to two legs helps to explain our persistent inability to get to grips with the origins of bipedalism. Because we, as offspring of our first bipedal ancestors, see their innovation as definitive and momentous, I have allowed this isolated event, plucked out of a long sequence of adaptive changes, to become the book's "core event." I try to mitigate such an anthropocentric bias by demonstrating that this particular adaptation, no differently from any other, must have been in response to the dynamics of behavior, ecology, and geography that drive all evolutionary change.

Speculations on the origins of bipedalism are often fascinating exhibitions of ingenuity—expressing, above all, that this is a theater for intellectual daring (15, 16). Early anthropologists thought that moving out of the forest, making stone tools, carrying food by hand, and walking upright were "decisions" that required peculiarly human intelligence! Such naivety became totally unsustainable once it was clear that the first bipeds were cranial apes, creatures with ape heads mounted on humanlike bodies. In spite of a vastly expanded theater of discourse, explanations locating postural change in the peculiarities of a specific ecological niche still tend to be neglected. Part of the explanation for this lies in the environment of students who first begin to grapple with the subject in the intellectual hothouses of universities far from Africa, where the raw materials of study are finger-worn plaster casts of fossils, dog-eared papers on evolutionary theory, and videos on popular natural history.

Why an argument over bipedalism should have become somewhat of an intellectual arena could take up many pages, as would the merest outline of hypotheses. Russel Tuttle (17) has conveniently summarized and labeled them with his own street-smart titles as aide-memoires. The following simplified list of some 13 distinct hypotheses is built on Tuttle's titles and illustrates what a diversity of possible explanations have arisen since Darwin.

1. Freeing the hands in defense of a terrestrial way of life (Darwin 1871) (1).
2. Brachiation responsible for the postcranial features we share with apes. Broken down into three phases: gibbonlike, chimplike, and bipedal (Keith 1923) (18).

3. "The upwardly mobile" hypothesis (also Tuttle's favorite): small-bodied arboreal apes modifying their vertical climbing to run bipedally along thick branches in the canopy (Tuttle 1974, 1975, 1981) (19–21).
4. Bipedalism emerging from the need to carry babies, food, and other objects back to base (Hewes 1961; Isaac 1978; Lovejoy 1981) (16, 22, 23).
5. The avoidance of predators: extra vigilance in the savannas with frequent peering over tall grass (Dart 1926) (24).
6. Phallic display directed at females (Tanner 1981) (25).
7. Intimidation displays directed at other or same species (Westcott 1976; Jablonski and Chaplin 1993) (26, 27).
8. An aquatic phase of foraging and avoiding predators in water (Westenhofer 1942; Hardy 1960; Morgan 1972) (28–30).
9. A thermoregulatory theory whereby savanna dwellers rear up to keep cool (Wheeler 1984) (31).
10. "Two feet better than four" hypothesis; energetic efficiencies in bipedalism (Rodman and McHenry 1980) (32).
11. A "gimmick" spread by imitation then favored by selection (Dawkins *in litt.*) (33).
12. Terrestrial squat-feeding—in grassland (Jolly 1970) (14) and on the forest floor (Kingdon 1997) (34).
13. Bipedalism explained by multiple factors (Napier 1964) (35).

Hypothesis number four, what Tuttle calls the schlepp hypothesis (*schlepp* is Yiddish for carry), has been elaborated into a theory of burdens carried as male bribes or gifts to females; this approach supposedly upstaged those of other, competing primates with "an unbeatable breeding package" (16). Treading water while searching for seafood may be more tongue-in-cheek. Intimidation of competitors and predators alike has been invoked as the origin for upright displays that somehow became two-legged walking. As discussed in later chapters, such displays may, indeed, have been significant for the survival of early hominins, but I cannot envisage them as the primary cause for an erect stance. An example of recently acquired faculties being packaged and projected back to much less plausible contexts is the supposition that bipedalism can be explained by the ancestral ape getting up to escape ground radiation and keep cool. This explanation amalgamates too disparate a bunch of separate faculties as well as making many assumptions about the habitat of the earliest hominins.

Many other efforts to understand the beginnings of bipedalism have been marked by the persistent tendency to lump together whole clutches of human or protohuman characteristics. In common with most biologists, I take the view that untangling the sequence of adaptive changes through which any evolving lineage passes is absolutely vital to understanding how the members of that lineage have arrived at their own unique permutation of traits (36). If the many staged adaptations that must have preceded getting up on two legs are to be understood, it will be important to identify such stages as discrete entities and then try to order them in a sequence that is biologically workable and theoretically plausible. For example, two-legged standing, in my view, preceded true bipedal walking and need not, perhaps should not, be lumped with it. Neither stance was necessarily synchronous with the acquisition of an erect back, nor with the ability to run.

The account that follows is an effort to translate the theoretical difficulties of explaining anatomical transformation and the abstractions of speciation into identifiable Time, Place, and Mechanism.

The question of timing for the emergence of hominins, long assumed to be a very ancient event, was revolutionized by Vincent Sarich and Allan Wilson in 1967 with their elegant demonstration of a "molecular clock" that could be applied to human origins (37). Since that time, geneticists and palaeontologists have tended to favor a relatively late date for the chimp-hominin divergence (7–5 mya). An enormous gap between this date and the first proven fossil bipeds (4.4–3.5 mya) has been dramatically closed recently with the discovery of a fossil biped dated to 6 mya.

The question of place is inextricably tied to the question of how the population that was to become bipedal and their four-legged parent population became separated. The questions are connected because the genetic isolation of populations is an essential prerequisite for speciation. There have been various suggestions for the isolation of vaguely eastern or southern ape populations (38), but none has identified a habitat both ecologically distinct enough to elicit an entirely new form of locomotion nor geographically separate enough to impose the necessary isolation. My own outline addresses both shortcomings.

It is not always appreciated that the broad character of today's major plant communities is not new. Boundaries may have fluctuated wildly, but the gross pattern of humid foci strung along the equator and arid hot-spots pulsing back and forth from the north and southwest would have been well established by the mid-Miocene. A humid focus in the east was identified by the botanist Frank White as a very peculiar "region of endemism" (39). As documentation of the peculiarity of this region's fauna and flora has improved, the relative isolation of the Indian Ocean littoral forest from more westerly forests has become ever more evident (figure 1.6). It is this littoral forest that I identify as the habitat in which the vital transformation took place.

As for mechanism, I have not sought global drought crises nor even fewer trees. Rather, I see an ape population adapting to a different, more deciduous kind of forest and see its isolation as of vital relevance. Unlike many of my predecessors, I have not looked for sparser resources; rather, I can point to a different, perhaps richer, menu. I have postulated a switch to more terrestrial feeding, but instead of actively pursuing fleet prey, east coast ground apes would have found a rich supplement of small, sessile animals and plant matter from the forest floor to augment crops of fruit; the latter being predictably less diverse and growing nearer the ground than in the high forests further west.

It is not immediately obvious how grubbing about for edibles on the forest floor could culminate in bipedal walking. The argument hinges on changes in the spine, pelvis, and head-neck junction (perhaps also in the heel) being necessary precursors to standing and balancing on two legs. It was during this phase that feet changed from being claspers to becoming platforms. In common with my colleague, Clifford Jolly, I hold that it was foraging, mainly on the ground, in a squatting position that demanded these necessary modifications. Hence the title, *Lowly Origin* (34).

The fossil record makes it certain that the particular type of ape bipedalism that gave rise to humans began in Africa, which is not to say that something like it never occurred anywhere else (40). Dating of the earliest fossils has implied that bipedalism began before 4.5 mya (41), even as early as about 6 mya. While a variety of Eurasian ape fossils from about 9 or 10 mya hint at the nature of hominin ancestors, molecular clocks (based on comparing the genes of humans with chimpanzees) suggest a common ancestry up to some 6 to 7 mya (42). If this statistic is correct, the timing for bipedal beginnings contracts to some time shortly thereafter. The location of all the earliest fossils makes it likely that it was African apes from the eastern side of the continent that first became erect (38).

Eastern Africa is a big area, but it has a relatively well-understood geological, climatic, and biotic history (43), which must invite a more specific context for hominin origins. I suggest that in-depth study of the biogeography of African fauna and flora quite literally narrows the most likely location down to forests of the eastern coastal strip. That these forests have suffered sustained ecological and physical separation from forests in central Africa is attested to by the distribution patterns of nu-

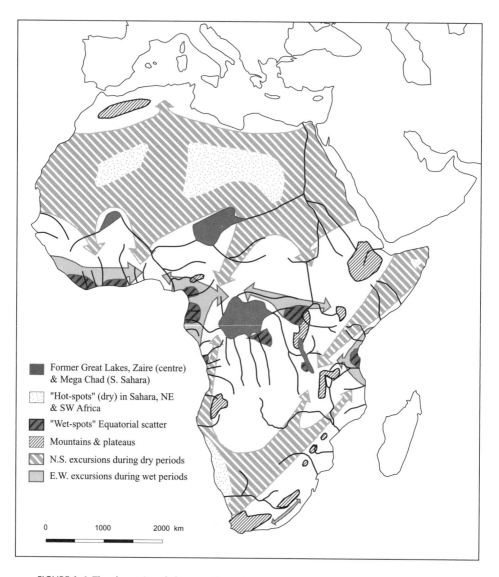

FIGURE 1.6 The dynamics of change. The pulsing of "wet-spots" (east-west, scattered along the equator), and "hot-spots" (north-south, in Sahara and northeast and southwest Africa). Former Great Lakes, Congo (center), and Mega-Chad (S. Sahara). (After Kingdon, J. 1990. *Island Africa. The Evolution of Africa's Rare Animals and Plants*. Princeton, NJ: Princeton University Press.)

FIGURE 1.7 Reconstruction of a foraging ground ape.

merous species of animals and plants (44). The separateness of these forests is of such long standing that it represents by far the most likely mechanism to have isolated incipient hominins from ancestral chimps. More important, the forest's distinctness as a habitat may help explain *why* the eastern apes became erect.

Reconstructing the ecology of the incipient hominin apes' habitat is a less hypothetical exercise than might be thought. These isolated forests still shelter many unique and ancient organisms, and the long-term climatic constraints on vegetation in eastern Africa are relatively well known (39).

As for the immediate impetus for getting up on two legs, there are good reasons to suppose that this posture followed and was dependent on an earlier adaptive phase of "squat-feeding" (41) (as I describe in chapter 5). Squatting would have induced the reorganization of the trunk that I contend was an essential precondition for balanced standing. Instead of simply assuming that standing was a brief balancing act that spanned the quadrupedal/arboreal adaptations of the common ape-hominin ancestors, I have isolated squatting as my topic. This lowly posture (figure 1.7) signifies not only the central subject matter and title of this book; it also exemplifies one component of my piecemeal approach to reconstructing human evolution.

The ability to stand without the expenditure of much energy requires good balance. Two-legged balance, in the brief waddling of apes or the performing of poodles, is a precarious artifice because there is too much weight concentrated at the top of the column and too little stability at the bottom. In other words, these quadrupeds are top-heavy. Easy, nonenergetic standing requires a downward displacement in the distribution of weight. If structures in the upper foreparts are no longer the largest and heaviest in the body, their smallest movements will no longer destabilize balance or threaten to topple the would-be walker. To achieve such shifts in the distribution of weight in living, functioning bodies, there must be substantial changes in the relative weight of muscles, bones, and organs. I contend that this necessary slimming down of the upper part of the eastern apes' bodies came about through ecological and behavioral changes that rendered exceptionally heavy, powerful fore-limbs redundant. I give the arguments, details, and rationale for this in chapter 5. It suffices here to assert, once again, that the single act of standing is inconceivable without such preliminaries. This phenomenon is commonly called "preadaptation," but I am, in the first place, trying to portray the ecological background for an "increment" that can be isolated as a manageable unit of evolution. My second objective is to highlight a conceptual approach that can enhance our ability to comprehend past events.

I contend that only an ape population that was able to exploit an intensive and reliable (rather than an extensive and irregular) food supply could permit a radical shift in the priority functions of their forelimbs. Hands and arms could, most frequently, and for long periods, be devoted to turning over leaf litter, selecting, processing, and handling foods; also, but rather less frequently, to vertical climbing and to some diagonal "propping" in the trees. As the incidence of bearing weight declined, there would come an identifiable point at which four-legged movement ceased to be as efficient as simple straightening of the legs. I contend that this point existed when the spine was balanced vertically.

The achievement of upright stance can be viewed as a moment of reorientation rather than action, a balancing act on a behavioral tightrope; the moment in which walking is still the unfulfilled potential of a standing "ground ape." Becoming fully erect can also be positioned symbolically on an ecotone between richly endowed rainforest and drier, less reliable and less homogenous "nonforests." It can also be positioned on a biogeographic boundary between the Indian Ocean coastal forests and the more diverse habitats of the eastern interior.

The incentives to get upright need have been no more than sporadic to begin with, but they would have had to be worth it. Worth could be measured in terms of extra food or compensatory food in places where and at times when it contributed toward survival (perhaps no more than seasonal gluts of milkwood or mustard bush fruit on the edges of a too-small home range).

It is only with hindsight that we can say that the ultimate worth of standing up, the hidden evolutionary prize, was the ability to find the way out of a sort of ecological and biogeographic cul-de-sac. Yet there are numerous other organisms in tropical Africa that have moved, in both directions, across the forest-nonforest boundary, and there are lessons to be learned from such species (these are explored in more detail in chapter 6). The most radical implication of "eastern ground apes" becoming transformed as they moved inland is that their occupation of a wide scatter of major river basins may have led to the evolution of more than one lineage.

There are numerous implications, not least for nomenclature, in the possibility that our own line of descent diverged at the ground ape level, not the level of "Lucy" or her kin. For many years, there has been widespread acquiescence to the inclusion of Lucies in the direct human lineage; indeed, *Praeanthropus* (*"Australopithecus afarensis"*) is frequently portrayed as the prototypical first ancestor (36, 45). We must now question our assumptions about these Lucies, who were once seen as very early hominins but are now recognized as relatively late players on the hominin stage (46). Long-held models of a single lineage must suffer still further erosion as the evolutionary tree gets ever more bushy (47–49). More significantly, new data show that some of the peculiarities of our own lineage were absent from the Lucy lineage yet must have evolved by the same time. This raises numerous new questions about the ecological and behavioral roots of our own specific line. The possibility of divergence at the ground ape level will, of course, precipitate more uncertainties about the ultimate roots of humanity. Even so, I hope it will invite much more discussion and research that is couched in terms of geography, ecology and behavior; until recently these tended to be rather subordinate parts of discussion (50).

On balance, I think it very likely that Lucies, once seen as very early but now recognized as relatively late hominins, are well off the main line of human evolution; nonetheless, they remain one of the best illustrations of an early biped because their fossils are so numerous. If Lucy-like traits creep into my self-portrait as an early hominin, that is partly because any illustration at this time is, of necessity, very broad-brushed. "Evolution by River Basin" could account for this high level of convergence, a parallelism that is already implicit in the anatomy of existing hominin fossils (51, 52). Indeed, parallel anatomical adaptations in different river basin populations could have been a natural feature of this crucial moment in the emergence of humans and, perhaps, other hominins. In any event, I contend that the fossil record already provides evidence that bipedal gaits were built on an erect back, no matter whether the evolution of walking was a single or a multiple development in the descendants of eastern ground apes.

It is possible that one of the reasons why the acquisition of erectness should have remained such a controversial and enigmatic problem is that we are hostages to an iconic history in which two legs are not only the mark of our uniqueness but an automatic antithesis to four (53). Darwin's reminder that we should include limbless, hagfish-like vertebrates in our ancestry may offer us an interesting slant on the problem and open to doubt the preferred categories of the debate: quadruped versus biped; bipedalism evolving from quadrupedalism.

The fact that most terrestrial mammals, including apes, carry their weight on four legs makes the word *quadruped* seem a rational enough category, but that could once be said of "quadrumana," or "four hands"—the now obsolete and anthropocentric term for monkeys. However, "four legs" misses a distinction within the category of limbs that could substantially alter the way we study bipedalism and its origins.

Suppose we question the assumption of four-leggedness as a base line? Suppose walking on four legs is in as much need of explanation or wonder as walking on two? Justifying such a counterintuitive argument depends on how far back we are prepared to go and how fundamental we choose to get.

Organic "bodies," whether single- or multicelled, are organized to respond to numerous challenges, such as extremes of temperature, humidity, saturation, chemistry, material textures, rays, or waves such as light, noise, vibration, or water. The simplest and most universal responses to such stimuli are simple stop-go or attract-repel actions. Having the ability to sense and then react to such stimuli has an intimate bearing on the way we are built as animals, given expression in that front-end concentration of sensory equipment that we call a head. Through it we receive, process, and respond to innumerable messages via sophisticated sensory and neural pathways. We are responding at this most primitive level when we recoil from a blast of heat or cold, or even from a loud noise or a bad smell.

Some invertebrates and effectively all vertebrates are irreducibly linear. Vertebrate bodies began as segmented, finned columns that were propelled forward by a series of rhythmic muscle contractions operating on either side of the tail end of their long, thin "backbones" (54). This train of cartilaginous discs enclosed a continuous thread of nerves and overlay a digestive tube that was protected by pronglike extrusions from the vertebrae: the beginnings of ribs. As a direct reminder of such primitive antecedents I, or you, have only to see or feel spine, back muscles, gut, and

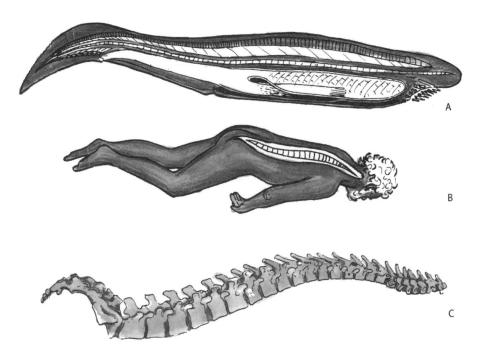

FIGURE 1.8 A. *Amphioxus*, a primitive chordate, in cross-section. B. My backbone, as felt. C. My backbone, as X-rayed.

ribs in our own bodies. As we squirm on the floor during back-strengthening exercises we can indulge the illusion of "remembering" our early vertebrate evolutionary infancy in the sensations emanating from each of these elements. We have inherited them from a pre-fish level of ancestry, together with their very necessary functions (figure 1.8).

Darwin was going back to just such elementary fundamentals when he wrote, in *Descent of Man*, that "we can see that the early progenitor of all the Vertebrata must have been an aquatic animal . . . more like the larvae of the existing marine Ascidians, than any other known form." For Darwin, the larva served to illustrate a phase, a single increment, in humanity's piecemeal emergence from its "lowly origin." Too much has been made of Darwin's supposed belief that human characteristics evolved "in concert," as a "package" (55). It is true that it needed Mendel, Crick, and Watson to show the particulate genetic mechanisms that underlie evolution and that Darwin may sometimes have smudged the boundaries between specific adaptations. However, the broad thrust of his arguments was unequivocally that evolution proceeded piecemeal, step by step, increment by increment.

It was vertebrae and associated bones toward the tail end of early verte-brates that first differentiated into hindfins and, eventually, back legs and pelvis. Like the rest of the rear end, their primary function was propul-sion. For the present argument, the significance of rear-end propulsion is that from the very beginning, tail-end limbs have had quite different ori-gins and functions from forelimbs.

The latter first evolved in close association with the head—indeed, so close that pectoral fins were actually tethered to the head of early fishes. When early terrestrial vertebrates, tetrapods, developed forelimbs from these fins, which became detached from the skull, one pair of gill arches was dragged away to become the shoulder blades or scapulae. Likewise, the anterior sources of forelimb nerve networks testify to very separate limb origins (figure 1.9). Subordinate to the brain, forelimbs still serve many vertebrates to alter and adjust their sense-driven decisions about the direction and pace of forward movements. This arrangement is par-ticularly true of aquatic animals such as frogs and fish. There are, of course, many seal and whale species in which the forelimbs play a minor role in providing thrust during swimming, but that function belongs to the hind limbs and tail.

The differentiation of limbs goes back to the very first limbed and fin-gered animals. That joined-up, light-weight, multiple-rayed fins are ele-gant solutions to the problems of fast, maneuverable swimming in open water is proved by the survival and perfection of fins over some 500 mil-lion years of fish evolution. The delicacy of fins renders them less useful when the water is full of obstructions and tangles, as happened when plants began to flourish on land and at water's edge at the start of the De-vonian, 400 mya. To progress through shallow forest swamp waters, larg-ish animal bodies must twist and turn, squirming or levering themselves over and around stems and branches, finding a purchase with something less fragile and slippery than a cartilaginous fin. Significantly, lungfish-like fossils, their fins borne on the end of blunt oarlike stumps, first ap-pear in the Devonian, But the major switch from fins to something more like true limbs comes with the appearance of the first, still aquatic, tetrapods after about 370 million years (56).

One of these, *Acanthostega*, illustrates both the piecemeal nature of evolution and also exemplifies the potential for fundamental differentia-tion between limbs at the front and back of an animal. The forelimbs of this newtlike tetrapod were similar to those of lobefinned lungfish, but the powerful hind legs were furnished with eight digits, not fin rays (fig-ure 1.10). Other species had five, six, and seven digits, but the number eventually stabilized into the five-fingered standard possessed by most

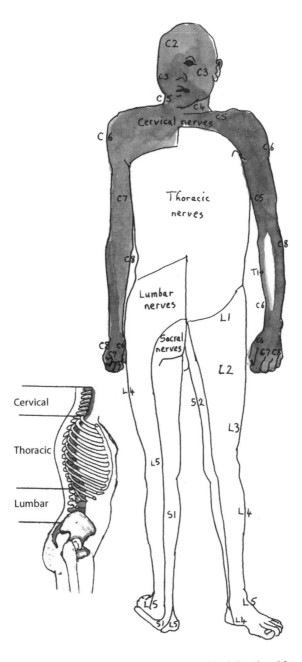

FIGURE 1.9 Cervical nerves serve the surfaces of both head and forearms.

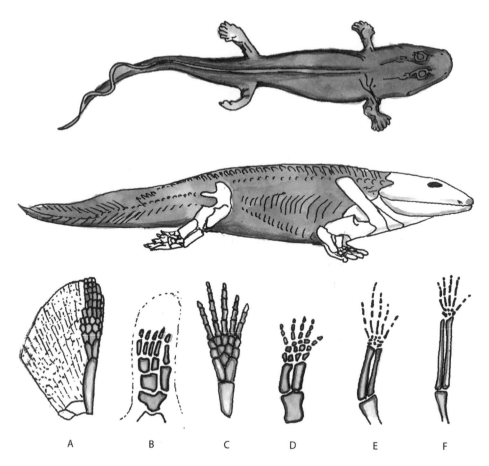

FIGURE 1.10 Top: The Devonian tetrapod *Ichthyostega*. (Note the close association of fore-limb bones with the head.) Bottom: Evolution of the forelimb. A. Breast fin of a shark (shaded section illustrates the components that persist in higher vertebrates). B. Lobe-finned fish. C. Generalized amphibian. D. Five-fingered early amphibian. E. Lizard. F. Human. (In part after Ahlberg, P. E., and A. R. Milner. 1994. The origin and early diversification of tetrapods. *Nature* 368: 507–514; Ankel-Simons, F. 2000. *Primate Anatomy: An Introduction.* San Diego, CA: Academic Press; and McLeod, M. 2000. One small step for fish, one giant leap for us. *New Scientist* 19 August 2000.)

modern amphibians, reptiles, and mammals (57). Another tetrapod, *Ich-thyostega*, which was broadly contemporaneous with *Acanthostega* and was also a large (>1 meter long) predator, had powerful front limbs that supposedly helped keep its air-breathing throat apparatus clear of the ground during forays out of the water. By 338 mya, tetrapods were fully terrestrial and highly diverse in form (58).

Because the fore-ends of primitive vertebrates were the first to en-

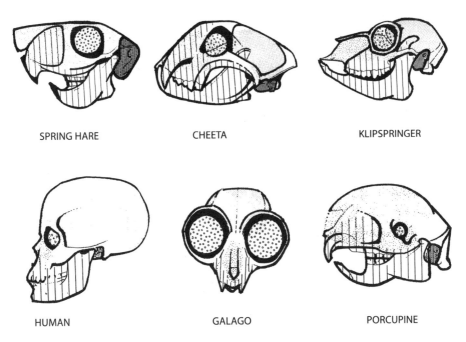

SPRING HARE   CHEETA   KLIPSPRINGER

HUMAN   GALAGO   PORCUPINE

FIGURE 1.11 Skulls of six mammals to illustrate the enclosure of functional activities, such as seeing, chewing, smelling, thinking, etc., in discrete capsules. A. Spring hare. B. Cheetah. C. Klipspringer. D. Human. E. Galago. F. Porcupine. (From Kingdon, J. 1971–1982. *East African Mammals. An Atlas of Evolution in Africa* [3 vols., 7 parts]. London: Academic Press.)

counter both food and obstacles, basic sense organs and a mouth developed there. The need to respond to light, chemical, electric, or vibrational signals led to the differentiation of cells until they developed into eyes, nose, and ears, each encapsulated in compartments that shared the upper part of what became the skull. The lower section of this structure became a hinged mandible, and eventually both jaws developed teeth. My skull still consists of a series of connected capsules (figure 1.11), and it is in the relative size and permutations of connecting bridges, struts, and welds that the species-specific differences among animal skulls become obvious.

The fact that my fore-end, including my forelimbs, encountered and "processed" the environment while propulsive force resided at my rear-end remained an inherited "given" long after my aquatic ancestral "self" had spawned land-dwelling mammals (indeed, even to the point where some had become tree-dwelling primates).

Perhaps the point was too indirect, or even too obvious, for Darwin to pursue, for while he emphasized that organisms retain general structures from their aboriginal progenitors and he explicitly described hands as acting "in obedience" to human will (at the same time noting "free, independent action in the arms and upper body" of *Homo sapiens*), *The Descent of Man* makes no mention of functional separation between forelimbs and hindlimbs predating the radiation of higher vertebrates. Yet it is in the context of deep Darwinian time frames that the evolution of standardized forelimbs and hindlimbs in terrestrial animals becomes no less a gravity-fighting contrivance than walking on two legs. It is somewhat of an exaggeration to portray our many quadrupedal ancestors as occupants of a mere interlude in our evolutionary history. It is no stretch to suggest that restricting propulsion to the back legs alone is as much a "return" as it is an innovation. When the primitive connection between head and forelimbs is remembered, the latter's emancipation from serving as supportive props becomes somewhat less revolutionary. The development of a way of life that rebalanced the tyranny of foursquare gravity becomes as remarkable for what it "restored" as for the novelties it unquestionably introduced. As for human two-leggedness being "revolutionary" (and supposedly uniquely beautiful) (59), human gaits scarcely compare for balance, grace, and speed with those of ostriches or emus!

This diversion into the nether regions of vertebrate history therefore serves to remind us that "quadrupedalism" is mainly a contrivance that land-living animals have evolved to overcome the many problems of gravity and the need for faster movement. Among the reasons for this reminder that forelimbs have historical neural connections with the head, one has been to demystify bipedalism; second, challenge the assumption that not using the forelimbs for propulsion was an event of absolutely unprecedented originality; and, third, to prepare the ground for discussing later head-hand linkages. A further purpose is that if you have been persuaded that the functional anatomy of a hagfish is even remotely relevant to an ape standing up, the next leap ahead, to the earliest primates (in chapter 2) will be that much easier to embrace.

Reconstructing some of the series of events that must have gone on before and after becoming erect may also help to underline the incremental nature of the evolutionary process itself (figure 1.12). Each such increment needs to be studied as a complex of interrelated changes, set in time, place, and environment. And I have already indicated that a historically plausible sequence of "increments" has provided the substance and the structure for each of the succession of chapters that follows. So why lift one increment out of its proper temporal sequence?

The reasons are twofold. One is to establish the central topic of the book. The other is to reinforce my descriptive technique of breaking "bipedalism" down into component parts. Thus, I interpret an erect back,

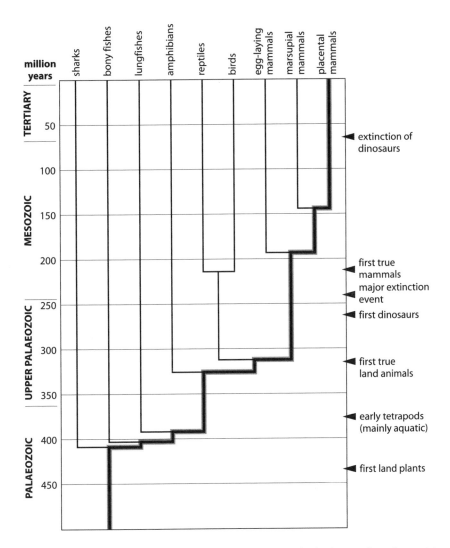

FIGURE 1.12 The radiation of vertebrates through time, with the human line of ancestry traced in bold.

balancing on two legs, walking, and running as four distinct entities. I acknowledge that they may have important structural connections, but each can be productively studied as a separate adaptive response.

To balance on two legs is an essentially static activity. Alone, it has not gotten you anywhere. However, if you were previously four-legged, it is an anatomical and physical feat; it is also a necessity for economic locomotion. Conceptually, it is a "moment"; chronologically, standing comes before walking and after quadrupedalism. Yet the realities of adaptive change in intelligent, active, versatile animals make it unlikely that any such "moment" could ever find such a neat, static niche. As more than one wag has remarked, what came between four legs and two? Three? The witticism is less outlandish than it may seem, as is explored in chapter 4.

In the context and format pursued in this book, standing is, above all, an evolutionary moment. Many living mammals have literal moments of standing up on their hind legs. These are easy to document, but the evolutionary moment in which an ape becomes habitually erect is a temporal unit that is more difficult to visualize, describe, or study. There are other important theoretical reasons for trying to reconstruct the series of events that must have gone on before and after becoming erect.

The difficulty of reconciling four-legged walking, especially knuckle-walking, as the immediate precursor of bipedalism has led to a polarization of views as to whether it was a short, sudden, instant adaptation or a protracted, sporadic response to particular stimuli—an activity that merely increased in its incidence. It is a polarization that has some theoretical importance because a sudden "flip" from one mode of locomotion to another supports the idea of "punctuated equilibrium" (60) and combined "packages" of adaptive change, whereas protracted, step-by-step change in specific domains is, by definition, more gradualist (61). In asserting that there was no sudden leap, no magic response to some cosmic change, I am supporting and augmenting Jolly's 1970 proposition that a (literally) lowly but important intervening stage of evolution should be registered. Furthermore, in laying out a road map of human evolution with a succession of wayside stops, I am suggesting that greater attention be paid in future to identifying and reconstructing the ecology of identifiable ecological "islands." This topic should be an essential dimension of evolutionary studies because the genetic isolation of populations is a key component of speciation.

In the pages that follow, I draw on several new lines of evidence to suggest places, mechanisms, ecological contexts, and times for what I regard as the plural beginnings of hominin evolution.

Becoming erect can be seen as a symbolic moment for the irreversible separation of the human lineage from all other animals; it has also become the rather arbitrary taxonomic criterion for separating all human-like hominins from all apes (62). I think this is a false duality. Standing was but a momentary pause in a long series of adaptive changes, but it was predicated on the accumulated benefits of earlier evolutionary phases: the primacy of vision, diurnal habits, taillessness, and the spatial

mobility of an ape with brachiating ancestors. These piecemeal preconditions for bipedalism were followed by no less piecemeal and extraordinary consequences for our own evolution as humans. An incremental
perspective on evolution suggests that the set of adaptations that culminated in erectness had no inevitable connection with adaptations for big
brains. That much is corroborated by the fossil record. We know that
brains enlarged in our own lineage, but the first really successful bipeds
seem to have had no need for brains larger than those of apes. In chapter
5, I propose that our intellectual beginnings may have started in an initially insignificant sidebranch of the ground apes.

Furthermore, the development of what we like to call cleverness must
have been equally subject to incremental change and equally tied in to
selection for a particular type of mental versatility. A step-by-step improvement in cleverness must have involved change in quite separate
sensory, motor, physiological, and behavioral, mental, and neural skills.
Each of these modifications must have had its own ecological and behavioral contexts. So while this book does not set out to examine the evolution of intelligence, I share with others the conviction that many parallels must have existed between physical and mental development (63).
Staged changes in ecology and behavior must have corresponded with
piecemeal changes in the way in which evolving humans constructed
mental models of the world or worlds in which they found themselves.
These would have included literal "models" (including the physical con-
structs we call tools) whereby they could eke out a living through their
own and their social group's actions.

The conviction that all our adaptive traits, including intelligence, derive directly from animal ancestors is still a very minority view. It is a
conviction that has profound implications for our worldview and for
how the only truly bipedal animal will have to conduct itself if it is to
survive. Matt Cartmill summed up both the simplicity and the magnitude of the personal choice we all have to make: "to seek to show that all
things human are prefigured or paralleled in the lives and adaptations of
our fellow animals—is at bottom to doubt the reality of the moral boundary that separates people from the beasts. Whether we fear or welcome
the dissolution of that boundary is the real issue" (64). It is an issue that
has as much bearing on long-term human survival as it has for new human moralities.

The intellect that Darwin celebrated "penetrating the movements and
constitution of the solar system" still bears the stamp of numerous increments of change. Each increment is embedded in a distant past, but identifying their sources and their transformations will be the task for future

self-portraitists seeking to chart the architecture of that mysterious amalgam that is a living human being.

My own self-portrait is more explicitly physical. So, having explored some of my rudimentary vertebrate dimensions and identified the evolution of bipedalism as a pivotal topic for the book, I now backtrack from the latter and, with apologies to neglected amphibian and reptile ancestors, leapfrog to sketch out my (and your) life as an early mammal and primate.

REFERENCES

1. Darwin, C. 1871. *The Descent of Man and Selection in Relation to Sex.* London: John Murray.
2. Kingdon, J. 1993. *Self-made Man. Human Evolution from Eden to Extinction?* New York: John Wiley & Sons.
3. Spoor, C. F., B. A. Wood, and F. Zonneveld. 1994. Evidence for a link between human semicircular canal size and bipedal behavior. *Journal of Human Evolution* 30: 183–187.
4. Kingdon, J. 1997. *The Kingdon Field Guide to African Mammals.* London: Academic Press.
5. Olson, T. R. 1981. Basicranial morphology of the extant hominoids and Pliocene hominids: The new material from the Hadar formation, Ethiopia, and its significance in early human evolution and taxonomy. In *Aspects of Human Evolution*, ed. C. B. Stringer, 99–128. London: Taylor and Francis.
6. Falk, D. 1988. Enlarged occipital/marginal sinuses and emissary foramina: their significance in hominid evolution. In *Evolutionary History of the "Robust" Australopithecines*, ed. F. E. Grine, 85–96. New York: Aldine de Gruyter.
7. Kingdon, J. 1990. *Island Africa. The Evolution of Africa's Rare Animals and Plants.* Princeton, NJ: Princeton University Press.
8. Brunet, M., A. Beauvilain, Y. Coppens, E. Heintz, A.H.E. Moutaye, and D. R. Pilbeam. 1996. *Australopithecus bahrelghazali*: une nouvelle espece d'Hominide ancien de la region de Koro Toro (Tchad). *C.R. Acad. Sci. Ser. IIa* 322: 907–913.
9. Cerling, T. E. 1992. Development of grasslands and savannas in East Africa during the Neogene. *Palaeogeography, Palaeoclimatology, Palaeoecology, Global and Planetary Change* 97: 241–247.
10. Kingdon, J. 1971–1982. *East African Mammals. An Atlas of Evolution in Africa.* (3 vols., 7 parts.) London: Academic Press.
11. Foley, R. A. 1999. Evolutionary geography of Pliocene African Hominids. In *African Biogeography, Climate Change, & Human Evolution*, ed. T. G. Bromage and F. Schrenk, 328–348. New York: Oxford University Press.
12. Wheeler, P. E. 1985. The loss of functional body hair in man: the influence of thermal environment, body form and bipedality. *Journal of Human Evolution* 14: 23–28.
13. Wheeler, P. E. 1993. The influence of stature and body form on hominid en-

ergy and water budgets: a comparison of *Australopithecus* and early *Homo* physiques. *Journal of Human Evolution* 24: 13–28.

14. Jolly, C. J. 1970. The seed eaters: a new model of hominid differentiation based on baboon analogy. *Man* 5: 5–26.

15. Morgan, E. 1982. *The Aquatic Ape. A Theory of Human Evolution*. London: Souvenir Press.

16. Lovejoy, C. O. 1981. The Origin of Man. *Science* 211: 341–348.

17. Tuttle, R. H. 1981. Evolution of hominid bipedalism and prehensile capabilities. *Philosophical Transactions of the Royal Society, London, Series B.* 292: 89–94.

18. Keith, A. 1923. Man's posture: its evolution and disorder. *British Medical Journal* 1: 451–672.

19. Tuttle, R. 1974. Darwin's apes, dental apes, and the descent of man: Normal science in evolutionary anthropology. *Current Anthropology* 15: 389–398.

20. Tuttle, R. H., ed. 1975. *Primate Functional Morphology and Evolution*. Mouton, (L. Junk) The Hague: 291–326.

21. Tuttle, R. H. 1969. Knuckle-walking and the problem of human origins. *Science* 166: 953–961.

22. Hewes, G. 1961. Food transport and the origin of hominid bipedalism. *American Anthropologist* 63: 687–710.

23. Isaac, G. 1978. The sharing hypothesis. *Scientific American* April 1978: 90–106.

24. Dart, R. 1926. Taung and its significance. *Natural History* 26: 315–327.

25. Tanner, N. M. (1981) *On Becoming Human*. Cambridge University Press: Cambridge.

26. Westcott, R. W. 1976. Hominid uprightness and primate display. *American Anthropologist* 69: 78.

27. Jablonski, N. G., and G. Chaplin. 1993. Origin of habitual terrestrial bipedalism in the ancestor of the Hominidae. *Journal of Human Evolution* 24: 259–280.

28. Westenhofer, M. 1942. *Der Eigenweg des Menschen*. Berlin: Maunstaedt & Co.

29. Hardy, A. 1960. Was man more aquatic in the past? *New Scientist* 642–645.

30. Morgan, E. 1972. *The Descent of Woman*. London: Souvenir Books.

31. Wheeler, P. E. 1984. The evolution of bipedality and loss of functional body hair in hominids. *Journal of Human Evolution* 13: 91–98.

32. Rodman, P. S., and H. M. McHenry. 1980. Bioenergetics and origins of bipedalism. *American Journal of Physical Anthropology* 52: 103–106.

33. Dawkins *in litt* to J. Kingdon.

34. Kingdon, J. 1997. Ecological background to the possible origins of bipedalism in early hominins. Ms. and lecture for J. Sabater Pi Festschrift "Ecological Background to Human Evolution." September 1997, Barcelona University.

35. Napier, J. R. (1964) The evolution of bipedal walking in the hominids. *Archives Biologie* (Liège) 75: 673–708.

36. Szalay, F. S., A. L. Rosenberger, and M. Dagosto. 1987. Diagnosis and differentiation of the order Primates. *Yearbook of Physical Anthropology* 30: 75–105.

37. Sarich, V. M., and A. C. Wilson. 1967. Immunological time scale for hominoid evolution. *Science* 158: 1200–1202.

38. Coppens, Y. 1991. L'évolution des hominidés, de leur locomotion et de leurs environnements. In *Origine(s) de la bipédie chez les Hominidae*, ed. Y. Coppens and B. Senut, 295–301. Cahiers de Paléoanthropologie. Paris: CNRS.

39. White, F. 1983. *The Vegetation of Africa*. Natural Resources Research 20. Paris: UNESCO.
40. Kohler, M., and S. J. Moya-Sola. 1997. Ape-like or hominid-like? The positional behavior of *Oreopithecus bambolii* reconsidered. *Proceedings of the National Academy of Sciences* 94: 11747–11750.
41. White, T. D., G. Suwa, and B. Asfaw. 1994. *Australopithecus ramidus*, a new species of early hominid from Aramis, Ethiopia. *Nature* 371: 306–312.
42. Sarich, V. M. 1968. The origin of the hominoids: An immunological approach. In *Perspectives on Human Evolution* 1: 94–121, ed. S. L. Washburn and P. C. Jay. New York: Holt, Rhinehart and Winston.
43. Russell, E. W., ed. 1962. *The Natural Resources of East Africa*. Nairobi: East African Literature Bureau.
44. Lovett, J. C., and S. K. Wasser. 1993. *Biogeography and Ecology of the Rainforests of Eastern Africa*. Cambridge: Cambridge University Press.
45. Johanson, D. C., and M. A. Edey. 1981. *Lucy: The Beginnings of Humankind*. New York: Granada.
46. Day, M. H., R.E.F. Leakey, A. C. Walker, and B. A. Wood. 1976. New hominids from East Turkana, Kenya. *American Journal of Physical Anthropology* 45: 369–436.
47. Strait, D. S., F. E. Grine, and M. A. Moniz. 1997. A reappraisal of early hominid phylogeny. *Journal of Human Evolution* 32: 17–82.
48. Tattersall, I. 2000. Once we were not alone. *Scientific American* January: 38–44.
49. Bayley, E. 2000. Only Human? We used to share the Earth with many other types of human. How did we come to survive them? *Focus* 94: 44–52.
50. Bromage, T. G., and F. Schrenk, eds. 1999. *African Biogeography, Climate Change & Human Evolution*. New York: Oxford University Press.
51. Turner, A., and B. A. Wood. 1993. Comparative palaeontological context for the evolution of the early hominid masticatory system. *Journal of Human Evolution* 24: 301–318.
52. Wood, B. A. 1993. Early Homo: how many species? In *Species, Species Concepts, and Primate Evolution*, ed. W. H. Kimbel and L. B. Martin, 485–522. New York: Plenum.
53. Cartmill, M. 1983. Four legs good, two legs bad: Man's place (if any) in nature. *Natural History* 92(11): 64–79.
54. Gray, J. 1953. *How Animals Move*. Cambridge: Cambridge University Press.
55. Leakey, R.E.F., and R. Lewin. 1992. *Origins Reconsidered*. London: Little, Brown and Company.
56. Ahlberg, P. E., and A. R. Milner. 1994. The origin and early diversification of tetrapods. *Nature* 368: 507–514.
57. Coates, M. 1996. The Devonian tetrapod *Acanthostega gunnari*; Postcranial anatomy, basal tetrapod interrelationships and patterns of skeletal evolution. *Transactions of the Royal Society of Edinburgh—Earth Sciences* 87.
58. McLeod, M. 2000. One small step for fish, one giant leap for us. *New Scientist* 19 August.
59. Cartmill, M. 1990. Human uniqueness and theoretical content in paleoanthropology. *International Journal of Primatology* 11: 173–192.
60. Eldredge, N., and S. J. Gould. 1982. Punctuated equilibria: An alternative to

phyletic gradualism. In *Models in Paleobiology*, ed. T. M. Schopf, 82–115. San Francisco: Freeman, Cooper and Co.

61. Williams, G. C. (1966) *Adaptation and Natural Selection: A Critique of Some Current Evolutionary Thought*. Princeton, NJ: Princeton University Press.

62. Tobias, P. V. (1998) Ape-like *Australopithecus* after seventy years: was it a hominid? *Journal of the Royal Anthropology Institute* 4: 2.

63. Byrne, R.W. (1995) *The Thinking Ape: Evolutionary Origins of Intelligence*. Oxford: Oxford University Press.

64. Cartmill, M. (1993) *A View to a Death in the Morning*. Cambridge, MA: Harvard University Press.

CHAPTER 2

On Being a Primate

From Gondwana to the Forests of Egypt

*O*n the nature of mammals. Primates and humans as the most mammalian of mammals. Africa the likely place of origin for anthropoid primates. Primate hands and feet, padded fingers, thumbs, and hand-brain coordination. The challenge of becoming diurnal. Egyptian and North African fossils from 36 to 31 mya main source of information on early primates.

*A*s darkness falls, in the vestigial rainforest on Mount Baldy, in northeastern Australia, a short, still silence is broken by a flutter of leaves that resembles the stirrings of an uncertain breeze. It is the waking movements of numerous possums, gliders, and other tree-living marsupials (figure 2.1) as hunger drives them out of their lairs and hollows. Most touch ground or are exposed to daylight only when they die, so complete is their divorce from the diurnal, terrestrial world of today's people and dingoes and yesterday's predatory reptiles.

Is this furtive rustling a last echo of the first mammals primal way of life? Did modern mammals begin as discrete tree climbers, escapees from the sunlit world of dinosaurs and other large reptiles? From the structure of their eyes, it is certain that all mammals went through a prolonged

spotted cuscus

FIGURE 2.1 Spotted cuscus, *Spilocuscus maculatus*, one of the many arboreal marsupials with clawed but very effective climbing hands.

phase of being nocturnal. They are also likely to have been tree dwellers. Some of the most ancient of living mammal lineages—not only possums but also bats, rodents, anteaters, pangolins, flying lemurs, and of course primates—have features consistent with such an evolutionary history. There are also remarkable resemblances between some possum and primate species in spite of possums being typical pouched marsupial mammals from Australasia whereas primates are generally envisioned as an advanced, mainly Old World order of placental mammals. The similarities seem to owe as much to retentions from a primitive way of life that characterized their common ancestors as to their contemporary occupation of equivalent ecological niches. If so, both possums and nocturnal primates can serve as illustrations of early representatives of very separate and distinct orders but may also illustrate key features of the common rootstock for all mammals.

Exploring our evolution in terms of long successions of major adaptive changes poses fundamental questions: why mammals? why primates? why apes? And why bipedal apes rather than bipedal possums or bipedal foxes? Familiar textbook definitions can only reveal so much about the

FIGURE 2.2 *Hadrocodium wui*, a 195 million-year-old fossil mammal from China. (After Luo, L. X., A. W. Crompton, and A. L. Sun. 2001. A new mammaliaform from the Early Jurassic and evolution of mammalian characteristics. *Nature* 292[5521]: 1496, 1535.)

intrinsic nature of any group of animals. This limitation exists partly because it is very difficult to recreate the conditions in which these major categories emerged. We can guess that mammalian orders, including primates, evolved in response to challenges posed by continuous change, but they were changes in a world so ancient and unfamiliar that we can scarcely imagine it. Even the timing for mammalian and primate emergence is largely conjectural. We need all the help we can get, so is there any fossil evidence for lineage histories and common ancestors?

It is frustrating that, for the greater part of their existence on Earth, modern mammal lineages have fossilized all too sparingly. The gaps are enormous, but recent fossil discoveries have provided some surprising indications of the timescales involved, and molecular evidence has tended to support much earlier origins and more ancient divergences for mammal orders than was once thought possible, even a decade ago (1). For example, the nearest fossil relative of all mammals was discovered in China only as recently as the year 2000 and has been reliably dated to 195 mya (figure 2.2). This minute animal, smaller than a pygmy shrew and named *Hadrocodium*, is classed as a "Mammaliform," the group from which all the surviving mammals (including the egg-laying monotremes) were derived. It has most of the key features that distinguish mammals from mammal-like reptiles, such as a single jaw, a stable jaw joint, and the three middle-ear bones that were jaw components in their reptile ancestors. Another key fossil illustrates the common ancestry of marsupial and placental mammals. These are Therians with tribosphenid (three-cusped, interlocking) molar teeth, and their fossil representative is a shrew-sized mammal from Madagascar (2). Its highly diagnostic teeth have been found in deposits that are 165 million years old, a period when today's island was almost certainly mainland deep within the fragmenting landmass of Pangaea. Although this discovery decisively locates the earliest example of a therian mammal in a southerly, equatorial region, *Hadracodium* shows that still earlier mammal ancestors existed in the north as well.

We know that early marsupial and placental mammals coexisted with a variety of other, rather similar animals. How and where the two diverged is still unclear, but that divergence has been estimated to have been in the region of 135 to 145 mya (2, 3). The very rarity of early mammals implies that they may have been ecologically or geographically localized, and present indications are that they were very small. In any case, for the greater part of their existence, they were swamped into insignificance by the sheer abundance, diversity, and size of reptiles, a dominance that came to an abrupt end only 65 mya with the cataclysmic extinction of the dinosaurs.

There are several explanations for a paucity of mammal fossils, among them that tiny forest animals, especially arboreal ones, rarely fossilize, even under the best of conditions. A scarcity, or near-absence of known deposits from the relevant periods in the Old World tropics might mean that they lived in parts of the world or in habitats that, so far, have left little or no trace. It may also amount to little more than the notoriously uneven distribution of palaeontologists and their study sites.

Facing up to this absence of fossils, can we use comparisons with other mammals, especially primitive ones, to reconstruct a template for the very first primates? To illustrate the difficulties, take the question that began this chapter: was nocturnal tree climbing a primary point of divergence or were some or all early mammals already arboreal? Attempts at an answer are inseparable from even more fundamental questions about the beginnings of mammals in general (including Australian marsupials) and placental mammals in particular (4).

Mammal emergence began during the reign of reptiles, which lasted 100 million years and generated a huge diversity of lifestyles and sizes (an abundance only sketchily reflected in the present dinosaur fossil record). This period was also a reign of the plants: not only did plants sustain such diverse, numerous, and sometimes huge animals as dinosaurs, the bodies and branches of trees provided a diversity of habitats. The evolving angiosperm (flowering) plants were more complex in structure and diversity than the earlier conifers (firs and pines). While it is possible that these plants helped create new sets of habitats for a new class of animals, the main diversification of angiosperms was too late to explain mammal or even, perhaps, primate origins (4). Some of these plant biomes would have been closed to reptiles; for example, many plants grow at cooler latitudes or altitudes (5). Lower temperatures (sometimes aggravated by deep shade or night) still pose metabolic, sensory, and physiological challenges, even for today's advanced reptiles, which are, for the most part, adapted to daylight. Their dependence on warmth is most clearly manifested in the declining numbers of reptile species with ascending latitude.

There is something of a consensus that four elements—temperature, light levels, metabolism, and trees (particularly light and temperature)—were relevant for mammal emergence (6). Today, mammals remain predominantly nocturnal and can maintain body temperatures of about 37°C under a range of fluctuating external temperatures, with metabolic rates that can be 25 times that of a reptile. Clearly, such internally generated heating systems are costly in fuel (7) and must have taken a long time to develop. However, the earliest mammals had mammal-like reptile ancestors that lived more than 100 million years earlier—a long research and development phase, by any standard!

If the niches avoided by reptiles have always tended to cluster around cooler parts of the world with cooler nights, cooler seasons, higher altitudes, and shadier habitats, it is most likely that these were the conditions that would have most favored the emergence of mammals. Before 160 mya, the terrestrial world consisted of a single supercontinent, Pangaea (figure 2.3), so latitude, altitude, and ecology rather than continental isolation would have been the relevant parameters, providing distinct ecological opportunities for the incipient mammals.

For further clues to where mammals might have found niches in a world of reptiles, we need to reexamine some of the well-known differences between living mammals and living reptiles. Mammalogists agree on a general description or diagnosis that includes the following generalities. Mammals are animals whose mothers have mammary glands. Except for monotremes, they are born, rather than hatched, and suckling is part of a general trend toward prolonged maternal care. They can process food more efficiently with the help of diverse and specialized teeth set in robust but maneuverable jaws that follow complex, eccentric chewing patterns. This eating technique allows a higher energy metabolism than that of reptiles. Mammals generate, control, and maintain high and relatively stable temperatures that are conserved against cold by fur or hair or, if subject to overheating, dissipated by sweating or panting. Higher

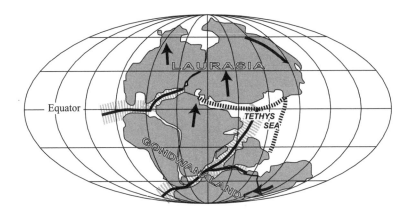

FIGURE 2.3 Approximation of the positions of the continents at the end of the Triassic, 180 mya. (After Deitz, R. S., and J. C. Holden. 1970. The breakup of Pangaea. *Scientific American* offprint 892, 12 pp.)

energy budgets involve the separation of oxygenated from deoxygenated blood within a double circulation. The efficiency of the lungs and breath control is boosted by a diaphragm in the chest, a secondary palate beneath uniquely scrolled bones and tissues in the nasal region. The latter are also associated with sensitivity to chemical signals, further expressed in elaborate systems of scent production and extraordinary olfactory powers (8).

Compared with reptiles and birds, mammals have eyes that are more extraordinary for what they lack than for what they possess. Their nocturnal ancestry is manifest in rudimentary or degraded muscles within the eye, simple lenses, and poorly developed internal reinforcements to control eye shape and keep the eye from going spherical. These are all structures that are very much better developed in birds and some reptiles (9). Hearing and tactile senses, instead, involve specialized structures. The mammalian inner ear has been elaborated from three reptilian jawbones that first declined in size, then detached themselves from the mandible, and finally were captured by the evolving ear capsule on the base of the mammalian cranium, where they were remodeled into functional components of the inner ear. To serve these refined senses, neural control and general coordination have been enhanced by enlargement and extension of the brain and neural networks. Maintenance of all these systems, including the brain, are dependent on a stable and fast turnover of energy in mammals (10).

This combination of strengths and weaknesses makes mammals able to outcompete reptiles in many (but not all) contemporary habitats.

When dinosaurs were about, it would seem that these diverse and abundant reptiles effectively put a cap on mammals getting very large or even very common.

The properties of mammals, especially the very small, early ones, may seem rather distant from the evolution of a bipedal ape. Nevertheless, an adjustable, efficient metabolism is a necessary prelude to the development of advanced mammal limbs. An energetic lifestyle, particularly an agile, arboreal one, demands flexible leverage, an efficient grasp, and speedy reaction times. To this end, mammals, most especially primates, have realigned limb joints and articulations and have developed ingenious swivel points in the limbs and neck and a firm pelvic girdle. Some specialization and refinement in the detailed architecture of these bony joints and junctions, especially in the region of the pelvis, not only distinguished mammals from their reptilian predecessors but accompanied the differentiation of primates from other mammals, of apes from monkeys and of hominins from apes.

Consider now the broader time frames of mammal beginnings. The mammal-like reptiles had generated at least ten recognizable lineages during the Permian and Triassic (290–210 mya), many of which are represented by well-preserved and diverse fossils. Mammals are thought to have arisen, toward the end of this period, from a particularly advanced group, the Cynodonts. During the Jurassic and Cretaceous (after 210 mya), early mammals, in turn, branched into at least eight major lineages, of which only placentals, marsupials and egg-laying monotremes survive. Whether the parting between placentals and marsupials had a continental separation at its beginnings remains unknown at present. However, more than one biologist believes that the global pattern of first appearances points to the African portion of Gondwana as the place of origin for placental mammals as a whole (11), although Asia has equally strong claims (12).

The preceding paragraphs have presented a very summary list of mammalian peculiarities and history. But what of primates, the central group of concern? At present, there is a huge temporal gap between the very tentatively dated first placental mammals about 145 to 135 mya (2, 3) and the first appearance of fossil primates that might, in any way, be ancestral to modern anthropoid forms (at about 60 mya at Adrar Mgom in Morocco) (13). If, as I suggested at the beginning of this chapter, primates do indeed conserve many of the most conservative traits of mammals, they could have been among the first lineages to differentiate during the Cretaceous (14). Suggestions of a 90 million-year-old branching between prosimians (properly the suborder Strepsirhini) and simians (properly the

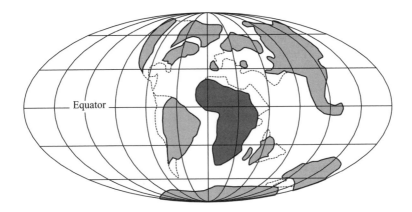

FIGURE 2.4 Reconstruction of plate positions approximately 85 mya, when Africa was show-ing lengthy isolation. (After Hedges, S. B. 2001. Afrotheria: Plate tectonics meets genomics. *Proceedings of the National Academy of Sciences* [USA] 98: 1–2.)

suborder Haplorhini) (1) could imply that the time of emergence of pri-mates and their splitting away from other basal orders could be well in excess of 100 mya. Such an early emergence could have meant that pri-mates existed thereafter in both the Asian and African land masses. The timing of this emergence needs to be determined because Africa suffered some 50-million-year isolation (about 100–50 mya; figure 2.4) during which time an entire superorder, "Afrotheria" (embracing elephants and five other orders), evolved in isolation from the other continents (15). If anthropoid primates are of African origin, they would have suffered a similar period of isolation from other primates.

The physiological and anatomical advantages that were listed for mammals are all possessed by primates, although for some in reduced de-gree. A weaker sense of smell in monkeys, with a corresponding decline in olfactory apparatus (i.e., the nasal portion of a "muzzle") partially ex-plains the general tendency in diurnal primates, especially anthropoids, to shorten the face with a corresponding reduction in the number of front teeth. As a whole, their teeth have tended to remain relatively un-specialized. The more obvious factors that might help explain the differ-entiation of primates from other mammals concern geography, habitat, physiology, and feeding habits. A reasonable working hypothesis is that primates arose from the already arboreal and nocturnal basal therian stock as a peculiar lineage of tiny visually oriented insect eaters, foraging at night, singly, and probably rather slowly, through the fine foliage of tropical rainforests (16, 17, 4). Early primates were unequivocally climbers. By the Eocene (53–34 mya), fossils showed evidence of blunt, flat-nailed fingers with opposable big toes or thumbs. These traits are possessed by all the most conservative living primates: "the last common ancestor of the extant primates, like many extant prosimians . . . subsisted to an important extent on insects and other prey, which were visually located and manually captured in the insect-rich canopy and undergrowth of tropical forests" (18). Some peculiarities of their limb extremities are discussed shortly and, like enlargement of the brain, are a recurrent topic in the pages that follow.

I take up the question of vision and bifocalism later, though it suffices here to remark that realignment of the eyes seems to have been an intrinsic part of the muzzle's shrinkage, an expansion of the brain, and a downward bending of the skull and shortening of the face (19). Less direct evidence supports the idea of a progressive improvement in visual discrimination during the evolution and diversification of primates. For example, some prosimians, like most mammals, are poor at color perception. Diurnal primates, instead, have differentiated the structures of the inner eye, thereby gaining excellent color vision and an enhanced ability to accommodate to bright light (20). In an evolutionary perspective, these improvements essentially redifferentiated and regained properties of vision that were never lost by birds, reptiles, even some fish! Among mammals, primates are better served by their eyes than their noses, but even "improved" simian eyes compare poorly, in structural complexity, with those of birds and even some reptiles (9). Among diurnal primates, rebuilding color/daylight vision is likely to have taken many millions of years, so its acquisition must have long predated the earliest fossils that have been inferred to have daylight vision (about 36 mya) (21).

When it comes to assessing what attributes gave our ancestors such an early edge over other animals, it is clear that the visual superiority of diurnal primates was among the keys to success. Jumping tens of millions of years ahead, it seems likely that good color discrimination was an essential component in the development of a discriminating and analytical intellect in humans.

Other mammalian traits are well exemplified by primates and endorse the qualification of primates as typical, perhaps archetypical, mammals. Of all mammalian peculiarities, those that enhance the quality, rather than the quantity, of offspring are most typically primate. Females have long gestations and long suckling and care periods for only one or two young at a time. Their young mature slowly and are late to breed but live relatively long lives (8). Later, it will be clear that these attributes were crucial for the development of specifically human characteristics, reinforcing the paradoxical assertion that humans are in some respects the most mammalian of mammals (22)!

Turning to the biogeographic context, primates are essentially tropical and subtropical animals, with the greatest diversity in Africa where there are some 85 currently recognized species belonging to 5 families and 23 genera (excluding Madagascar). Although there is a broad resemblance between Eurasian and African primates today, and past exchanges are known to have flowed both ways, the deeper roots of certain groups could well be in tropical Asia, but the origins of anthropoids are more likely to be African (23).

Before the breakup of Pangaea, about 160 mya, fossil species of all sorts, animals and plants, seem to have had very wide distributions. This pattern would be consistent with the absence of serious barriers between populations. Because they subsequently appear in both parts of the former Pangaea (Laurasia in the North, and Gondwana in the South), placental mammals could either have originated before Pangaea was split by the Sea of Tethys (a modern vestige being the Mediterranean), or they could have spread later through chance raftings over a still quite narrow Tethys. As Tethys widened, a very long period of isolation began for the southern continents (which have continued to break up and drift up to the present). The earliest placentals, too, would have suffered similar lengthy isolation but probably remained small and insignificant parts of the fauna until the end of the Cretaceous. With the sudden worldwide extinction of dinosaurs and massive ecological change at about 65 mya, mammals seem to have emerged from obscurity and radiated rapidly, in many cases from already distinct but presumably small-sized ancestors, many of them originating from Africa (24).

The primates must have been one of these burgeoning groups. However, before exploring the primate radiation, we need to return to deferred questions surrounding tree dwelling as a fundamental characteristic of primate or mammal beginnings (and an essential prerequisite for the very much later development of bipedalism).

In 1880, T. H. Huxley (25) initiated a century of speculation on the nature of mammal origins by pointing out the abundant evidence for arboreal origins for all known marsupials. Mathew (26) thought all placentals also had arboreal origins, and Lewis (27) concluded that the built-in versatility of early placental hands and feet implied movement between trees and the ground. These speculations suggest that the ancestral stock of both marsupials and placentals was mainly arboreal and nocturnal. If so, my vignette of Queensland possums will serve well enough as a contemporary glimpse of the first therian mammals' nocturnal way of life.

There is another reason why possums are relevant to both mammal and primate beginnings—a reason that allows me to portray an attribute that links humans with a much wider family of animals. The sketch of hands that follows is one of the earliest increments in a composite portrait. As more parts are drawn in, the picture should get more and more recognizably human. Possums and primates have similar hands and feet that are "primitive" in the sense of retaining five digits and a complement of the primitive tetrapod and reptilian palm and wrist bones that were illustrated earlier. This "primitivism" must be qualified by their possession of strong, flexible grasping fingers with opposable "thumbs" or big toes and bare pads on the digits and palms. These pads have important tactile and mechanical functions that are served, in both orders, by complex tissue and nerve structures (28). These physical features are, almost certainly, not just the convergence of similar structures to serve similar ends in different lineages but rather the common retention of one of the very earliest refinements of climbing mammals. The most striking resemblances concern whorled ridges in the naked skin of palms and digits (best known as the impressors of "finger prints" at scenes of crime!) and clusters of nerve fibers that underlie these ridges (29). The latter are known as Meissner's corpuscles, which are the highly sensitive tactile end organs that enable blind people to read Braille. They are restricted to primates and some marsupials and clearly evolved in the ancestors of these tree climbers to enhance and augment other sources of information about their precarious and changeable environment (18). Hypersensitive fingertips are demonstrably useful when their owners live in thorny vegetation, and I have often marveled at the ability of bushbabies to run and jump, pell-mell, through the thorniest acacia bush. When the interface between an animal and its living environment consists of soft, porous skin, special problems arise. Thinly skinned, easily damaged pads (liberally perforated with skin pores) are more susceptible to infection than feet protected by thick, horny, or scaly surfaces. Although this hypothesis has yet to be supported for all digit-padded mammals, it is likely that these animals all benefit from secretions (primarily from eccrine skin glands) that possess antibiotic properties. If this protective action can be shown to operate on the palms of possums and primitive primates, such as bushbabies, we may be better able to understand a unique feature of modern humans, namely the "spread" of eccrine glands from the palms of hands and feet to virtually the entire body surface.

There is no doubt that both fore and rear "grasping organs" are primarily adapted to climbing on stems and branches. The simple property of clinging to a support is commonly known as a prehensile capability, but when it comes to "hands"—as distinct from "feet"—there are extra considerations. Food clasping and feeding techniques clearly shaped the evolution of hands, especially during the very early phase of nocturnal insectivory. Virtually all living primates possess more versatile but weaker hands versus less versatile but stronger feet. Behind these obvious differences is the influence of striking or grabbing (something that is almost exclusively a property of forelimbs) a skill that is particularly dependant on close hand-eye coordination. For herbivorous species, manual skills are less obvious than in predatory primates for whom the need to catch live prey, often with one hand, puts a premium on the speed and accuracy of a hand/arm strike (18). Both omnivorous and more strictly vegetarian primates must owe part of their manual dexterity to their tiny nocturnal insectivorous primate ancestors. This evolutionary history makes it impossible to explain the properties of hands and feet only in terms of direct adaptation to the diets and ecologies of contemporary species.

With such qualifications in mind, it can be seen that soft finger pads, relatively short palms, and long digits provided climbing (or rather "clasping") primates with firmly gripped bases for movement in any direction. The downside of these features would have been restriction to a narrower range of stem diameters because broader, flatter surfaces offer less purchase.

Hence open, large-trunked woodlands would not have favored the earliest primates so much as continuous, moist, twiggy, and liana-hung forest canopies. Woodlands would have suited a different guild of mammals, including some possums, gliders, squirrels, and other arboreal mammals. If bats derive from early mammals that resembled primates, as seems possible, more open, woodland microhabitats could have favored more persistent leaping (and ultimately gliding and powered flight) while more conservative types, including the early primates, kept to the more continuously dense habitats. Most trunk climbers possess sharp claws, but few have truly prehensile hands in that they are unable to habitually secure objects within the firm grasp of a single hand. The absence of true prehensility is evident in the two-fisted feeding of clawed squirrels and other arboreal rodents.

In the great majority of climbing mammals, both extinct and living, legs provide the main motor force for climbing and so are more heavily muscled than the arms; the spread and power of their grip is also greater than in the forelimbs. Both possums and many primates commonly rely on well-clamped feet to cantilever the entire weight of their forequarters out to reach another branch or to stretch out for an item of food.

There is an important distinction to be made here between faster and slower climbers. The latter, like chameleons, rely on cautious deliberate movements and a firm grip on supports. Their preferred routes through the trees tend to involve less exposure because of their reliance on smaller branches and twigs. Typically, they minimize predation by freezing or crouching at the slightest disturbance; more rarely, they brazen things out by being aggressive or producing noxious smells. Slow climbers tend to have exceptionally powerful grips, with relatively long, thick, and sinewy digits. Fast climbers, instead, are more reliant on the momentum of their own bodies; firm, sticky grips can become encumbrances. Their digits tend to be shorter and more lightly built. Living examples of the two modes of locomotion are the African pottos and some of the galagos.

In species that leap or jump, the propulsive power of frog-like, elongated hindlimbs is obvious. But even in species with well-matched limbs and great similarities in hand and foot structure, a clear separation of functions between the fore- and hindlimbs is evident. All four extremities function as clamps, but there is a premium on both strength and ankle flexibility in the rear clamps whereas "exploratory skill" and deft, multiple use distinguish the fore clamps. These differences between forelimbs and hindlimbs are neither trivial nor recent. As I pointed out in chapter 1, propulsion has always been the main function of hindlimbs, while elements of subordination to the head and its senses have remained perceptible, even in the most footlike of forelimbs, as is occasionally demonstrated by elephants in subtle movements of their forelegs.

Watching Australasian possums has convinced me that basic manipulative skills evolved earlier, rather than later, in our mammal ancestry. Most striking of all is the insectivorous striped possum, which, like a piano player, flutters its sensitive fingers over resonant timber—grub-laden deadwood or ant-sheltering bark that echoes its secrets beneath featherlight drumming. Once detected and exposed by rodentlike incisors, the insects are extracted by equally precise and sensitive finger (and tongue) work. Other species of possum show quite different but equally impressive skills. Although such species-specific skills may have been refined relatively recently, the fact that they have been elaborated in several ancient and very different marsupial lineages denotes that all are built on a long established common dexterity. Furthermore, while the relative length, strength, or proportions of digits and the form of nails may readily alter during evolution, the hands also have to establish complex linkups with the brain to serve specialized diets or modes of locomotion. In this respect hands serve all the senses as an animal interacts with the many elements of its environment. It must identify and correctly assess food versus nonfood, friends versus enemies, and viable versus dead-end branches. All such interactions depend on information taken in through eyes, nose, ears, and whiskers. As for the many decisions that follow such scanning, species vary widely in reaction times and the velocity of their movements, but both locomotion and food-getting behaviors have to be implemented by hands that respond efficiently to what the brain and senses dictate.

Watching the abilities of marsupials in hand-brain coordination certainly forced me to revise my primate-centered ideas about the novelty of manipulation in hominin evolution. Connecting nocturnal possums and their capabilities to the earliest mammals and primates has been a necessary detail in my "self-portrait" as a first mammal and primitive primate. When it comes to later, larger primates, manual potentials may well have been more dormant than absent. For example, the clumsiness of ape hands is, perhaps, a mere by-product of enlarged body weight and the extra loads imposed on both limbs and hands. Perhaps the bipedalism of the first hominins released some extremely ancient skills that had become compromised in apes by the crude function of fighting gravity.

So, when I watched possums expertly finger-drumming bark to locate larval burrows or maneuvering witchetty grubs out of holes and into mouths, I suspect that I was witness to some of the most ancient skills that distinguish not just possums or primates but, maybe, archaic mammals as a whole. Touching, gauging, gouging, probing hands and fingers that are so closely coordinated with smelling, seeing, hearing, and tasting that their refinements have as much to do with serving senses and feeding appetites as with clambering through branches. If, as most evolutionary biologists would contend (30, 31), human anatomical history is made up of successive increments, perhaps we need look no further than our hands to discover a legacy that could stretch back some 140 million years (figure 2.5).

The smooth, continuous skin covering the palms of our hands and the soles of our feet may seem rather different from the blobby, spatulate pads on the digits and palms of primitive climbing mammals, but these same pads are perceptible on the hands and feet of small human fetuses, demonstrating that our own sensory surfaces are merely modifications of a very primitive condition. Later, it will be apparent that manual skills played a central part in the emergence of bipedalism. Nonetheless, in the broader context of this book, it is only proper that discussion of the ori-

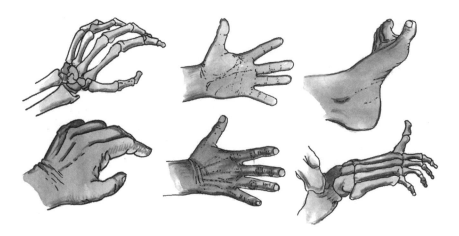

FIGURE 2.5 My five-fingered hand differs only in its proportions from those of many other primates. My foot can still separate the big toe from the other four toes, in spite of being tied into a single structure.

gins and age of the "hand increment" should come close to its start, while big brains and little teeth come toward the end.

In this "history of increments," we still cannot be absolutely certain whether primate dexterity had its first tentative beginnings in the early Jurassic with very primitive climbing mammals or its proven existence with primates of the late Cretaceous; but my bets are on the former. This 100-million-year gap between the possible and known age of clasping hands and feet is a humbling margin of uncertainty. All the more so when we remember how central piano-playing, pencil-pushing, scalpel- or spanner-wielding fingers are to a contemporary human's definition of self, let alone to the dignity these skills confer on those who practice such "manual" professions.

If Huxley, Mathew, and Lewis (25–27) prove to have been right about arboreal origins for all mammals, this conclusion places primates close to the central core of what it is to be a mammal. In this perspective, both primates and arboreal marsupials would be remarkable less for their various specializations than for retaining and refining a common, profoundly conservative lifestyle. When it is remembered that some climbing frogs and reptiles have more than passable "hands," the fundamental question about the first mammals is not *whether* they had hands but rather what *sort* of climbers they were. Hands were a necessary adjunct of being arboreal. The basic model of grasping hands and feet has been reorganized in a limited number of highly derived orders, notably the ungu-

lates, whales, sea cows, seals, and bats, in which five fingers have been reduced to four, three, two, or even one digit, enclosed in flippers or all five
elongated into webbed "hand-wings."

The structure of grasping hands and feet comprises an array of tightly
bound bones, with complex, interlocking, articular surfaces that are attached, at the wrist or ankle joint, to the limb-bone ends. These are
bound, equally tightly, to the four long bones of the palm, or meta-
carpals, and a single divergent thumb metacarpal. Four closely aligned
digits, or fingers, protrude from this palmar complex. Each of these consists of three elongated, slightly curved, and tapered phalanges. A fifth
digit, the thumb, has only two, rather shorter phalanges. The four fingers
vary greatly, from species to species, in their proportions. On their own
they can form relatively weak hooks around stems, but when the digits
curl around, into the palm, stems or twigs are gripped in a double-locked
hook grip that is strong and firm. The thumb comes into play during
more complex manipulations, the strongest being a power grip in which
the thumb opposes, overlaps, or buttresses the hold of the fingers. The
hand can adjust the digits and the angle of attachment, to accommodate
cylinders, spheres, or irregular objects, and exert force to various ends.
The fingers and thumb can also be opposed in various ways to grip,
pluck, pull, or twist small objects with much greater precision or delicacy
but also, necessarily, with less strength. Finally, the fingers can be flut-
tered or rubbed over objects to assess their tactile properties or to stimu-
late reactions. Of course the outer, nonprehensile surfaces of hands can
also be used as sensors or, as blunt or clenched instruments, to push,
punch, or club.

That hands and fingers are hallmarks of mammal beginnings may still
be speculation, but no such doubt surrounds the early primates, partly
because such an extraordinary diversity of primates still survives, all of
which have highly developed and dexterous hands.

Lemurs have long served as models for primitive primates, and there
has been a large measure of agreement that the earliest primates, like
many living lemuroids, were both nocturnal and arboreal. As for size,
many scholars (32, 1) maintain that primates began as energetic twig
climbers—in other words, tiny, mouse-sized animals not unlike pygmy
galagos or mouse-lemurs (29). The claws of numerous climbing animals
(notably squirrels and tree shrews) demonstrate that a small animal's
progress over large branches and trunks is well served by sharp grappling
hooks. By contrast, clinging, spatulate fingers (such as those of climbing
frogs) confirm that soft-tipped fingers, on well-padded hands and feet are

better for clinging to fine stems and moving through twigs and twiglets. Even so, pads on the tips of digits are not incompatible with lightening-fast swats by the spread-eagled fingers of a single hand to catch insects.

The outermost extremities of trees and bushes, whether in the canopy, understorey, or shrub layer, constitute a significant habitat for small foli-vores, frugivores, and predators. The finer branches and their foliage com-prise, by volume, a high proportion of the trees' occupation of space and the greater part of any forest habitat. Fruit and flowers are most often borne here, and the foliage also attracts and sustains great numbers of small invertebrates. This is a difficult but potentially rewarding habitat for any animal that can subsist off small fruit, flowers, buds, or invertebrates. It requires small size, a relatively energetic lifestyle, and the ability to clamber, scurry, crawl (and occasionally leap) through layers of twigs and foliage. The main competitors, as well as predators, are flying animals.

If we seek contemporary models, the night dwellers would have dif-fered but little, in superficials, from the aptly named Malagasy mouse lemurs. Indeed, they would have had much in common too with pygmy possums, which are equally agile in trees and just as nocturnal. As can be partially deduced from the big ears of nocturnal lemurs, hearing is excep-tionally well developed and was likely to have been equally critical in the Palaeocene. Pointed noses and the extensive use of gland and body odors in lemurs and possums betrays that scent is still a very important sense. At night, noses and ears are obviously exceptionally well suited to finding well-hidden prey. Scents are also well suited to marking out small, well-traveled home ranges in dark trees. In confirmation that this is so, most prosimians seem to live in a continuous cloud of scent, feverishly rescent-ing themselves, their neighbors, and their surroundings after every rain-storm and at every new meeting (33). Nonetheless, in relation to other primitive mammals, it is clear that visual superiority was a primate trait and a specialization that was functional, even in poor light.

The prepossession of better vision might have tipped the balance in fa-voring an early adaptation to daylight living in the ancestors of modern advanced primates, whereas the preeminence of other senses might have deterred other mammal groups from becoming diurnal. Early mammals had clearly adapted so completely to very low light levels, and over such a substantial period of time, that structures for managing exposure to full sunlight had become atrophied (9).

The comparative anatomy of vertebrate eyes (with some indirect sup-port from fossil orbits) shows that day-adapted retinal double cones, rapid pupil control, refined lenses, inner eye muscles, and other mecha-nisms for adjusting and refining vision in full daylight had probably been

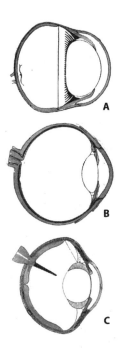

FIGURE 2.6 Cross-sections of the eyes of (A) galago (*Galago* sp.), (B) human, and (C) reptile (lizard). (After Walls, G. L. 1942. *The Vertebrate Eye and Its Adaptive Radiation*. New York: Hafner.)

evolved by the common ancestors of reptiles, birds, and mammals (figure 2.6). Yet these features were in effect severely degraded during the mammals long, long restriction to night living (9). For tree-living mammals to become truly diurnal, some of these degraded faculties would have had to be revived or rebuilt from what had become a night-specific visual apparatus, but it would seem that primates made a very early start in this process so they might have had less atrophy to reverse. Even so, it is unlikely that this was a rapid development, and there is ample evidence that living diurnal primate species still vary quite substantially in visual acuity (20).

Several contemporary primate species in Africa, Madagascar, and Asia still seem to occupy very ancient niches. Indeed, the 45 million-year-old *Tarsius eocaenus* from China is indistinguishable at the generic level from contemporary tarsiers (34). Likewise, a common Madagascan lemur, the Sifaka, *Propithecus verreauxi*, has striking resemblances with a 50 million-year-old fossil, *Smilodectes gracilis* (1). In monkey-free Madagascar (the biogeographer's "Lemuria"), lemurs have evolved into both nocturnal and diurnal species. A similar demarcation by activity period seems to lie at the root of the primates' primary subdivision into prosimians (Strepsirhini) and simians (Haplorhini).

We know that primates had already differentiated into at least three major lineages well before 54 mya. By that time, lemurlike "strepsirhines" were already well differentiated from "haplorhines," while the two constituents of the latter category, anthropoids and tarsiers, were also likely to have been distinct at the time of the dinosaur extinction. Today, the tarsiers are true "living fossils," a relict group confined to tropical Asia but once present in Europe and North America (as well as Africa, where an Oligocene Egyptian fossil has been named *Afrotarsius*). They are among the most conservative of mammals and are of special interest for illustrating how anthropoid primates might have evolved from the haplorhine prosimians (35). Although they have become secondarily nocturnal, they lack certain night-adaptive prosimian features and share with higher primates the distinction of retaining "daylight-type" eyes (albeit vastly enlarged to capture light). These are encircled by eye sockets that look like nothing so much as paired bony acorn- or egg-cups.

Because tarsiers are already recognizable as secondarily nocturnal animals 45 mya, it is clear that their ancestors and the ancestors of all anthropoid primates must have become diurnal long before that time. If Martin's calculations are of the right order, that event could date to as early as 90 mya. Furthermore, the tarsiers' eye sockets also serve as a medium for the exploration of a landmark in our own evolutionary history because the biology, anatomy, and physiology of this living fossil can be used to extrapolate key diagnostic details of real fossils. These grotesquely huge orbital cups allow us to understand some of the implications contingent on our ancestors' emergence into daylight (figure 2.7).

For a start, the Strepsirhine-Haplorhine split (or lemur-anthropoid-tarsier differentiation) broadly corresponds with a nocturnal-diurnal divide that may be rooted in a very early division within the most ancient fossil primates into Adapids and Omomyids. In spite of being known only from relatively late (Eocene) deposits, the latter can serve as models for the once nocturnal ancestors of anthropoid primates (including tarsiers) because of similarities in their dentition and skull structure (19). Lemurs, instead, seem to be closer to the Adapids (figure 2.8) (1, 14).

Lemurs differ from tarsiers and anthropoids in having eyes that are bounded by mere bony struts and, like many other nocturnal mammals, possess a light-collecting modification of the eye that is known as the "tapetum lucidum" (which reflects back brightly when a torch is shone into the eyes of night-living animals). By contrast, humans, together with monkeys, apes, and tarsiers, lack a tapetum but are the only mammals

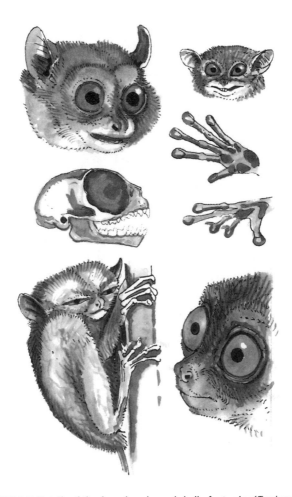

FIGURE 2.7 Details of the face, hands, and skull of a tarsier (*Tarsier* spp.).

with that peculiar dimple in the retina known as the fovea (Latin for "pit") (9). The fovea consists of a bowl of concentrated photoreceptors that are known to increase the resolution of images that are transmitted to the brain (the fovea's structure may also serve to enlarge those images) (19). As a significant enhancer of visual acuity, the fovea is shared with most birds and reptiles (and even some predatory fish), and its proper performance is dependant on a stable eyeball sitting within its own bony socket. All Haplorhines have evolved just such eye sockets by expanding three strut bones surrounding the eye.

This peculiarity—bony walls or septa that separate eyeballs from the jaws and chewing muscles—is the anatomical solution to a problem that

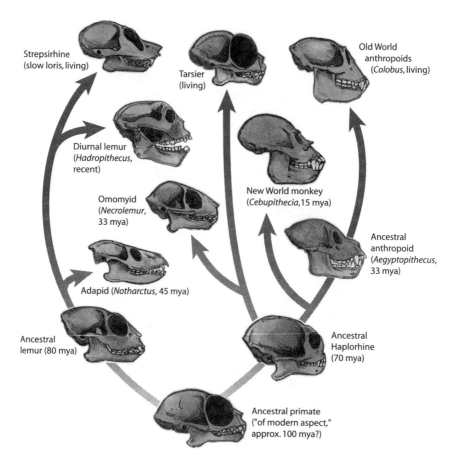

FIGURE 2.8 Outline of primate phylogeny, with tentative date lines, showing five major groups of living species and six extinct groups. (After Martin, R. D. 1990. *Primate Origins and Evolution. A Phylogenetic Reconstruction.* Princeton, NJ: Princeton University Press; and Ross, C. F. 1996. Adaptive explanation for the origins of the Anthropoidea [Primates]. *American Journal of Primatology* 40: 205–230.)

was probably aggravated by changing angles of mastication as the protodiurnal primate skull bent its axis, reduced its olfactory apparatus, and shortened its jaws. With less need to keep a good lookout while feeding, the main problem for most prosimians and all their earliest ancestors was to see in nearly total darkness. The tapetum, being a light-gathering device, is not dependent on eyes being stable, so there is no need for eyes to be isolated in their own self-contained compartments. In these circumstances, the primitive mammalian ring of bony struts and soft tissues that encircle lemuroid eyes have been under no selective pressure to change.

Daylight, on the other hand, would have exposed early primates to

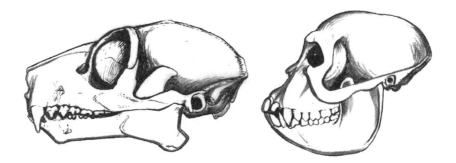

FIGURE 2.9 Skulls of a lemuroid bush-baby and mangabey monkey. Simple rings surround the bush-baby's eyes, whereas the mangabey's eye sockets are typical of all anthropoids and tarsiers in enclosing the eyes in bony cups. (From Kingdon, J. 1971–1982. *East African Mammals. An Atlas of Evolution in Africa* [3 vols., 7 parts]. London: Academic Press.)

many more visually based predators (both reptiles and birds), and these dangers would have been augmented by competition from animals with superior eyesight. The development (or, more likely, the *re*development) of a fovea in a stable eyeball was probably the primary selective pressure to develop a wall between the eyeball and chewing muscles (figure 2.9). In terms of visual function in the biology of primitive primates moving into daylight, there would have been intense selection for them to become simultaneous lookers and feeders (16). Later it will be apparent how the ability to scan and feed at the same time must have been crucial for the vulnerable phases of evolution that anticipated and accompanied bipedalism.

For nocturnal primates that were to become diurnal, changes in the structure of the eye and its sockets were technical details. The threatening challenge of predators, particularly sharp-eyed hawks and eagles, would have been particularly acute for solitary foragers. Having the benefit of more eyes to keep a lookout would have been a prime incentive to becoming more social; at the very least, there would have been obvious advantages to moving about in family groups. When social mammal species are compared with solitary relatives, the former show enlargement in the size of the brain's neocortex (a development that probably involves larger amounts and a qualitatively more complex storage of information, more social memory (36). A corresponding difference in brain size is evident when nocturnal prosimians are compared with diurnal anthropoids. Furthermore, an enlarged neocortex, in the view of some authorities, helps to explain greater frontality for the eyes because expansion at the front of the brain would have tended to push the orbits into a more closely aligned and vertical position (19).

Thus, the invasion of primates into daylit niches can be predicted to have involved new social habits and behavior. But what of the beginnings of this shift? What of the incentives?

It might be thought that the prior existence of birds and reptiles would have been sufficient disincentive to becoming diurnal, yet some birds, bats, and reptiles were probably already nocturnal, already sources of significant competition or predation, even at night. Yet three factors could have been influential for the night-day crossover. First, night primates had had many millions of years to become proficient foragers for invertebrate prey that typically would have belonged to such types as were vulnerable to certain arboreal mammalian traits. These assets might have included the primates' relatively stealthy movements (followed by fast strikes); thorough visual, auditory, and olfactory inspections; and the predators' peculiarly good coordination between fingers, eyes, and specialized teeth. Second, primates would have been at their best in settings that were particularly difficult for birds—say, dense, thorny vegetation in which two-legged, flyers were less maneuverable. Third, and perhaps most important, nearly all predatory types tend to specialize in peculiar methods and techniques to exploit the particular vulnerabilities of their preferred prey. Wherever that prey type goes, their predator will follow, and that pursuit might just as well take prey and predator across the day-night divide as across a boundary between vegetation types.

Thus the beginnings of this ecological, anatomical, and behavioral revolution might have started with a climatic shift or other changes that favored an extension in the activity of the primates' preferred prey. Any change in climate or habitat that expanded opportunities for a specialized invertebrate prey would have been tracked by their well-established primate predators. Alternatively, some small advance in primate hunting techniques might have allowed them to expand their menu or extend their foraging period. Night-dwelling animals could have begun by becoming active earlier in the evening or by prolonging activity into the morning, or both. Becoming active by day during seasons or eons of cool, overcast weather was a likely by-product of climate change, and a step-by-step expansion of the activity period might also have mitigated the initial difficulties of coping with superior contestants.

The fossil record is consistent with the idea that the mass extinction of dinosaurs 63 mya would have opened up many new niches for diurnal species. It is also generally acknowledged that large, diurnal ground-dwelling mammals could have come into their own only after the demise of the dinosaurs. However, the night-day switch among primates almost certainly concerned minuscule animals and so must have been broadly independent of the giant animal theater. The supposition that primates had a substantial start on other mammals in becoming diurnal is given some credence by the fact that most other day-living mammals have poor color discrimination and tend to be more generally myopic than is good for them. It is significant that the main other group of mammals with proven color vision and good acuity—the squirrels—is small and arboreal. Other rodents are thought to see in something approaching monochrome or are generally poor sighted. Furthermore, an early start for squirrel diurnalism is implied by their being, in many other respects, among the more conservative of rodents.

Wherever and whenever it happened, one or more primate species eventually adapted to daylight, a switch in activity that has been replayed, much later, by several lemurs in Madagascar. In that isolated realm, invasion of the day has involved rather modest changes in the body build and behavior of surviving species of lemurs. But the living forms have already been greatly depleted, and some recently extinct diurnal lemurs were much better models for the process of turning a lemur into a monkey than today's survivors. The skull of one species in particular, *Hadropithecus*, was so astonishingly monkeylike that a leading biologist, at the time of its discovery, actually thought it was a unique type of Malagasy monkey (37).

Useful as such mimics are to illustrate the late, provincial island replay of an older pageant, we must turn to broader continental stages to see what wholesale transformations were involved in becoming diurnal.

Central to our understanding of the geography and anatomy of primate beginnings has been the excavation, by Elwyn Simons, of fossil beds rich in primates, in the Fayum depression southwest of Cairo. Any comparison of these fossils must take account of the miniature scale of the players. Even as late as 36 mya, the earliest known primate fossils are still relatively small. Thus, reconstructed weights of Fayum anthropoid primates are as follows (19):

300 g (*Qatrania wingi*)
600g (*Apidium moustafai*);
600–900g (*Catopithecus browni*)
700–1,300g (*Oligopithecus* spp.)

This diversity of genera and families illustrates the success of anthropoid primates in Africa and supports the supposition that they evolved there (38). Although a slightly earlier fossil, *Eosimias*, is known from China, its claim to anthropoid status is contested, and it postdates the Eocene land bridge between Africa and Eurasia that allowed faunal exchanges to take place in both directions.

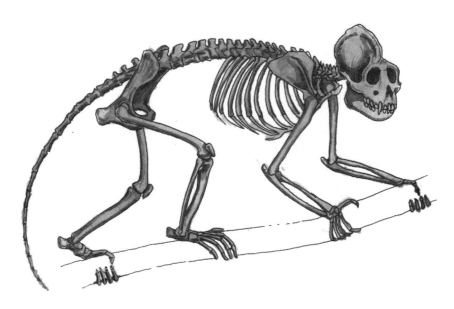

FIGURE 2.10 Skeleton of *Aegyptopithecus*. (Courtesy of E. Simons.)

Skulls from the Fayum Depression, in Egypt, are of Eocene (36 mya) and Oligocene ages (31 mya). Weathering out on the northernmost margins of the subtropics, they are, for the present, the best sample of what might have been going on in the evolutionary furnace of equatorial Africa. The earlier species, sometimes described as "basal anthropoids" are best exemplified by *Catopithecus browni*, a tiny monkey that was probably already diurnal and, of contemporary equivalents, shows most resemblance with South American marmosets (39). The sexes in this species were markedly different in size. They also showed the "postorbital closure" of the space between eyes and cheeks with a bony wall that is a diagnostic peculiarity of all simians. *Catopithecus* had forward-facing orbits that had converged sufficiently to allow good bifocal vision. There was also sufficient brain enlargement to cause some upward extension of the braincase behind and above the orbits. Perhaps as a direct result of this reorientation of the eyes, the muzzle was also displaced or tilted downward, bending the axis of the skull's base. These alterations in the architecture of the skull effectively built a "monkeylike" head from the older, point-muzzled, "lemurlike" one.

For an illustration of the next phase of anthropoid evolution, the 31 million-year-old "Egyptian monkey," *Aegyptopithecus zeuxis* (40), also from Fayum, still has hints of its prosimian origins but was unequivocally diurnal. It combines resemblances with both Old World and New World

FIGURE 2.11 Reconstruction of *Aegyptopithecus*.

higher primates (as illustrations of skull, skeleton, and reconstruction demonstrate). More recently discovered fossil monkeys from southern Africa also reinforce the supposition that the original stock from which all South American monkeys are descended must have rafted across a very narrow Atlantic from Africa well before the Oligocene. Furthermore, African locations for these higher primates imply that the parental stock for anthropoid apes and, ultimately, hominins was more likely to have been African rather than Asian.

Aegyptopithecus (figures 2.10 and 2.11) provides a nice illustration of higher primate evolution. Although there are features unlike those of any living primate, the shape of some bony structures in its skull and proportions of its skeleton anticipate or, perhaps, parallel those of protoapes (commonly called "dental apes" because of their blunt, unspecialized teeth) that flourished in East Africa during the Miocene, more than 10 million years later. Today, we know that conservative species can survive in marginal enclaves long after more successful descendants or cousins have come to dominate mainstream habitats. It is therefore possible that the Egyptian monkey might have been a local Egyptian offshoot or survivor of a stock so old that it came from a point close to the ancestry of all surviving simian primates. In any event, it is one of the earliest and best preserved of Old World primates and illustrates a level of organization that typifies a common ape-monkey stem.

Were we able to see evolution speeded up, the shrinkage of muzzles that transformed prosimian faces into simian ones would probably seem quite comical. If this wholesale suppression of the snout is caricatured as a sort of sink-hole in the center of the skull, it may be easier to visualize how some adjacent features became remodeled, almost as by-product effects. To imagine hard bones as a sort of pliable clay conjured by evolutionary forces is not easy, yet we know, from watching children grow up, bone is far from rigid and inflexible; relative proportions swell and shrink during periods of growth (41). In the evolutionary workshop, conjuring seems constrained only by what already exists. Existent genotypes must always provide the raw material for change, but limits to the swelling and shrinkage of any feature, proportion, limb, or faculty can have surprisingly generous margins of latitude.

How animal bodies or body parts arrive at their particular shape is best understood by identifying functional zones or "activity compartments" such as brain cases, eye sockets, ear capsules, abdomens, or limb levers. These zones or compartments occupy spaces that directly reflect the relative importance of the activity (i.e., thinking, seeing, hearing, digesting, or running). Each activity or function must also make a whole slew of accommodations with its neighboring activities and with the neural, circulatory, and structural components that bridge or tie disparate parts into the wholes that we recognize as "head," "chest," or "hindquarters." They must also respond to physical forces such as gravity or temperature.

When circumstances change for a particular species, its accommodations are not just behavioral. The decline of one faculty (such as scent) is often compensated for by another (such as sight); likewise, enlarged molar teeth demand corresponding realignment and enlargement of the jaws and muscles that grind them. The bone on which such muscles are anchored also reacts, and that infinitely malleable whole, the skull, alters its shape and proportions with every such change.

It is not just heads that change. All animal forms, in all their diversity, reflect the accumulation of hundreds of millions of years of structural accommodation to all the vicissitudes that have been thrown at them. In a speeded-up cartoon of the process, primitive mammal morphs into primitive primate, early monkey morphs into protoape and on to the far-from-wholesale changes that built a bipedal ape.

There are enough fossil and living primates to draw a reasonably accurate cartoon of this remodeling process. (The cartographic technique for demonstrating transformations was first developed by artists, notably Albrecht Durer, and elaborated in a biological context by D'arcy Wentworth Thompson (42) as "cartesian coordinates.") The end products, the artifacts that emerge from this technique, are gridironed faces or profiles that almost look as though they have been punched, bent, and stretched by some brutal sculptor. However, coordinates have the special virtue of highlighting which sensory areas and which functional components (along with their buttresses) enlarge, shrink, or get remodeled.

It is possible to use cartesian coordinates to compare the elongated profile of the Egyptian monkey with that of a prosimian and an early ape. The comparison confirms that the Egyptian monkey has a reduced muzzle but reveals that so much has been sucked away in the ape that the "nasal sinkhole" almost drains empty, causing a sort of "collapse" in the center of the face and a wholesale flexing of the skull. This technique serves to illustrate how the face tilts downward while the eyeballs shift in to face forward (figure 2.12).

The appearance of complete bifocalism in the evolution of ape and monkey eyes is sometimes portrayed as if it were a direct response to defects in the three-dimensionality of the ancestors' vision. Recall that long-muzzled lemurs are bifocal enough to pursue fully functional lives in the trees. Possums and, presumably, innumerable extinct primitive arboreal mammals were too (although most were likely to have been a lot slower in the trees than modern monkeys, partly because checking out sounds and smells is a less instantaneous process than making visually based decisions). The head posture of all these long-nosed mammals depresses the muzzle, while their orbits tilt upward enough to bring varying degrees of overlap to the two fields of vision. The mechanisms assisting realignment of two partially overlapping eyebeams into an effectively single, wholly bifocal eyebeam involve a significant change in the way retinas send their messages to the brain. In mammals with laterally directed retinas, the visual neurones send their output to the opposite side of the brain. Bifocal primates instead have rerouted the left eye's connection from the right side of the brain to the left side; this is clearly an adaptation that allows instantaneous comparison between the output of *both* eyes, a neatly logical solution but one that must have involved quite substantial reorganization of the visual cortex and its neural pathways. In this connection, it is interesting that the partially bifocal cat retains much of the earlier two-channel system but has achieved partial rerouting for the central region of visual overlap (19).

Shrinkage of the olfactory apparatus, especially in the apes, is clearly due to progressive reduction in the importance of the sense of smell. That

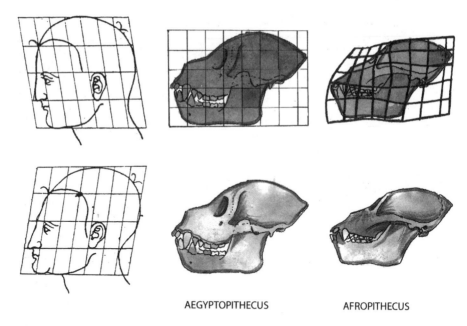

AEGYPTOPITHECUS   AFROPITHECUS

FIGURE 2.12 Left: Two drawings from Durer's sketchbooks (1514) illustrating his use of co-ordinates. Right: The skulls of *Aegyptopithecus* (33 mya), *Afropithecus* (17 mya), *Sivapithecus* (13 mya), *Dryopithecus* (11 mya) and a gorilla (recent) with cartesian coordinates to illus-

this has affected the neighboring orbits is obvious, illustrating a recurrent theme in evolution: that alterations in one realm of activity must have an impact on other activities and structures. In this instance, the reorientation of the orbits must owe as much to changes in the muzzle as to an independent, self-contained, vision-improving migration of eyes.

The interrelationship between apparently separate structures or properties goes beyond skull proportions. When constraints on body size were lifted, as seemed to be the case some time after primates became diurnal, a dietary shift toward larger, often harder fruits demanded deeper jaws with a strongly fused symphysis on the chin and harder, blunter, more resistant teeth. As it happened, flexure of the skull (influenced, in part, by the decline in olfaction) would have helped the jaw muscles find new angles of mechanical advantages. This is not to say that innovations in vision and mastication were merely incidental effects of shrinking muzzles, but the latter process must have prompted change in other parts of the head. Not only that, but shifts in one part of the skull would have actively favored one particular structural solution over its alternatives when it came to accommodating change in separate functional areas.

It is this sort of proliferating feedback from each and every change,

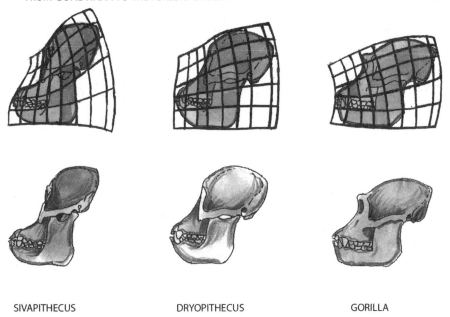

SIVAPITHECUS   DRYOPITHECUS   GORILLA

trate buckling of the skull mainly due to reduction of the olfactory function and more powerful chewing.

however minor, that helps sculpt the specific shape of every species. Major reorderings of senses, such as the rise of vision and decline of the
sense of smell in higher primates, are externally obvious through simple
comparison between different classes of living primates. Further demonstration that the sense of smell in apes and humans has degenerated has
been confirmed by the Human Genome project, which found that more
than half of the 1000 genes involved in olfaction have been effectively
lost in humans (43). Such gross swapovers in primary senses provide useful illustrations for our understanding of evolutionary processes and especially help us to visualize continuous changes in form and function
within complex three-dimensional structures.

While genetics can now confirm a loss of olfactory powers that had already been noted in the anatomy of living and fossil primates, there are
other less obvious evolutionary changes that may be less amenable to genetic analysis. Take the primate-marsupial soles, palms, and fingerpads
that were mentioned earlier. As the literal interface between an arboreal
mammal and its habitat, the skin of hands and feet must have multiple
properties. The olfactory potential of arboreal scent trails is evident in the
ability of scent-oriented lemurs to track one another along branches and

trunks. The effectiveness of such trails is often augmented with urine or deliberate gland rubbings, but highly sensitive mammalian noses can apparently follow an individual animal along nothing more than its points of contact with the environment.

Skin secretions are the main sources of such scent trails, and pads and palms (in common with other areas of skin) are punctuated by scent-producing pores. On the grasping surfaces of primate skin, these microscopically small pores consist of eccrine glands that channel a sweatlike secretion up to the surface. Any olfactory dimension that this secretion might once have had is undetectable by today's apes and humans; yet both, particularly humans, have vastly increased the number of eccrine glands they possess. Why? I am deferring the specifics of this discussion to the profile of modern humans; however, understanding the larger role of eccrine glands in apes and humans is inseparable from exploring the varied functions of skin glands in mammals as a whole. The main sources of scent signals in mammals are specialized skin glands, many of them deriving from the apocrine glands that are attached to hair follicles. Indeed, it was "sweat" similar to that produced by the apocrines that became modified into milk, and it was clusters of such glands that gave rise to the breast, that most fundamental of mammal attributes. We know that breast milk serves more purposes than nutrition alone. The most notable are regulatory or antibiotic properties that both resist infection and help to ensure a healthy passage from ingestion to excretion. Likewise, waxy secretions in the ear and the skin glands—eccrine, sebaceous and lacrimal—include certain regulatory and protective functions in addition to the more obvious ones of scent production, coolant, lubrication, and so forth. Furthermore (as illustrated by their modification into mammae), skin glands can be restructured to perform new functions. With the olfactory channel nearly obliterated in higher primates, we should examine skin glands (as the premier scent-producing organs) for signs of atrophy or rejigging for other purposes.

The apocrine glands, which are the most widely used for scent production, are common and well distibuted over the bodies of most mammals. They are scarcer in African apes, where they are interspersed with eccrines; whereas humans, the least olfactory of all primates, do without them until puberty (when the glands develop, very selectively, around the genitals, nipples, armpits, and ears). By contrast, the eccrine glands that are now so typical of humans are absent, rare, or confined to foot pads in the majority of mammals. It seems that the decline of olfaction in higher primates has led to the replacement of smelly, oil-based apocrine "sweat glands" by salty, water-based, and explicitly antibiotic eccrine "sweat glands."

In addition to this inversion of gland frequencies within the primate order, a third class, the fat-secreting sebaceous glands (like the apocrines) is attached to hair follicles and functions to oil and protect hair shafts. Like the apocrines, they develop in modern humans under the influence of sex hormones: they emerge at puberty and become larger and more active in males. Sebaceous glands are densest in the areas that are most prone to infectious diseases, notably around the anus, the mouth, and the eyes.

While a decline in apocrines corresponds with a decline in olfaction within primates as a whole, an increase in the function of eccrine and sebaceous glands specifically distinguishes humans. I think this difference between ape and human glandular physiology signifies developments that are linked to the late emergence of modern humans. Hence my deferral of further discussion.

It is sufficient here to point out that while scent glands and scent signals have declined, skin glands remain important in the regulation of hormonal balances in sexual and social life and in the maintenance of healthy skins.

We have seen that the Egyptian monkey *Aegyptopithecus* has less muzzle than a lemur but more than an ape. Its teeth, too, are somewhat intermediate, so it can conveniently illustrate evolutionary changes in olfaction and dentition. Less directly, comparing the Egyptian monkey with lemurs and apes suggests that wholesale contraction in a major sensory system is symptomatic of early and radical responses to the challenges of "going diurnal" and "going visual." Furthermore, anatomical changes that involve the greatest perturbation of established axes and masses probably "force" the direction of change on less influential activities. Thus, the sequence in which "increments" of change appear may be important by both limiting and directing further options in the remodeling of other structures. This continuously constrained feedback may find its clearest manifestation in evolving skulls, but it must also operate on less tangible structures, such as glands and behavior. This process, analogous to the sculptural remodeling of an already established statue, informs much of what follows and provides the components for the step-by-step emergent self-portrait that is a major purpose of this book.

During the protracted process of becoming diurnal, primates embarked on a replay of much earlier radiations. They could diverge into bigger/smaller, faster/slower, and so on, each difference signifying an alternative technique of making a living. Many animals, even closely related ones belonging to a single lineage, may adopt opposite approaches to finding food. Where one species or group of species forages widely for easily found items, others become expert gleaners, combing carefully through more restricted but better-known home ranges: expansive versus intensive foraging techniques (22). The energy budgets and lifestyles of such species are typically very different, and it can be predicted that when an adequate fossil sample of early diurnal primates has been found, there will be evidence of such a basis for evolutionary divergence. This is a theme that is taken up in later chapters.

I have already noted that the abundance and variety of species that have been found in a single site in Egypt strengthens the proposition that primates first evolved in Africa. However, the siting of the Fayum deposits raises other important issues for our understanding of the biogeography of primate evolution. Egypt's relative proximity to past Africa-Eurasia land bridges (running between today's Anatolia and the Straits of Hormuz) and its relative distance from the equator indirectly supports an African origin for anthropoid primates because the Fayum fauna is more continental than immigrant. There is too much ecological uniformity within a corridor, particularly if it is open only for a short while, to permit more than a very few (sometimes only one) rather widely successful species to become the founder of a new radiation. Older, indigenous lineages instead display much greater heterogeneity. Egypt's Fayum fauna conforms better with the latter situation.*

The diversity of Fayum monkeys not only supports an African origin for the higher primates, it also reinforces the argument for New World primates deriving from Africa. South America is known to have had monkeys by the late Oligocene, but the Atlantic was already wide by then and the monkeys must have been well established to turn up as fossils in Argentina; rather far south for a primate. The continent's original detachment from Africa, over 100 mya, would have been too early for a diurnal monkey connection. However, given that floating islands of living debris commonly drift down large tropical rivers and out to sea, the rafting of "floating forest islands" over relatively narrow equatorial Atlantic straits

---

*To finally settle primate priority for Africa or tropical Asia will require more fossils as well as more certain dating for connections between Africa and Eurasia. The very earliest could have been close to 60 mya (44), but northern Arabia (then part of the African land mass) was more certainly and more continuously sutured with Eurasian Turkey and Persia between about 44 and 35 mya, to be followed by long periods of isolation (45). Later reconnections between Africa and Eurasia are known at about 18, 15, 10.5, and 6 mya.

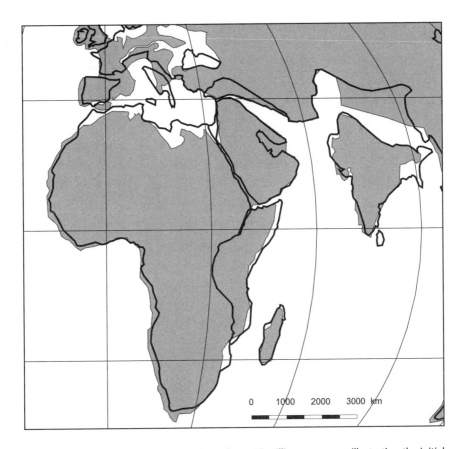

FIGURE 2.13 Position of Eurasia and Africa about 45 million years ago, illustrating the initial suturing of the continents after Africa's long isolation. (From a best-fit amalgam of Dercourt, J., L. Zonenshain, and B. Vrielynck. 1993. *Atlas Tethys Paleoenvironmental Maps*. Paris: Gauthier-Villars; Smith, A. G., D. G. Smith, and B. M. Funnell. 1994. *Atlas of Mesozoic and Cenozoic Coastlines*. Cambridge, UK: Cambridge University Press; and Hempton, M. R. 1998. Constraints on Arabian Plate motion and extensional history of the Red Sea. *Tectonics* 6: 687–705.)

might have been possible up to about 50 mya. An African source for New World monkeys is not only conceivable at about this time, there is no evidence of any sort for a Eurasian connection via North America (figure 2.13) (12).

With larger size, there are further developments of the primate brain, arboreal agility and versatility in the use of limbs, hands and feet; but all these developments are subordinate to that great crossing of the day-night borderline (46). When our ancestors abandoned a furtive existence in the forests of the night, a way of life that had lasted for many tens of millions of years, the curtain rose on a new chapter in primate evolution. By invading the day, they thrust a new mammal presence into the teeming theater of birds, lizards, insects, and millions of flowering, fruiting, sun-fed plants.

Even today there are countless night-to-day successions that mirror, or at least symbolize, that extraordinarily prolific and significant evolutionary event. I remember a particular dawn in Uganda. A blush of pink had begun to suffuse the eastern sky, a delayed fruit-bat winged urgently back to its communal roost, but the red-tail monkey troop that I had risen so early to be with was already on the move. Animals followed one another out of their sleeping tree in the valley bottom and, in an untidy procession, moved up the slope, through a broken canopy. Before the sun was up, they were into the red milkwood trees, stuffing their cheek pouches with sweet orange cherries. By the time parrots swept in with fast, braking swoops and the hornbills arrived, braying, low over the treetops, the red-tails were half ready to go, having creamed off the ripest fruit from the richest clusters. By mid-morning the trees were alive with pigeons and barbets, turacos, still more hornbills, gentle monkeys and mangabeys; but the red-tails were gone. In hastening to make the most of another dawn in their ever-changing forest, I envisioned those monkeys as triumphant modern successors to a procession of long-extinct primates, among them my own direct ancestors. As I lumbered through the undergrowth, I could find some solace in the thought that I, too, once traversed branches high in a dawn-lit canopy, close on the heels of my fellow troop members, all of us drawn by the anticipation of savory cockchafers or sweet cherry pulp. It is only now, with an imagination that is informed and inspired by Darwinian (or Mendelian) insights, that we can treasure and value such moments, snatched from the rich texture of primate life in equatorial Africa. We can connect, today, with some of the most vital and vivacious expressions of life that have ever flourished on this planet: primates, here for some 100 million years, diverse, constantly changing, and source of our own existence.

REFERENCES

1. Martin, R. D. 1990. *Primate Origins and Evolution. A Phylogenetic Reconstruction.* Princeton, NJ: Princeton University Press.
2. Wyss, A. 2001. 195 MY old *Hadrocodium* is a mammaliaform. *Nature* 292(5521): 1496–1497.
3. Goodman, M., A. Romero-Herrera, H. Dene, J. Czelusniac, and R. E. Tashian. 1982. Amino acid sequence evidence on the phylogeny of primates and

other eutherians. In *Macromolecular Sequences in Systematics and Evolutionary Biology*, ed. M. Goodman. New York: Plenum Press.

4. Susman, R. W. 1991. Primate origins and the evolution of angiosperms. *American Journal of Primatology* 23: 209–223.

5. Zimmer, C. 1999. Fossil offers glipse into mammals' past. *Science* 283: 5410.

6. Romer, A. S. 1962. *Vertebrate Paleontology*. Chicago: University of Chicago Press.

7. Gordon, M. S. 1968. *Animal Function: Principles and Adaptations*. New York: Macmillan.

8. Young, J. Z. 1957. *The Life of Mammals*. Oxford: Clarendon Press.

9. Walls, G. L. 1942. *The Vertebrate Eye and its Adaptive Radiation*. New York: Hafner.

10. Hammel, H. T. 1968. Regulation of internal body temperature. *Annual Review of Physiology* 30: 641–710.

11. Storch, G. 1984. Die altertiare Saugetierfauna von Messel—ein palaobiographisches Puzzle. *Naturwissenschaften* 71: 227–233.

12. Hoffstetter, R. 1974. Phylogeny and geographical deployment of the primates. *Journal of Human Evolution* 3: 327–350.

13. Gheerbrant, E., H. Thomas, S. Sen, and Z. Al-Sulaimani. 1995. Nouveau primate Oligopithecinae (Simiiformes) de l'Oligocene inferiur de Taqah, Sultanat d'Oman. *Comptes Rendus Academie Science Paris* (Series IIa) 321: 425–432.

14. Martin, R. D. 1994. Bonanza at Shanghuang. *Nature* 368: 586–587.

15. Hedges, S. B. 2001. Afrotheria: Plate tectonics meets genomics. *Proceedings of the National Academy of Sciences* 98: 1–2X.

16. Cartmill, M. 1970. The orbits of arboreal mammals: A reassessment of the arboreal theory of primate evolution. Ph.D. Thesis, University of Chicago.

17. Cartmill, M. 1992. New views on primate origins. *Evolutionary Anthropology* 3: 105–111.

18. Cartmill, M. 1974. Rethinking primate origins. *Science* 184: 436–443.

19. Ross, C. F. 1996. Adaptive explanation for the origins of the Anthropoidea (Primates). *American Journal of Primatology* 40: 205–230.

20. Jacobs, G. H. 1995. Variations in primate color vision: Mechanisms and utility. *Evolutionary Anthropology* 3: 196–205.

21. Simons, E. L. 1992. Diversity of the early Tertiary anthropoidean radiation in Africa. *Proceedings of the National Academy of Science U.S.A.* 89: 10743–10747.

22. Kingdon, J. 1997. *The Kingdon Field Guide to African Mammals*. London: Academic Press.

23. Simons, E. L. 1995. Egyptian Oligocene primates: A review. *Yearbook of Physical Anthropology*. 38: 199–238.

24. Gheerbrant, E., J. Sudre, and H. Capetta. 1996. A Palaeocene proboscidian from Morocco. *Nature* 383: 68–70.

25. Huxley, T. H. 1880. On the application of the laws of evolution to the arrangement of the Vertebrata, and more particularly of the Mammalia. *Proceedings of the Zoological Society of London*. 1880: 649–661.

26. Mathew, W. D. 1909. The arboreal ancestry of the mammalia. *American Naturalist* 38: 811–818.

27. Lewis, O. J. 1983. The evolutionary emergence and refinement of the mammalian pattern of foot architecture. *Journal of Anatomy* 137: 21–45.

28. Winkelmann, R. K. 1965. Innervation of the skin: notes on a comparison of primate and marsupial nerve endings. In *Biology of the Skin and Hair Growth*, ed. A. G. Lyne and B. F. Short. New York: Elsevier.

29. Cartmill, M. 1979. The volar skin of primates: its frictional characteristics and their functional significance. *American Journal of Physical Anthropology* 50: 49–51.

30. Darwin, C. 1871. *The Descent of Man and Selection in Relation to Sex*. London: John Murray.

31. Dawkins, R. 1982. *The Extended Phenotype*. Oxford: W. H. Freeman.

32. Martin, R. D. 1980. Adaptation and body size in primates. *Z. Morph. Anthrop.* 71: 115–124.

33. Jolly, A. 1966. *Lemur Behaviour: A Madagascan Field Study*. Chicago: University of Chicago Press.

34. Beard, K. C., T. Qi, M. R. Dawson, B. Wang, and C. Li. 1994. A diverse new primate fauna from middle Eocene fissure-fillings in southeastern China. *Nature* 368: 604–609.

35. Cartmill, M., and R. F. Kay. 1978. Craniodental morphology, tarsier affinities, and primate suborders. In *Recent Advances in Primatology*, vol. 3, ed. D. J. Chivers and K. A. Joysey, 205–214. London: Academic Press.

36. Dunbar, R.I.M. 1992. Neocortex size as a constraint on group size in primates. *Journal of Human Evolution* 22: 469–493.

37. Forsyth Major, C. I. 1896. Preliminary notice on fossil monkeys from Madagascar. *Geology Magazine* 3: 433–436.

38. Simons, E. L., and D. T. Rasmussen. 1996. Skull of *Catopithecus browni*, an early Tertiary Catarrhine. *American Journal of Physical Anthropology* 100: 261–292.

39. Simons, E. L. 1995. Skulls and anterior teeth of *Catopithecus* (Primates: Anthropoidea) from the Eocene and Anthropoid origins. *Science* 268: 1885–1888.

40. Simons, E. L. 1967. The earliest apes. *Scientific American* 217(6): 28–35.

41. Huxley, J. 1932. *Problems of Relative Growth*. New York: Dial Press.

42. Thompson D'Arcy, W. 1942. *On Growth and Form*. 2nd ed. Cambridge, UK: Cambridge University Press.

43. Weiss, G., and A. von Haeseler. 1996. Estimating the age of the common ancestor of men from *ZFY* intron. *Science* 272: 1359–1360.

44. Hempton, M. R. 1987. Constraints on Arabian Plate motion and extensional history of the Red Sea. *Tectonics.* 6: 687–705.

45. Dercourt, J., L. Zonenshain, and B. Vrielynck. 1993. *Atlas Tethys Paleoenvironmental Maps*. Paris: Gauthier-Villars.

46. Charles-Dominique, P. 1975. Nocturnality and diurnality. An ecological interpretation of the two modes of life by an analysis of the higher vertebrate fauna in tropical forest ecosystems. In *Phylogeny of the Primates: A Multidisciplinary Approach*, ed. W. P. Luckett and F. S. Szalay, 69–88. New York: Plenum Press.

CHAPTER 3

On Being an Ape

Excursions to Asia and Back

Self-portraits in the faces of apes. Origins of protoapes dated to over 20 mya, the most significant being *Morotopithecus bishopi*. Ape excursions into Eurasia 18 to 20 mya. Development of separate African and Asian ape lineages. Rise of modern monkeys a likely influence in the decline of apes. Emergence of Eurasian tree apes (Dryopithecines) before 12 mya and probability of their emigration into Africa about 10.5 mya. Arid rainshadow through eastern Africa may explain separation of chimp and human ancestors possibly as early as 7.8 mya.

There is a famous cartoon of Charles Darwin, drawn during his lifetime, of his bearded face looking out from the quadrupedal body of an ape (figure 3.1). Intended to mock his claim to apelike ancestors, the image epitomizes Victorian rejection of such an obscenity. Not just an obscenity, but for many a blasphemy; after all, the Italian philosopher Lucilio Vanini was burned alive in 1619 for the sacrilege of suggesting that humans originated from "apes" (1), and the institutions and opinions responsible for his murder were still influential more than 200 years later.

FIGURE 3.1 A caricature of Charles Darwin that appeared in 1871, the year in which *The Descent of Man* was published.

To be fair to the cartoonist and his times, it must have been difficult for a closeted Victorian to visualize any part of our ancestral self-portrait in a chimp or an orangutan. Anyone can see that ape arms, fingers, and ears are almost identical to ours, as are those most private of human parts, the breasts and genitals. Yet these similarities are offset by differences in the proportions of limbs and facial features that have served to reinforce the traditional stereotype of apes as monstrous aliens, not close relatives. What has changed is that genetic literacy has taught us that human genes once inhabited bodies very different from their present dwellings and given us an appetite to translate jaw-breaking pithecine names into comprehensibly sentient ancestors—ancestors we can acknowledge, not reject (2, 3). That knowledge means that our vain psyches and aesthetic traditions must entrust our ancestry not to mythic images of Greek gods or an invisible aerial sculptor, but to the unglamorous physicality of fragmentary fossils and apes as hairy modern analogs.

Apes, politicians, religious leaders, and Darwin are all fair game for the cartoonist's pencil, but the mentality that idealizes humans while demonizing apes will be ill-equipped to find a self-portrait in the face of an ape. Yet see ourselves in that face we must if we are to make any progress in grasping the realities of our own evolution. I see it as essential to stare, unflinching, into the mirror and look, long and hard, into the eyes of apes as it is to document our common history. In any case, I have reached a point in this serial self-portrait where seeking realities in the subtleties of my cousins' expressions has become the necessary prologue to a chapter on ape evolution, dispersal, and diversification.

Seeing myself in an ape face, rather than an alienating cartoon, depends on the way in which I analyze and process what I see. Start with the eyes, ears, and mouth, viewed as receivers and transmitters of messages. Go on to consider what sort of visual impression is made by the uneven distribution of naked and hairy patches. These differ by sex and age, so do they make a different impression on males, females, or the young; and can faces and hair tufts send signals to other species? In this respect there is an excellent model for comparison in the faces of orangutans. In bald-faced baby orangs, each of these features is surrounded by a ring of bright pink skin that helps emphasize every pucker of lips or eyes (skin pigment tends to become more uniform with age). The resulting clarity in these infant expressions is clearly designed to signal to and elicit responses from the mother and, perhaps, a few orang neighbors. Judging from the responses of female zoo-goers, captive orang babies force strong responses from humans as well. This must be evidence for an exceptionally well designed cross-species signal system; or, more simply, ape and human baby faces are sufficiently alike to appeal to any hominin mother.

Self-portraiture becomes rather more problematic with the very hairy adults, but return to the concept of features as receivers and senders of messages. It is obvious that to see and be seen, eyes need to be free of obstruction (ears are less critical). Were these apes to have hair all over their heads, not only would they look like inexpressive red haystacks, their hair would get contaminated by food and their senses inoperable. As seen by a viewer, the naked or seminaked face of an adult stands out as a clearly visible, isolated structure. As males mature, they develop platelike cheek pads that serve to enlarge the signal value of a naked face and transform it into a single circular target (figure 3.2). In spite of its circular design, the orang face breaks down into two subunits: the eyes and the mouth, each expressing emotions by different means. Both sexes grow beards and moustaches the hair of which is differently colored and textured from that of the rest of the body (especially among Sumatran

FIGURE 3.2 Bornean and Sumatran orangutans compared, showing great sexual and individual variation. Above left, a mature Bornean male. Above center, a Sumatran female. Above right, a self-portrait. Of the heads below, the five on the left are Sumatran, the seven on the right, Bornean.

orangs). These adornments are skimpy in females and some males but well developed in most older males. The beard's color, together with a tendency to have the skin around the mouth differently colored or textured, marks out the mouth region as a distinct entity from the eyes.

The facial signaling system of humans has kept many of these apelike traits. My eyes and their immediate setting of naked facial skin scarcely differ from those of an orang except that I have very hairy eyebrows. Even my beard seems to conform to a similar principle, that of enhancing the separation of eye and mouth transmitter systems into visually discrete regions of a singly perceived mass. I suspect that I am only marginally, if at all, more expressive facially than an orang. Then again, I am usually talking when I grimace, and although I am more conscious of my facial expressions when I am with nonhuman animals, much of this self-consciousness disappears whenever I am with fellow humans because my expressions become subordinate to words. By contrast, orang expressions are, in our terms, mute (see figure 3.2), yet these animals clearly read and respond very attentively to their own and our faces, whereas our facial language has been vastly augmented and to some degree overlaid by words. When, for whatever reason, an ape (or any other visually oriented animal) has fully focused its attention on a human face, it is quite clear that very small changes in human expression are registered and responded to. This is most evident in pets and domesticates and was well exemplified in the celebrated exhibition horse "Clever Hans," who created quite a stir in turn-of-the-century Germany by correctly answering numerous verbal questions concerning numbers, musical chords, or the qualities of objects. Hans nodded his head or tapped his hoof in response to almost imperceptible twitches or tensions in the faces of his interrogators whenever the correct answer came up (4, 5). Seeing bureaucrats' or poker-players' faces might persuade one that the acquisition of speech has actually atrophied human expression, but a Meryl Streep or Lawrence Olivier film is a quick antidote to any such notion. The artifact of speech has enlisted an apelike face, and the ape's close attention to facial expressions, to link one of our highest and most distinctive faculties with a truly primitive yet wonderfully subtle facial anatomy.

Although apes and humans are genetically equidistant from their common ancestor, chimps, gorillas, and orangs are the closest and most complete approximation to ancestors that still survive. Yet that recognition has, until quite recently, remained the prerogative of a few scientifically educated people and some remote African tribes. Before such insights can be more widely shared, that most natural of impulses, seeing ourselves as the measure of beauty, "uprightness," and "rightness" (even "righteousness") will have to be inverted. We will have to consciously disown projecting primates as comical copies of people—even, perhaps, develop an apelike sense of humor in seeing ourselves as aberrant, sometimes comic, versions of *them*.

To find something of our "older" selves in apes and to retrace our common emergence and dispersion in and out of Africa, this chapter must return to pick up the genetic thread of ancestry left with the Oligocene monkeys of Fayum. A big problem is that no matter how real that thread must have been, there still remains a gap of 10 million years between the fossils of *Aegyptopithecus* (31 mya) and those of the earliest known "protoapes" or "dental apes" (21 mya). Modern apes, like us, have carried on evolving, and a complete image (let alone a complete skull or skeleton of our common ancestor) remains effectively unknown. Even so, there are sufficient fossils to tell us that it was much more like them than us. The fossils, fragmentary as they are, point to a tropical African beginning and the emergence of at least one form, possibly more, that were so versatile and abundant that they were able to spread out of Africa.

But this is to leap too far ahead; the first task is to try and bridge that 10-million-year gap. My own first acquaintance with the world of Miocene protoapes will serve to illustrate the nature of the evidence and the challenge to our imaginations. It may also suggest how contemporary landscapes and their animal or human inhabitants can be transformed by the eyes of a palaeontological detective, especially when those eyes are sensitive to the drama of countless lives and deaths played out in the vast evolutionary theater of Africa.

The plains of Karamoja, in northeastern Uganda, have swallowed up the debris of several once-gigantic equatorial volcanoes, of which a few resistant fragments remain in the form of oddly shaped mountains or hills. As former ash-falls and streambed bone-traps erode away, they release rare Miocene fossils. These fossils provide precious clues about the fauna and flora that lived and died 20 mya on the fertile, forested sides of what were then periodically active volcanoes.

Clambering over these long-dead screes with pioneering field palaeontologists Bill Bishop and Sonia Cole was to time-travel in visionary company. I remember Bill stopping beside a fire-scorched acacia bush on the screes of an ancient volcanic plug (locally known as Lolim) and gesturing up into a blue sky: "The summit, at about three miles high, was up where that wisp of cloud is; run your eye down to the left: those low hills are all that remains of the western base. But look to the right and you can pick up a lot more of the original mass; that's the original talus contour, preserved in Napak, over there, covered in forest then."

While camping in Karamoja, our search for fossils had none of the character of a grim forensic science; it was recovering events, lives, and ecologies as full of vitality as anything happening today. Holding a toothy bit of jaw on his left palm, Bill's right hand swept over the invisible contours of a chimp-sized skull that had once inhabited a living primate yet, at this moment, existed only in the imagination of a single person (6). It was an imagination that searched not only for more bits of a 20 million-year-old protoape in the cinders of long dead volcanoes but also sought to encompass the environment and evolutionary meaning of extinct hominins in eastern Africa. Big Bill, the geologist, puzzling over what rocks and formations to search next for that meaning. The feeling that we were witnesses to our own evolutionary history began to take hold there, on the gravel of Lolim, but I was impatient to clothe the bones with flesh and life. I found myself trying to match these long-lost animal and plant communities with the fast-disappearing ones that I was studying at that time. I knew that every species, whether its extinction was ancient or contemporary, once had its own unique geography, but I had little conception of how different Miocene environments and their inhabitants must have been. It was difficult to see dry, mainly flat Karamoja as a wet, mountainous region.

Now, many years later, time-traveling imaginations have been led down trails that wind along some very unexpected routes. As apparently disconnected facts accumulate, I've been joined by new generations of puzzlers seeking patterns from numerous new sources of evidence—not only more fossils, but a whole new genetics and new Darwinian insights from the ecological and behavioral sciences (7).

In the midst of all this novelty, the tattered fossil that I first saw lying on Bill Bishop's palm poses some important questions about the course of primate evolution, provoking controversies very far from Karamoja. Later expeditions have found many more pieces of the same individual, more parts of the face, backbone, legs, and shoulder (8). Now named after him (and the provenance of the fossil, Mt. Moroto), *Morotopithecus bishopi*, "Bishop's ape," at 20.6 million years old, is acknowledged as one of the earliest substantial fossils of a protoape found so far. However, its position within a plethora of possible primate trees is contested. Also involved are fundamental questions concerning the very nature of apes, their origins, their habitat preferences, and their history after the Miocene. To learn just how wet or how forested Karamoja was 20 mya is a matter of some significance because, among other things, it has a bearing on whether Bishop's Ape or its close relatives were plausible candidates for emigration *out* of Africa. The corridor between continents, when it opened up (sometime between 20 and 18 mya), was unlikely to have been dense rainforest, so were there apes of that period living in seasonal woodlands?

This question introduces a further, strongly contested debate as to whether human evolution has been a continuously African story or whether there was a Eurasian dimension. There have long been protagonists for Asian rather than African origins for the primates as a whole, but this position is now rather less credible, at least for the anthropoid primates. What *is* undeniable is that orangutans, gibbons, and an assortment of non-African fossil apes are collateral relatives belonging to the same broad lineage as gorillas, chimps, and people. Enter here the "molecular clock" that converts differences in two sets of DNA into the time since their lineages split. This technique dates the divergence of African and Asian great apes (chimp/gorilla versus orang) at about 14 mya (9) and the greater and lesser apes (Asiatic gibbons) at about 18 mya. Such dates (supposing that the clocks have been calibrated correctly: a disputed assumption) force the issue of *where* these bifurcations took place.

There is a limited number of possibilities. The greater-lesser ape split can be confidently located in Asia because true gibbon fossils have never been found in Africa. So was Asia the original ancestral home of all apes? Did the Miocene African fossils derive from still earlier emigrants and modern hominids from later emigrants out of Asia? This is unlikely, because, as we have already seen, Oligocene Egypt was host to numerous and diverse anthropoid primates at a time when connections between the two continents were still relatively tenuous. (Comparable higher primates have yet to be conclusively proven from much more numerous Eurasian fossil beds.)

A second option, that the orangutan lineage differentiated in Africa and its ancestors then migrated into Asia, requires that they left behind the relatives that would later become gorillas, chimps, and people. This option founders on the absence of any traces of oranglike fossils in Africa; they are exclusively Asiatic. Whereas orangs, for fairly obvious reasons, now demonstrate an exceptionally poor ability to disperse beyond tropical rainforests, fossils attest to much more versatile relatives living in drier habitats. If the ancestors of gibbons and orangutans had been able to colonize Asia, could not some of their relatives have retained equivalent mobility and ecological plasticity? This is a rhetorical question, because the Eurasian tree apes, or Dryopithecines, belonged to the same emigrant ape stock and became a diverse and successful group in Eurasia between 14 and 3 mya. They occupied a huge geographic area and ranged into quite northerly, decidedly seasonal regions of Eurasia (10). The closest contemporary analog for these Miocene ecological communities is Indian subtropical woodlands, where winter dry seasons are followed by warm summer rains.

The third, most realistic model of ape dispersal acknowledges that the hominid family tree must accommodate the orang as well as movements between Eurasia and Africa. It would seem that at least one ancestral ape left Africa before 18 mya. (Remember that, at that time, the early "dental" or protoapes of Africa filled many, if not most, of the niches that today are typically occupied by the more recently evolved monkeys, in a climate that was more extensively benign.) For this immigrant stock, Eurasia would, on current evidence, have been virgin territory, apparently devoid of any competing anthropoid primates. Apes therefore found themselves, for a while, in a very large primate vacuum. They became very successful and evolved a wide variety of species (including a giant form, bigger than a gorilla, as well as little gibbons). In the middle of this spectrum of sizes were the smallish, versatile Dryopithecine tree apes that, like their colonist ancestor, could live in drier, more seasonal forests and woodlands. In these northerly European latitudes, primate diversity was clearly lower than in tropical Asia and Africa, and it was rare for more than one species of tree ape to occur at any one time (11). Proponents of the third option think that it was just one of these highly adaptable *Dryopithecus* species that found its way back to Africa and that its modern descendants are people, chimpanzees, and gorillas (9).

It is in this controversy that Bishop's ape plays a role: it is more like modern apes than any other early Miocene protoape. In this it could be seen to support African ape continuity (12); yet considering how very early on it lived, it may just as well anticipate the sort of ape that made it to Eurasia (perhaps a million or so years after Bishop's specimen died on Mt. Moroto).

Bishop's ape could be held to represent an interesting example of continuity in anatomy and lifestyle within Africa. Nonetheless, Eurasian apes are actually more plausible "main-liners" than the numerous fossil protoapes that are known from African Miocene deposits. I therefore favor the excursion to Eurasia because it seems to conform best with a very broad range of biological and biogeographic patterns and parameters (13). (Even so, many of my proposals for and portrayals of an incremental progression could still be reconciled with a wholly indigenous ancestry for the surviving African apes and humans.)

At the time of its discovery, the *Morotopithecus* jaw was allocated to *Proconsul*, an abundant and well known type of early Miocene primate that came in several sizes; the new teeth were similar but quite the largest known. With the discovery of more body parts, this allocation has had to

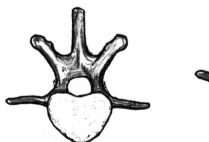
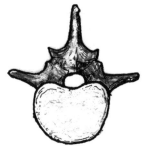

FIGURE 3.3 Lumbar vertebrae of monkeylike "Proconsul," on the left, compared with the more apelike vertebra of the Bishop's ape, *Morotopithecus*. (After Ward, C. V. 1997. Function and phylogeny in hominoid trunk hindlimb evolution. *American Journal of Physical Anthropology* [Suppl. 16]: 176–204.)

be revised. The most immediate difference concerns its lumbar vertebrae, which resemble those of a modern ape in being short and without drawn-out flanges or processes (figure 3.3). *Proconsul* species, on the other hand, have long-bodied, more complex vertebrae that more closely resemble those of modern monkeys such as colobuses (14). Among the implications of long and short backs is that of the two "dental apes": *Proconsul* may have retained some of the characteristics that now typify monkeys, whereas *Morotopithecus* shows more features of a true ape.

It is still a difficult proposition to accept that "monkeys" have evolved from any sort of an "ape." Such a possibility turns things on their head: traditionally, apes are supposed to have evolved from monkeys! Even the name *Proconsul* is loaded with assumptions; Consul was the name given to a London Zoo chimpanzee, so the first fossil name was explicit in describing a "primitive ape." Yet the fossil record is unambiguous in exhuming modern monkey remains from more recent geological formations—a finding that is in close accord with molecular data, which also suggest a relatively recent monkey radiation (although their root stock may be over 20 million years old). *Proconsul* is not on a direct line of ancestry to modern monkeys, but there is now general agreement that it is well off the direct line of hominid descent (figure 3.4) (9).

This inversion of older ideas hinges on the observation that diurnal primates may have diverged into "expansive foragers" and "intensive gleaners." The former could have ranged over a wide area and moved through the branches with a fast and easy gait. The latter, by contrast, made slow, careful movements, using their strong, clamplike hands and feet on even more dextrous joints. The distinction may be crucial for our understanding of ape emergence and hence the birth of our own lineage.

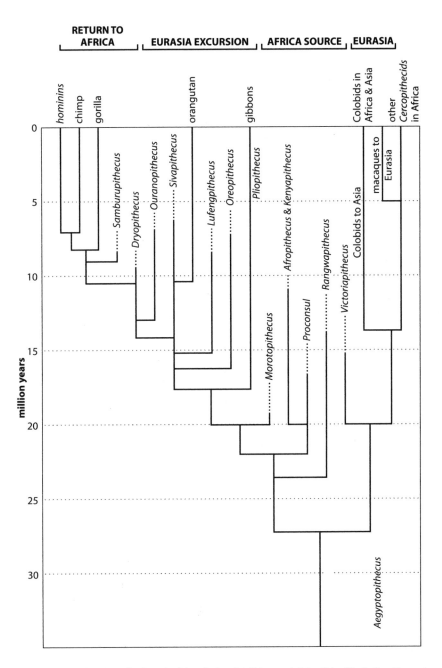

FIGURE 3.4 Phyletic tree for hominoid radiation in Africa and Asia. (Modified after Stewart, C.-B., and T. Disotell. 1998. Eurasian ape as great ape ancestor. *Current Biology* Jul/Aug 1998.)

The early Miocene primates are often referred to as "dental apes" because their teeth retained blunt primitive cusps while other features were as much monkey as ape. The elongated dorsal vertebrae of *Proconsul* betray monkeylike, flexible backs, and their narrow shoulder blades flanking deep chests imply a monkeylike, "arboreal quadruped" gait. While this gait could be taken to resemble those of many other running animals in aligning the limbs and shoulders for long fore-aft strides by both legs and arms, *Proconsul* joints show an ability to swivel and rotate, a combination that is not found in living African higher primates. Unlike modern monkeys, proto-apes were not built to be agile runners or leapers. Furthermore, at least one species of *Proconsul* is known to have been tailless, and tails are of demonstrable utility when it comes to running and leaping through the canopy.

At issue is the whole nature of the original, very ancient divergence between apes and monkeys. Was it led by differences in locomotion? Did body size play a role? Bishop's ape could provide vital clues here. For example, the shoulder suggests an accomplished climber, albeit a heavyweight one (confirmed by very thick thigh bones, also typical of modern apes). More of the skeleton might confirm whether a more conical ribcage was already developed before the Eurasian excursion.

There are still rather few leads on the topic, but one of the most obvious points of difference between most monkeys and all apes is the presence or lack of a tail. Because some monkeys, notably terrestrial ones, have very short or vestigial tails and others show signs of secondary lengthening, there has been a general assumption that tail lengths are trivial and too easily modified to be treated as a phylogenetic trait. If a mainline, true ape could be shown to have a tail, that assumption might be justified, but I suspect that the absolute loss of a tail could be linked not only with the divergence between apes and monkeys but also to related questions concerning fore- and hindlimb dominance. Effectively, changes in tail length may be a relatively trivial variation among actively evolving monkeys (particularly among species that are habitually terrestrial). However, irreversible taillessness in apes might have been a diagnostic detail of their very earliest evolution as well as becoming an ape-specific limitation in their later history. In confirmation of this irreversibility is the apes' specialization of a curled-up vestigial tail, known as the coccyx, which has been amalgamated into a wholly internal structure. Because tails are of demonstrable utility to most tree dwellers and the surviving apes are all to some degree arboreal, why lose them?

Long tails do not just signify "balancing organs"; rather, they imply quite distinctive ways of interacting with trees, branches, the ground, and other members of the same species and hence with the resources of the forest. It is true that tails provide better balance, not just during arboreal running but especially during leaps or steered drops between distant branches, a life-saving faculty in forests with broken canopies or gaps of any sort. Tails are, by the same token, linked with faster movements and the limbs' precarious, point-to-point momentary contact with the branches or ground that they run over. The advantages of fast movement are numerous: they facilitate traversing gaps, escaping predators and rivals, encompassing a larger home range, and moving rapidly through the trees from one patch of food to another. These benefits seem to greatly outweigh disadvantages, so what ecological context could negate so many advantages?

Given that tails have some sort of ultimate precedence, why should they have atrophied in the first place. Even more puzzling, why, if "apes" are, in any sense, monkey ancestors, should one or more branches of these descendants develop an almost infinite variety of tail forms and lengths? The answer to the first question seems to be that they may (like birds' wings on small, predator-free islands) be physiologically expensive to maintain in the absence of any compelling reason to retain them. Excluding exposure to extreme cold, there are at least two plausible explanations for an entire lineage suffering atrophy of the tail. The choice hinges on the body size of the parental stock. If the ancestors were terrestrial, they were more likely to have been large animals. This has been the preferred interpretation for many years and, without the need for balance, is generally cited as the primary reason for ground dwellers dispensing with tails; yet many terrestrial mammals still find tails sufficiently useful not to lose them.

An ancestral stock that was arboreal was more likely to have been small in body size. If comparisons between small, arboreal tailed lemurs and equally small near-tailless pottos and lorises are a guide, redundancy is most likely to be correlated with slower, more cryptic movement through consistently dense vegetation. Not all fruits are easily found or harvested, especially the smaller, more scattered types of berries, and many nutritious invertebrates, such as scale insects, are abundant but minuscule or very well concealed. Living off such resources, as pottos and lorises do, seems to confine such inconspicuous species to foraging using a slow, systematic technique, to being small, and to living in restricted home ranges in evergreen and consistently productive forests. These animals are "intensive gleaners" rather than "expansive foragers." Today, most tailless or nearly tailless primates are small and nocturnal, but there are some, a few lemurs and New World monkeys, that are both diurnal and moderately large.

A choice between small and large body sizes in the ancestry of apes need not be entirely mutually exclusive, in that the primary (and total) loss of a tail could have occurred during a lengthy period as small arboreal animals. A subsequent increase in body size could have preceded or followed more terrestrial habits.

Gibbons provide an interesting perspective on the problems posed by irreversible taillessness. Note first that they feed on fruit, buds, leaves, and animal matter, mostly garnered from the extremities of outlying branches in dense evergreen tropical rainforests. Instead of a slow, cautious approach, gibbons arrive at this precariously positioned resource by a unique form of arm-swinging "suspended ricochets" that rivals branch-running for speed and efficiency (15).

Apart from birds, the ancestors of gibbons would have faced their first direct competitive challenge from monkeys, newly arrived out of Africa, perhaps some 10 mya. In the face of faster competitors with more specialized teeth and digestions, there were few options open to an irrevocably tailless primate adapted to slow, intensive foraging. Yet gibbon species exist today within a similar range of (medium) sizes to those of the many highly competitive monkey species with which they coexist. This specialized mode of locomotion has involved very extreme modifications in the anatomy of their limbs and the acquisition of a gait that operates only in bursts. Momentarily faster than the arboreal running of monkeys, and probably several orders of magnitude faster than anything achieved by ancestral apes, gibbon arm-swinging seems to have been part of their solution to mounting competition.

Another route to outcompeting monkeys was to enlarge beyond the point at which "optimal size" and fast movement offer decisive advantage. Large size can ensure priority of access and a greater measure of security from predators. Less expenditure of energy permits a less selective diet, which in turn allows bulk feeding. One obvious cost is restriction to thicker supports or more frequent travel on the ground, so it is interesting that apes seem to have made an earlier start in combining tree and ground dwelling than other primate lineages. This initiative is so fraught with risk that it may well have been favored by the possession of better brains. Here again, *Proconsul* may mark an early start for this long-term ape specialization with an estimated brain-body ratio that exceeded today's average for living simian primates (16). In sum, rotatory joints and large brains in *Proconsul* suggest that slower, more calculated foraging strategies among ancestral apes could explain the initial divergence between apes and monkeys. There is some controversy as to whether all species were tailless, but most species seem to have resembled modern apes in this respect.

With living Asian apes (and less certainly with Bishop's ape), there is a demonstrable shift away from hindlimb to forelimb dominance that is part of an extreme specialization for suspensory locomotion. As I emphasized earlier, this is no trivial change in emphasis. It concerns a deeply entrenched protocol in how animals move and interact with their world. Furthermore, the absence of any gibbonlike primates, fossil or living, outside Eurasia probably reflects much more severe limitations on innovatory evolution within primate-crowded Africa, where competition from monkeys seems to have progressively narrowed niches, especially for apes. Very narrow corridors between the continents must also have discouraged the traffic of extreme specialists both ways and certainly helps explain the inability of late-evolved gibbons to colonize Africa.

The anatomy of Bishop's ape, even incompletely known, suggests that the first migrant(s) out of Africa may have been less specialized than any of today's apes. Lumbar differences are now a diagnostic feature that separates short-backed apes from long-backed monkeys. The functionality of a short back can be interpreted in various ways. It may have no direct link with the postulated slow clamping movements of the very earliest "apes," but there are obvious constraints when the body becomes large and heavy. Thus, leverage (especially during suspension) is best exerted on a short, relatively stiff body rather than on a long, floppy one. If loss of flexibility in the thorax was linked with "suspensory brachiation" (movement hanging *beneath* branches, rather than progressing *above* them), then the first expression of this mode of locomotion in a fossil may lie in the bones of Bishop's ape. Today, a compacted trunk can be functionally linked with enhanced mobility in the shoulders (figure 3.5), and that mobility became a vital characteristic of the pre-bipedal lineage that led to humans. It remains an open question just how advanced shoulder mobility or compaction of the torso was in African Miocene protoapes. Once again, more bits of Bishop's ape could help resolve these questions.

The Eurasian excursion is also consistent with the dynamics of competition among primates on both continental masses. In Africa, apes of any sort must face formidable competitors in the many and varied descendants of "Victoria monkeys." These first appear in the fossil record between about 20 and 15 mya. They are the immediate ancestors of cercopithecoid monkeys, baboons, guenons, and colobuses. Many of their descendants are highly successful today, mainly because each species has

FIGURE 3.5 Left: Chimpanzees climbing slender hanging vine. Right: Swivel joints and leverage of a compact body in orangutans.

evolved a specialized technique for extracting a living. They have penetrated and are major mammals in almost every one of Africa's varied habitats.

Not only do these animals compete with apes, they define one another's roles and niches in very precise and interesting ways. For example, among a community of 12 higher primates in Bwamba, western Uganda, the gray-cheeked baboon-mangabey is strictly limited to narrow strips of swamp forest, living and feeding very largely in palm trees. Further east, in Mabira, where there are only four higher primates, the same species ranges widely through the forest, feeding on many more types of food than in Bwamba (17). Other examples could be given of monkeys specializing in an area or zone of the forest linked with their use of highly specific techniques to harvest a species-specific range of foods. In effect, the larger the primate community, the more resources are partitioned by place, time, and technique (table 3.1).

Fossils of African protoapes disappear from the record after about 12 mya. Why? It could just be an omission in collections or a coincidental accident of geology, but it is more likely to reflect a real decline, even extinction, for the slower, more conservative apelike primates. They could

TABLE 3.1. Diurnal primate communities in two Ugandan forests.

Semliki (10-km radius of Mongiro Hot Springs) 12 Species	Mabira (10-km radius of Njeru) 4 Species
Forest Types	Forest Types
Lowland Forest	*Lowland Forest*
Chimpanzee, *P. troglodytes*	Olive baboon, *P. anubis*
Red colobus, *P. foai ellioti*	Grey-cheeked mangabey, *L. albigena*
Guereza colobus, *C. guereza*	Red-tailed monkey, *C. (c.) ascanius*
Dent's monkey, *C. (m.) denti*	
Red-tailed monkey, *C. (c.) ascanius*	
Swamp Forest	*Swamp Forest*
Grey-cheeked mangabey,	Olive baboon, *P. anubis*
L. albigena	Grey-cheeked mangabey, *L. albigena*
De Brazza's monkey *(C. neglectus)*	Red-tailed monkey, *C. (c.) ascanius*
Montane forest	No montane forest represented
Angola pied colobus, *C. angolensis*	
Gentle monkey, *C. (n.) mitis*	
L'Hoests' monkey, *C. l'hoesti*	
Edge habitats	*Edge habitats*
Olive baboon, *P. anubis*	Olive baboon, *P. anubis*
Savanna monkey, *C. (a.) tantalus*	Savanna monkey, *C. (a.) tantalus*

have been crowded out by a proliferating suite of fast-moving African monkeys that progressively perfected a variety of ever more specialized teeth, digestions, and locomotions.

Eventually, at about 10 mya, leaf-eating monkeys found their way to Asia. Later still, at about 5 mya, so did pouched monkeys, represented by the macaques (9, 18). In a minor replay of their radiations in Africa, both monkey groups diversified and flourished, especially in tropical areas. As a result, their forerunners, the Asian apes, must also have faced unprecedented competition. However, because apes were the primate pioneers and the monkeys arrived and proliferated so much later, the asian apes' decline came much later than it did for the protoapes in Africa.

Although the divergence between apes and monkeys has been a proper preoccupation, it is important to remember the two groups' undeniably common origin. A crude suggestion of common roots between apes and

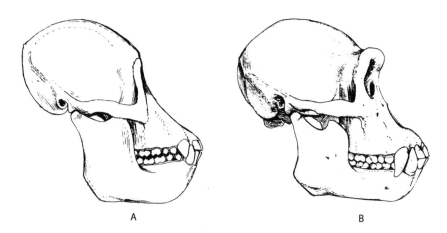

A   B

FIGURE 3.6 A. *Dryopithecus* skull reconstructed from Spanish and Hungarian fossils. B. Comparison with a chimpanzee skull.

monkeys is suggested by some gross facial similarities. For example, some gibbon and colobine monkey species retain skulls that were once thought to be conservative in general architecture (19) (while differing radically in their dental and postcranial specializations). Both have survived by becoming extreme specialists—colobines in diet, digestion, and dentition; and gibbons in a mode of locomotion that gives them easy access to fruit at the ends of long, unstable fronds. Both groups are anatomical mosaics of specialized features, together with more conservative ones. Their compact, short faces may be one or the other. It is this mosaic character in so many organisms, ourselves included, that makes tracing evolution so difficult and controversial (20), but it also makes the process more interesting (and, incidentally, helps justify the incremental, piecemeal structure of these chapters).

By about 12 mya, Eurasian tree apes, or Dryopithecines, had adapted to drier seasonal forests or woodlands and colonized the warmer parts of Europe; they were particularly abundant around the Mediterranean (21). As it happens, the first fossil discovery, of the lower jaw of *Dryopithecus fontani*, was described just before the publication of Darwin's *Origin of Species*. It represents, together with another tree ape, *Graecopithecus freyburgi*, a plausible model of the sort of Eurasian ape that might have found its way back into Africa (figure 3.6).

If it is assumed that a Eurasian ape population gave rise to the African great apes and people, rather than an indigenous stock, the founder species must have been in possession of adaptations sufficiently generalized to have allowed it to enter Africa at a time when benign climate co-

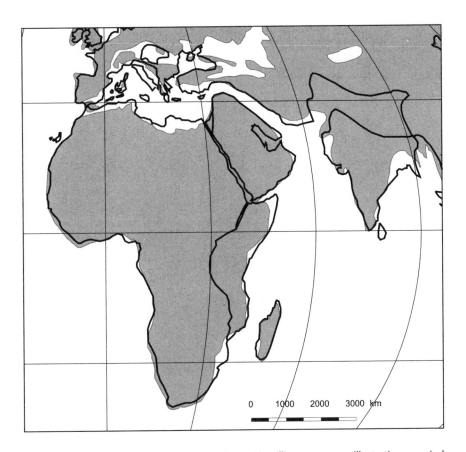

FIGURE 3.7 Position of Eurasia and Africa about 20 million years ago, illustrating a period when apes left Africa. (From a best-fit amalgam of Dercourt, J., L. Zonenshain, and B. Vrielynck. 1993. *Atlas Tethys Paleoenvironmental Maps*. Paris: Gauthier-Villars; Smith, A. G., D. G. Smith, and B. M. Funnell. 1994. *Atlas of Mesozoic and Cenozoic Coastlines*. Cambridge, UK: Cambridge University Press; and Hempton, M. R. 1998. Constraints on Arabian Plate motion and extensional history of the Red Sea. *Tectonics* 6: 687–705.)

incided with a good land bridge. Such a land bridge is now known to have existed at about 10.5 mya (figures 3.7 and 3.8) (10). Nonetheless, it needs to be remembered that however benign conditions were, in terms of rainfall and vegetation, the land bridge habitats would have fallen far short of being rainforests. Some guide to the Dryopithecines' environmental preferences can be inferred from the ecological settings of *Dryopithecus* fossils that have been found in temperate latitudes of Europe (22): seasonal semideciduous forests and woodlands might have been prototypical habitats for the putative Eurasian invader.

However, this distinction between drier and wetter forests does not al-

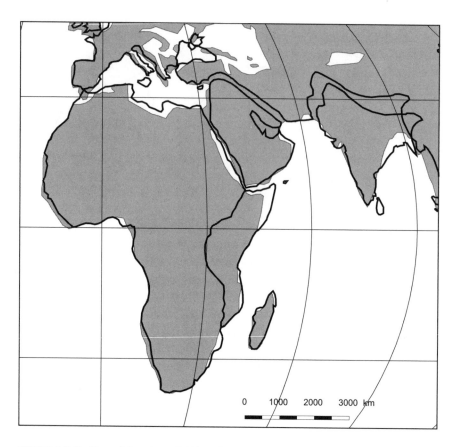

FIGURE 3.8 Position of Eurasia and Africa about 10.5 million years ago, illustrating an important period of intercontinental exchange—notably, Eurasian tree apes entering Africa. (From a best-fit amalgam of Dercourt, J., L. Zonenshain, and B. Vrielynck. 1993. *Atlas Tethys Paleoenvironmental Maps*. Paris: Gauthier-Villars; Smith, A. G., D. G. Smith, and B. M. Funnell. 1994. *Atlas of Mesozoic and Cenozoic Coastlines*. Cambridge, UK: Cambridge University Press; and Hempton, M. R. 1998. Constraints on Arabian Plate motion and extensional history of the Red Sea. *Tectonics* 6: 687–705.)

ter the fact that whatever the precise nature of their environments, drought and fruitlessness would always have restricted the apes' overall range, both at the global and local levels. Eurasian immigrant apes in the late Miocene would always have depended on the presence of trees, a yearround supply of suitable food, especially fruit, and a ready availability of water. Massifs in the Sahara, even the Ethiopian Dome, would have become uninhabitable for most primates during peak glacial periods because of a lethal combination of aridity and intense cold. These limitations remain central to any evolutionary model because truly arid (often very cold) areas have always been extensive, especially in northern Africa, periodically closing off large tracts of the continent to ape survival or penetration. Ecological, as much as physical, isolation has always played a major part in the speciation of primates in Africa, as is still evident in the distribution of living species.

The great biotic divides in Africa are not always obvious, something in which I once had an early lesson. My mother's albums, together with her sketches and photographs, have lain around, on an old Zanzibar chest, for as long as I can remember. The 1933 album recorded a walking safari, with porters carrying their tentage, that took her and my father west of Kasulu, in today's Tanzania, toward Lake Tanganyika. One day, as their little caravan trudged over the green grassy hills, they approached an enormous fig tree, and she recorded them disturbing a troop of feeding "baboons." However, the accompanying sketch unambiguously depicted chimpanzees. As a precocious juvenile naturalist, I took her to task on the mismatch of caption and sketch: she remembered them resembling "big black fruit," falling out of the tree in their haste to get away to the safety of the nearby valley forest. New to the region and to its fauna, neither of my parents had anticipated chimpanzees east of the lake at that time. Their friend, the local game warden George Rushby, subsequently wrote a short article about this unexpected population of chimps; this caught the attention of Louis Leakey and led on, many years later, to Jane Goodall's celebrated studies on chimps living here on the outermost eastern margins of their range, hard up against a settled, pastoral countryside (23).

Today, the remaining chimps have retreated to the lakeshore itself, to Gombe National Park, a last enclave where they are surrounded by settlement and cultivation. It is tempting to project this retreat back in time to a point where the chimps might have been progressively displaced by humans from a range that once stretched from the Atlantic to the Indian Oceans. Such a projection is certainly false for the living species of chimpanzees, but it is not entirely fanciful to envisage such an extensive range for a Miocene tree ape during periods of high rainfall. At such times, extensive gallery forests and moist woodlands could have webbed across the highlands and valley bottoms of eastern Africa, creating a tenuous but continuous habitat for chimp-sized apes from one side of Africa to the other. In confirmation that this may indeed have been so, an upper jaw with teeth that combine human and gorillalike features has been found in arid central Kenya, many hundreds of kilometers to the northeast of Kasulu, from deposits that have been tentatively dated at 9 mya (24).

If both molecular and geological clocks are right, this was a period when there were no gorillas, no chimps, and no people; they had not evolved yet. So what was this animal doing so far from the contemporary ranges of apes? Was this lucky find just a Late Miocene ape, or could it represent something close to the common hominin ancestor? This specimen, named *Samburupithecus koptalami* or Samburu ape, consists of a jaw fragment that shows some resemblances with the dentition of Eurasian apes that have been recovered from various subtropical seasonal habitats around the Mediterranean. So the ecological setting need not surprise, particularly because this was a very humid phase in which woodlands would have been at their most expansive. On the one hand, this arrangement would have favored a continent-wide dispersal throughout the now very widespread woodlands, but it might have denied the rainforest core of Africa to tree apes if they were still essentially woodland, not forest, animals. However, in terms of what is so far known of the African fossil fauna of the time, both the ape and its setting are rather startling, even without precedent. Allowing for a near total absence of forest archaeofauna, protoape fossils cease to be found in eastern and southern Africa after about 13 mya. Even up to that time, they were not animals that showed Eurasian affinities. It is difficult not to conclude, therefore, that the Samburu ape descended from one of the exchanging travelers known to have crossed an Africa-Eurasia land bridge about 10.5 mya (25, 9).

How many other species could have evolved from the supposed Eurasian Tree Ape invader? New possibilities have been revealed with the recent recovery of a nearly complete fossil hominid skull from Chad. *Sahelanthropus tchadensis* was discovered by Ahounta Djimdoubmalbaye in July 2001. Nicknamed "Toumai" ("hope of life" in the Goran language) this 6- to 7-million-year-old hominid demonstrates that a relatively flat, short face, small, blunt molars, and short, tip-worn canines were real traits in at least one of the African descendants of the Eurasian Tree Apes.

Michel Brunet (the leader of the discoverer's team) and his colleagues described *S. tchadensis* in admirable detail in the science journal *Nature* (July 10, 2002). The skull resembles that of a small chimp with an ape-like, narrow dental arch, but it has a shorter face, more inflated brow ridges, and widely spaced eyes. It resembles gorillas in having a thick neck and well-developed chewing muscles (figure 3.9). So far, no body parts have been found, but the short face and hominin-like teeth have encouraged hopes for bipedality and claims for direct human ancestry. This very significant fossil comes from a period when the ancestors of humans, chimps, gorillas, and other, now extinct, hominids were in the process of diverging. The skull of *S. tchadensis* hints at what their common ancestor might have been like. But how close was it to the main stem? My own hunch is that Toumai is the representative of its own

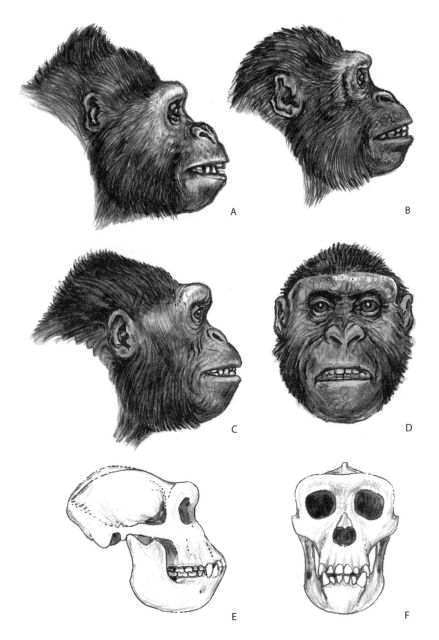

FIGURE 3.9 Reconstructions of *Sahelanthropus tchadensis*: profile and frontal view of supposed appearance (C, D) and profile and frontal view of reconstructed skull (E, F), with gorilla (*Gorilla gorilla*, A) and chimpanzee (*Pan troglodytes*, B) profiles for comparison.

branch but one that is only a short distance (in time, place, and morphology) from that early radiation to which we owe our own existence. If so, this implies that there may have been regional populations or species that exhibited unique combinations of characteristics. Before Toumai no fossil had ever shown such a unique combination of ape and hominin features, particularly from such an early time.

It should never be assumed that the incoming Tree Apes would have made an indiscriminate dispersal through the many habitats of tropical Africa without adapting and speciating. Lowland tropical rainforest would have been particularly resistant to invasion, primarily because it was inhabited by competitors and diseases to which these northern invaders would have been poorly equipped to resist. If rain forests were invaded would that have led on to speciation?

The probabilities depend on whether a continuous flow of genes would have eventually broken down as a steadily expanding range and diversification of habitats began to subdivide a single, far-flung population. This, in turn, hinges on the precise sequence of events by which immigrant ape genes found their way into the rainforest. There are two main possibilities. If invasion was gradual and across a broad front, gene flow could have been maintained. Conversely, any subpopulation that became wholly or largely separated from the parent gene pool in a locality that was especially favorable to adaptive change could soon become genetically distinct. The most likely mechanism inducing such isolation would have been a fluctuation in climate that changed forest boundaries.

Over many millions of years, intensely dry periods have affected the extent and margins of the equatorial rainforest belt in predictable ways. From the two northern desert foci (the Horn of Africa and the Sahara), dry polyps have extended south and, along two "corridors" (one in central equatorial Africa, the other through eastern Africa), have met up with similar polyps probing up from the Namib and Kalahari desert region in the southwest. In both cases, the corridors cut through along "rainshadows" behind humid coastal belts that have conserved some moisture, and hence forests, through almost all climatic vicissitudes, even the most severe (17, 26, 27, 28).

For the many species that are adapted neither to true forest nor to true desert, the intervening country provides a diversity of habitats: woodlands, savannas, bush, and steppe.. Although it was the moister of these that would have been the preferred habitat of incoming Eurasian apes, their boundaries would have fluctuated ever more widely during the climatic swings that began during the Miocene. During the warm, moist period that peaked at 9 mya, a sizable proportion of the continent could have been prime habitat for woodland apes. By contrast, during the cold, dry period that followed, peaking at about 7.8 mya, apes would have retreated as substantial areas became totally uninhabitable.

There are numerous anomalies in Africa's forest fauna and flora to suggest that these dry corridors have repeatedly introduced nonforest biota deep into the rainforest zone (26); the Okapi, a beautifully striped forest giraffe with known nonforest antecedents, is a favorite example. With each swing back to more rain, forests would have expanded again from a string of persistent equatorial "wet spots," or refuges. These expansions effectively engulfed nonforest elements living within the corridors, forcing a wide variety of species either to adapt or die out.

This mechanism could explain why and how the gorilla came to be the first species to diverge from the common ape-human stock. This conclusion can be reached from more than one line of evidence or reasoning. First, assuming a nonforest origin, the ancestors of those types that, today, show the most complete adaptation to moist rainforest could be supposed to have been the most likely pioneers of this recolonization (figure 3.10). The gorilla indubitably fits that description, and it may be relevant that recent molecular studies point to the gorilla budding off the chimphuman lineage sometime in the region of 7 to 8 mya (29).

To reconstruct the geography and chronology of that divergence, suppose first that a *Samburupithecus*-like population had spread to whatever parts of Africa could be reached or were specifically habitable. Remember that lakes, large rivers, very high mountains, waterless regions, and habitats saturated with competitors or diseases could all have marked boundaries and slowed or halted that spread. Right up to the present, the western "dry corridor," in the hinterlands of Cameroon and Gabon, seems to have acted as a persistent "trap" for nonforest organisms each time the forests have advanced after a prolonged dry period (18, 26). The abundance of gorillas in this region and their persistence, in the face of climatic changes in the past and intense persecution in the present, is consistent with this being the locality in which they have lived longest and to which they are best adapted. The very restricted overall distribution of gorillas also makes more sense when the animals are interpreted as the descendants of Eurasian immigrants rather than vestigial survivors of an indigenous ape with a formerly universal forest range. Thus, the combination of restriction to permanently humid forests and an inability to cross large rivers is sufficient to explain why the gorilla's range is bounded by the River Niger to the west and the Congo (Zaire) river to the south. So far, there is no evidence that they ever broke these barriers, even in the most distant past. If they were aboriginally indigenous forest

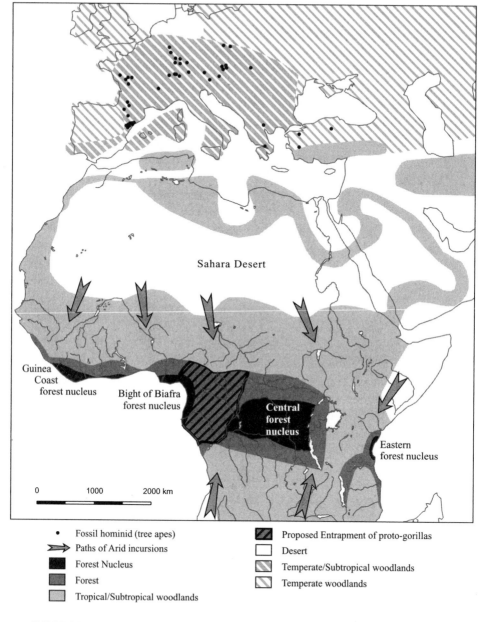

• Fossil hominid (tree apes)	Proposed Entrapment of proto-gorillas
Paths of Arid incursions	Desert
Forest Nucleus	Temperate/Subtropical woodlands
Forest	Temperate woodlands
Tropical/Subtropical woodlands	

FIGURE 3.10 Map of northern and central Africa and western Eurasia in the Miocene illustrating the range of Eurasian tree apes. (Fossil hominid locales indicated.) Possible preferred woodland habitats of tree apes in Africa differentiated from equatorial rainforest. Area of proposed "entrapment" of protogorillas lies between outer and inner margins of fluctuating rainforest. Paths of arid incursions indicated by arrows. (Modified from Nagatoshi, K. 1987. Miocene hominoid environments of Europe and Turkey. *Palaeogeography, Palaeoclimatology,*

apes, it would be necessary to invoke widespread extinction, for which, again, there is no evidence.

A subtle and persuasive argument in favor of the gorilla's precedence in the Cameroon-Gabon region lies in the fact that gorillas take a very much wider range of foods than chimps in the Lope forest of Gabon: 203 types of food for the gorilla to the chimpanzee's 142 (30). The longer a species has inhabited a specific region, the more time it has had to adapt to the chemistry of local foods and to learn what foods are edible.

As for timing, changes in global sea levels and the study of oxygen isotopes provide a crude guide to climatic change and suggest that dry periods peaked at 9.9 mya and again at about 7.8 mya, both followed by wet peaks at 9 mya and 6.9 mya (25, 31). For the incoming tree apes, the first wet phase peaked about a million years after their putative arrival. At that time, a corresponding expansion of forest can be supposed to have engulfed one small portion of a single, wide-ranging tree ape population. I envisage this engulfed subpopulation subsequently differentiating into the ancestors of modern gorillas.

Observers of the patterns of distribution and abundance have proposed that the Cameroon-Gaboon region must have been the "Area of Gorilla Origin" (32, 33). Although this hypothesis is almost certainly true, it has also been assumed that an established forest population simply speciated in response to the forests fragmenting "three ways" during a prolonged period of drought. Believing that chimps, gorillas, and hominins diverged in a single tripartite isolation event, primatologist Adriaan Kortlandt (32) described central-west Africa as the area of origin for gorillas and allocated far west Africa to the ancestors of chimps. He saw the drier land east of what he called the "Berlin Wall" of the Rift valley as the ancestral home of hominins.

Comparisons of the genetic differences between gorillas, chimpanzees, and humans suggest that their divergence was not simultaneously three ways but, rather, a staged event with the human ancestors budding off about a million years later than the gorillas.

Recall here the second genetic option for a nonforest, *Samburupithecus*-like ape population (before isolation of the putative gorilla). This option was to maintain a large gene pool while accommodating to increasing ecological diversity. Assuming the chimp-hominin ancestral ape to have been of Eurasian origin, many of the wooded riverbanks within the

Palaeoecology 61: 145–154; and Pickford, M. 1991. What caused the first steps towards the evolution of walkie-talkie primates? In *Origine(s) de la Bipedie chez les Hominides*, ed. Y. Coppens and B. Senut, 275–294. Paris: CRNS.)

spreading network of gallery forests that surrounded the main rainforest belt would scarcely have differed from those that had permitted its predecessors to colonize Africa in the first place. So long as rainfall and a reticulated, interconnected environment were sufficient to maintain genetic contact across wooded Africa, the parent population could have remained a single species until that genetic continuum was broken. I suggest that the mechanism separating future ancestral hominins from ancestral chimpanzees was climate change, just as it had previously separated the ancestral gorilla. Under the influence of shifting climatic and vegetation belts various hominid sub-populations might have arisen, much as they have in many contemporary mammal groups. The 6–7 million-year-old fossil hominid, "Toumai," *Sahelanthropus tchadensis* from north of Lake Chad had peculiarly enlarged brow ridges and widely spaced eyes that might have signified the existence of just such a local sub-population. Nonetheless its age and combination of chimpanzee and hominin features suggests that its immediate precursors could have been fair models for the common ancestors of chimps and humans. If dry conditions had permitted the newly arrived tree apes to penetrate south, deep into the equatorial zone, it may have been equally severe periods of aridity that isolated the protohominins, again, in a relatively benign but localized forest region. In both instances, protochimpanzees would have remained the most widely distributed, most adaptable, and least modified type of ape.

This conclusion, that the chimpanzee's ancestors originally occupied a broader spectrum of habitats than those of their modern descendants, was first put forward by Kortlandt and Kooij in 1963 (34), who envisaged the protochimps occupying semiopen habitats. (Their suggestion was further elaborated into a "dehumanization hypothesis.") Even more important, Koortlandt has consistently reminded us that the likely influence of humans, protohumans, and hominins in general should not be excluded in our efforts to interpret chimpanzee history. For much of that history, their ancestors may have been increasingly constrained by their hominin cousins. We may have shaped their ecological range and their physical evolution by gradually making their occupation of open woodlands untenable. If correct, such a history has important implications for exactly how chimpanzee biology is understood. For example, their ability to climb tall rainforest trees with apparent ease may be a skill that has been superimposed during the last 5 million years or so over an older, less specialized arborealism.

There is another way of interpreting the chimpanzees' specific ability to exploit the spaced-out tall emergents (which often tower above the

FIGURE 3.11 Structural similarities between spaced-out woodland trees and high rainforest emergents as sources of food for apes.

main canopy of shorter trees). This is to posit an analogy between the actual ground-level floor of a seasonal woodland (with its scatter of large fruit-bearing trees) and the forest canopy which, from the air, resembles a leafy carpet beneath a scattered "woodland" of emergents (figure 3.11). We can safely assume that equatorial rainforests would have represented a substantial expansion in food supplies and territory for incoming apes. Without being too fanciful, we can also see their new habitat interpolating an extra three-dimensional substrate, a second "floor" into the edifice of their earlier environment. Seen in this way, the emergents can be visualized as taller equivalents of their long-established sources of food in true woodland. Such equivalence means that the protochimps' anatomy and behavior might have needed minimal modification. In the forest, as in the woodlands, these apes were climbing up into well-spaced trees, but they were having to climb a lot higher to reach the fruits of these emergents. Such food, because it was perched above clean, clear boles or was more exposed to eagles, was either less accessible or less safe for the chimps' main competitors: monkeys.

Such an interpretation is consistent with the exceptional forelimb power that chimps bring to climbing tall, isolated trees, and it suggests that a faculty that was already well developed in a woodland setting could only get more marked when they entered forests. If such a history effectively reinforced an already well-developed anatomical advantage in chimps, we can safely surmise that it would have closed off any possibility of chimps becoming bipedal. For that outcome, an opposite trend would have been necessary: reduction, not amplification, of the forearms' climbing function.

The anatomy of ape limbs has been sufficiently plastic to allow the arms and shoulders of chimps and gorillas to accommodate to levering heavy bodies over the ground or to haul them up tall trees. It is this heavy-duty haulage that makes a striking contrast with the relatively light work that typifies human armwork. On present evidence, such plasticity most plausibly derived from an adaptable, generalized, and well-traveled Eurasian ape lineage. The African protoapes, sequestered for many millions of years in the forests of Africa, seem to have had more limited quadrupedal gaits. In any case, they appear to have become extinct before the arrival of the Eurasian tree apes.

Part of the true ape's spatial plasticity resides in the mobility of their shoulders, and if there is one structure in which apes and humans display startling similarity, it is in their shoulders. These prominent, hinged junctions between upper arms and body allow three-dimensional movement by attaching onto the shoulder blades or scapulae, a pair of exceptionally mobile, free-floating plates of bone on the back. The forelimbs are strengthened and stabilized by maintaining ties to the front of the rib cage by means of collarbones, or clavicles (35). Although collarbones can be seen as rigid bony anchors, numerous muscles attached to the trunk and neck tie and maneuver the shoulder blades (and with them the forelimbs). Shoulders are made up of a complex interleaving of muscles and ligaments that join trunk to scapulae and scapulae to humeri and provide leverage in all dimensions.

The mobility of ape and human shoulders contrasts with the more limited arcs of movement in more strictly quadrupedal species of mammal (figure 3.12). The latter have slab-sided chests that "hang" within a scapular "cradle": long, narrow shoulder blades are designed to maximize the stride, or fore-aft swing, of the forelimbs while restricting their scope for lateral movement. This arrangement is well exemplified by modern

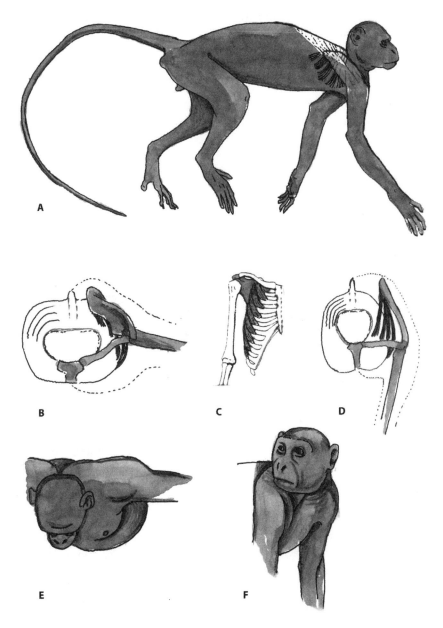

FIGURE 3.12 The vertical position of the *serratus* muscles and shoulder blade in a quadrupedal primate (A, D, and F) compared with a more lateral position in an ape (B, C, and E). (From Kingdon, J. 1971–1982. *East African Mammals. An Atlas of Evolution in Africa* [3 vols., 7 parts]. London: Academic Press.)

monkeys but is also perceptible in the fossil bones of some Early Miocene protoapes of the *Proconsul* group.

Mobile shoulders above a more or less conical rib cage are the hallmark of modern apes and humans. How early that structure first developed in primates is still unknown, but the arrangement found in fossil and living Eurasian apes is essentially homologous to that in African apes and humans, reinforcing the likelihood of a common ancestor. Thus, although their earlier ancestry is unequivocally African, the forebears of all surviving ape species look more likely to have been Eurasian between 18 and 10.5 mya. In Africa, it is the chimpanzee that may show the greatest morphological continuity with that ancestry, even if its contemporary near-restriction to rainforest represents a relatively recent retreat from the drier parts of its original woodland-forest range (34).

To return to the geography of ape speciation in Africa: if the habitats of early gorillas and early hominins were peculiarly different in each of their respective areas of isolation, each population, in quite different ways, might have had to make more adaptive accommodations than early chimps. Furthermore, the resources in both their respective enclaves were unlikely to have been poorer than those enjoyed by the protochimps. However, being more localized, they were, necessarily, more site-specific: a precondition for island-like evolution.

If the bifurcation within a common chimp-hominin ancestral population took place in eastern Africa, it is essential to explore how habitats in central Africa might have related to those in the east at the same time. In the Horn, recurrent cycles of drier climate have repeatedly interpolated an ecological barrier between east-central Africa and the humid coast. In 1971, I drew attention to this biotic disjunction which, to my surprise, has come to be dubbed "Kingdon's Line" (36). Disjunctions in the distributions of numerous, closely related sibling species suggest that this climatically controlled barrier, rather than the physical "Berlin Wall" of the Rift Valley, has most deterred the eastward expansion of forest biota (as well as reciprocal westward flow). Pointing out that proximity to the warm Indian Ocean must have allowed these forests to escape the rigors of arid periods, I inferred their isolation over a great period of time. Likewise, I suggested that the montane forests in south and east Tanzania were also likely to have been separate from those of central Africa for a substantial period. Noting that some of the eastern coastal endemics must have evolved there, I suggested that a north-south "Wallace Line" ran between the recent Mount Kilimanjaro and the much more ancient Pare/Usambara Mountains.

As with the gorilla, a first requirement for speciation was genetic isolation. For the chimp-human ancestral lineages to split, this prerequisite would have been met by a corridor of dry country that has sometimes been sufficiently extensive to connect northeast and southwest Africa. I contend, therefore, that it was climatic change operating through this fluctuating Somali arid zone, rather than the Rift Valley, that was the mechanism that permitted protohominins to become a new, highly distinctive and aberrant primate. Acknowledging the relevance of the "Kingdon Line" immediately shifts the theater of events further east, to the seaside forests of the east coast.

So long as hominins were assumed to have indigenous African forest origins, the degradation of what had been supposed to be a formerly continuous equatorial forest was thought to be the relevant phenomenon. In that tradition, the Kasulu chimps feeding in their savanna fig tree would have been seen as a perfect illustration of adventurous forest dwellers responding to the exigencies of a retreating forest. Up to now, speculation has consistently insisted on forest apes progressively accommodating to poorer resources as forests "dried out" in the savannas beyond the "Berlin Wall." This progression has been a dominant assumption underlying virtually all discussion on bipedal origins, and I contend that it has represented a major obstacle to comprehending this crucial event.

Eurasian origins instead invite an opposite model in which the expansive, parental population occupies the *drier* (not the wetter) end of the spectrum of possible (woodland/forest) habitats and both gorilla and human ancestors initially find themselves in *richer* (not poorer) enclaves.

In the model presented here, the humid Indian Ocean littoral becomes a distinctive ecological island where organisms can pursue their evolutionary destinies effectively divorced from other forest areas of Africa. One measure of the eastern forests' isolation is that as many as 25 percent of the larger trees (some 274 species) are endemic to the "Zanzibar-Inhambane Regional Center of Endemism" (38, 39). Only 62 percent also occur west of the Rift valley, confirming that some trees, like some animals, have had difficulty crossing the "Kingdon Line."

The human-induced degradation of today's eastern forests does not conceal the certainty that moist pockets of forest have provided refuges for a wide range of forest-adapted biota, including a significant number of true forest, tree-living endemics. The very conservative nature of so many of these species suggests that these refuges go back to the Miocene or still earlier times. These forests would have fluctuated in extent, like all tropical habitats (40), but the Indian Ocean is thought to have maintained more even temperatures than the Atlantic, thereby mitigating the severity of arid periods. In these circumstances, apes could have found vi-

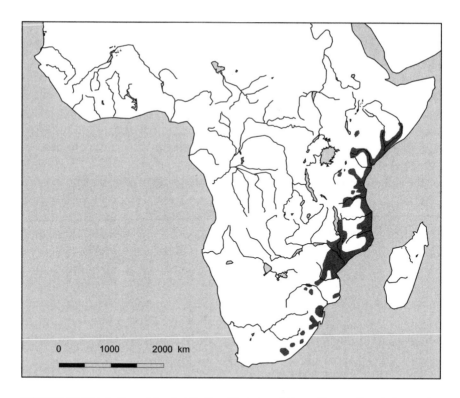

FIGURE 3.13 Distribution of the "white-throat" cluster of *Cercopithecus albogularis* complex of the Gentle Monkey superspecies. (After Kingdon, J. 1997. *The Kingdon Field Guide to African Mammals*. London: Academic Press.)

able refuges, even during the very driest climatic cycles. It could be objected that the common chimp-human ancestor, as a woodland-adapted animal, should have preferred similar habitats on the eastern littoral and therefore eschewed true forest. On a narrow coastal strip, that distinction in habitat preference might have had less meaning, because both food and water were likely to have been concentrated close to the banks of rivers and on coastal hills and a few nearby mountain ranges, such as the Usambara and Pare mountains. Prime primate habitats would have tended to be riverine, and the spectrum of plant species growing in the eastern coastal forests, even today, is as typical of woodlands as it is of forests.

Today, the extent of coastal forests (including many very degraded areas) corresponds with the distribution of white-throated gentle monkeys of the *Cercopithecus (n) albogularis* group (figure 3.13). This habitat is so narrow and the forests in a densely settled landscape so vestigial as to be effectively ignored on most older vegetation maps. At a continental level, the distributions of many plants and animals have a broad correspondence with less than a dozen well-defined zones. The Indian Ocean littoral forests comprise the only contemporary vegetation type in Africa that is longitudinally aligned (39). Elsewhere, latitude rules. As I show in later chapters, the almost unprecedented peculiarity of an extremely narrow, unstable, exceptionally long but ecologically continuous habitat may have been central to the evolution of bipedal apes.

One dimension of the gorilla-chimp-human radiation that has enjoyed massive research has been diet (41). Everybody knows that chimps prefer fruit (clearly the closest to the ancestral choice), gorillas take more herbage (42), whereas humans are more omnivorous. Before leaving the living apes behind, consider some peculiarities of gorilla and chimpanzee diet (43) that have an immediate bearing on one of the themes of this chapter: competition.

In their almost exclusive vegetarianism, gorillas have departed from an older dietary pattern. Partly related to the energetics of becoming very large, adaptation to eating much more herbaceous growth has not only made gorillas less "energetic" and limited them to regions of continuous rainfall and continuous herb renewal, it has fundamentally altered their relationship with competitors. The abundance and easy availability of their food does not mean that they tolerate other animals on their pastures. Indeed, they commonly see off other large browsers, such as giant hogs and bovids, an intimidation assisted by their imposing size and startling roars and screams. What they do not do is harm, kill, or eat other animals. In this they are unlike chimpanzees, people, and perhaps even their common ancestor. Nonetheless, unlike most ungulate herbivores they often display an intensely observant and apparently contemplative interest in other animals as different as chameleons and humans.

The presence and competition of other primates and other animals represents an important part of the environment in which any species must survive. It is of special interest that many competitors are also potential prey. So long as there are adequate staples, such as fruit, an economy of effort scarcely merits the extra energy and special skills required to take on the role of predator. Predatory habits could change, perhaps cyclically, wherever and whenever levels of competition rise above some easily tolerable optimum. This is not to say that there is any awareness on the part of apes of intolerable levels of competition from other primates; indeed, intolerance may be a latent force at any time. It may be as simple as encounters with monkeys rising with density to the point at which it becomes easier to kill them; and once the practice has caught on, it may be sustained. For the most part, monkeys are far too alert and agile to allow themselves to be caught: the more agile the species, the less likely it is to fall prey, thus guenons are rarer victims than colobus monkeys.

Most populations, predators or prey, fluctuate in numbers, so it may be significant that chimpanzees have been recorded going through bouts of intense interest in hunting—for monkeys or other animals (sometimes fellow chimpanzees too). Contemporary studies have provided some mainly behavioral correlates for bouts of hunting, but such immediate explanations need not invalidate longer-term phenomena. Relatively slow or cyclic alterations in levels of competition are much more difficult to measure or assess.

Whatever the proximate or ultimate reasons for hunting, a tendency to become "strategic," periodic predators could have been a recurrent feature in many hominin species. The ability to observe, learn, and anticipate the behavior of "competitor/prey" species could be strongly reinforced during such bouts. The fact that most such episodes are highly social in apes could have enhanced the ability of their descendants to make functional cross-connections between the observation of another species (generally potential prey) and sharing fast, coordinated responses with other group members to the cues elicited by prey behavior. In modern chimpanzees, predatory behavior tends to appear during the dry season and at times when one or more females in the group are in estrous. The number of successful kills increases with the number of hunters (a high proportion of them adult males), and the capacity to coordinate activities seems to be assisted by hoots, some of them quiet but "urgent." Temporary setbacks in the hunt can be followed by rapid regrouping. Successful hunters share the meat very selectively with estrous females and "friends." Unlike foods that are gathered without the assistance of group fellows, hunting has become socialized; this is clearest in the denial of spoils to non-allies. Such socially charged behavior is relatively infrequent but implies dormant skills in communication that depend on the excitement of a hunting emergency to find expression. The participants in every hunt presumably improve their ability to follow cues and interpret their fellows' signals (even when these are arbitrary personal quirks). Because they have the incentive of sharing meat, they also have an incentive to memorize tactics and remember the particular signaling systems of their more successful hunting partners. In later chapters, I suggest that the more frequent these "emergency" responses were, the more complex they might become, leading to a more structured and less instinctive set of responses to foraging. This response could have been a significant factor in withstanding the competition of other species.

For an incoming Eurasian Miocene ape, the keenest competition was likely to have come from monkeys (and, in different ways, pigs). Indigenous monkeys were already well diversified in Africa by 10.5 mya, whereas advanced monkeys seem to have been still absent or only newly arrived in Asia. Today, African monkey species can be seen to have diverged into fruit-, nut-, leaf-, and animal-eating specialists; small, medium, and largish forest monkeys; canopy, floor, woodland, swamp, mountain, and savanna specialists; slow, systematic, and thorough gleaners versus wide-ranging, fast but superficial foragers. With the passage of time, the end effect of all this specialization has been to make it progressively more difficult for generalized omnivorous species to find niches that were not more effectively exploited by a suite of smaller-sized specialist monkeys (44). In this respect it is interesting that the Eurasian tree apes, living in a smaller primate community, were substantially smaller than living African apes. An increase in the body size of living apes and humans may well represent an evolutionary response to competition from monkeys.

Our shared inheritance with chimpanzees embraces all the vicissitudes of our common history as primates. It includes our migration out of Africa, into Asia, and back again, and it includes the shared problem of outwitting much faster, much more numerous and diverse monkeys and baboons. A shared strategy for dealing with that challenge diverged some 7 mya and has gone on diverging ever since. If chimpanzees and most hominins have a long history of intolerance of other species, the particular form in which that self-interest is expressed in modern humans has, in turn, had a profound influence on other animals, especially the chimpanzee.

My parents' early experience of fleeing chimps left them with little more than an image of black bundles in hasty retreat. Over many years of my own pursuit of chimps and other primates, its frequent repetition has not dulled my own perception that I, my family, and my species continue to be relentless intruders into the territory of an older family, an older species. In my books, maps of the contemporary distribution of chimpanzees delineate boundaries that are far from "natural"; they are the cumulative outcome of billions of encounters between our two sibling lineages. (Many have been more hostile and less fleeting than my mother's.) It seems probable to me that chimpanzees slowly followed gorillas into the rain forests as a direct result of hominin preeminence in more open habitats.

Humans and chimpanzees are too closely related not to have been long-term competitors. It is a competition that would have begun the moment our respective lineages reestablished contact in central Africa after our ancestors' million years or so of segregation in eastern Africa.

Chimpanzees fleeing from a party of humans are not merely a motif; rather, their symbolic retreat and the whole nature of competition between primates has numerous dimensions and consequences for human evolution. In the following chapters I explore how our own intolerance of other species has been integral to the development of bipedalism and to the elaboration of a unique relationship with nature.

REFERENCES

1. Thomas, H. 1995. *The First Humans. The Search for Our Origins.* London: Thames and Hudson.
2. Jones, S., R. Martin, and D. Pilbeam, eds. 1992. *The Cambridge Encyclopedia of Human Evolution.* Cambridge, UK: Cambridge University Press.
3. Jolly, A. 1999. *Lucy's Legacy.* Princeton, NJ: Princeton University Press.
4. Pfungst, O. 1911. *Clever Hands: A Contribution to Experimental Animal and Human Psychology.* New York: Dial Press.
5. Cartmill, M. 1990. Human uniqueness and theoretical content in paleoanthropology. *International Journal of Primatology* 11: 173–192.
6. Allbrook, D., and W. W. Bishop. 1963. New fossil hominoid material from Uganda. *Nature* 197: 1187–1190.
7. Bromage, T. G., and F. Schrenk, eds. 1999. *African Biogeography, Climate Change, & Human Evolution.* New York: Oxford University Press.
8. Gebo, D. L. 1996. Climbing, brachiation and terrestrial quadrupedalism: Historical precursors of hominid bipedalism. *American Journal of Physical Anthropology* 101: 55–92.
9. Stewart, C.-B., and T. Disotell. 1998. Eurasian ape as great ape ancestor. *Current Biology* Jul/Aug 1998.
10. Bernor, R. L. 1983. Geochronology and zoogeographic relationships of Miocene Hominoidea. In *New Interpretation of Ape and Human Ancestry*, ed. R. L. Ciochon and R. S. Corruccini, 21–64. New York: Plenum Press.
11. Andrews, P., and L. Humphrey. 1999. African Miocene environments and the transition to early hominines. In *African Biogeography, Climate Change, and Human Evolution*, ed. T. G. Bromage and F. Schrenk, 282–300. New York: Oxford University Press.
12. McCrossin, M. L., and B. R. Benefit. 1992. Maboko Island and the evolutionary history of Old World monkeys and apes [abstract]. *American Anthropological Association* 226.
13. Begun, D. R. 1997. A Eurasian origin of the Hominidae. *American Journal of Physical Anthropology* Suppl. 24: 73.
14. Walker, A. C., and M. Pickford. 1983. New postcranial fossils of *Proconsul africanus* and *Proconsul nyanzae.* In *New Interpretations of Ape and Human Ancestry*, ed. R. L. Ciochon and R. S. Corruccini, 325–351. New York: Plenum Press.
15. Andrews, P. and C. Groves. 1975. Gibbons and brachiation. *Gibbon and Siamang* 4: 167–218.
16. Radinsky, L. B. 1977. Early primate brains: Facts and fiction. *Journal of Human Evolution* 6: 79–86.

17. Kingdon, J. 1971–1982. *East African Mammals. An Atlas of Evolution in Africa* (3 vols., 7 parts). London: Academic Press.

18. Kingdon, J. 1997b. *The Kingdon Field Guide to African Mammals*. London: Academic Press.

19. Delson, E., and P. Andrews. 1975. Evolution and interrelationships of the catarrhine primates. In *Phylogeny of the Primates: A Multidisciplinary Approach*, ed. W. P. Luckett and F. S. Szalay, 405–446. New York: Plenum Press.

20. Delson, E., and A. L. Rosenberger. 1980. Phyletic perspectives on platyrrhine origins and anthropoid relationships. In *Evolutionary Biology of New World Monkeys and Continental Drift*, ed. R. L. Ciochon and R. S. Corruccini. New York: Plenum Press.

21. Moya-Sola, S., and M. Kohler. 1993. Recent discoveries of *Dryopithecus* shed new light on evolution of great apes. *Nature* 365: 543–545.

22. Kelley, J. 1997. Palaeobiological and phylogenetic significance of life history in Miocene hominoids. In *Function, Phylogeny and Fossils: Miocene Hominoid Evolution and Adaptations*, ed. D. R. Begun, C. V. Ward, and M. D. Rose, 173–208. New York: Plenum Press.

23. Goodall, J. 1986. *The Chimpanzees of Gombe: Patterns of Behavior*. Cambridge, MA: Harvard University Press.

24. Ishida, H., and M. Pickford. 1997. A new Late Miocene hominoid from Kenya: *Samburupithecus koptalami* gen. et sp. nov. *Earth and Planetary Sciences*. 325: 823–829.

25. Thomas, H. 1985. The early and middle Miocene land connection of the Afro-Arabian plate and Asia: A major event for hominoid dispersal? In *Ancestors: The Hard Evidence*, ed. E. Delson, 42–50. New York: Alan R. Liss.

26. Kingdon, J. 1990b. *Island Africa. The Evolution of Africa's Rare Animals and Plants*. Princeton, NJ: Princeton University Press.

27. Kingdon, J. 1981. Where have the colonists come from? A zoogeographical examination of some mammalian isolates in eastern Africa. *African Journal of Ecology* 19: 115–124.

28. Kingdon, J., and K. Howell. 1993. Mammals in the forests of eastern Africa. In *Biogeography and Ecology of the Rainforests of Eastern Africa*, ed. J. C. Lovett and S. K. Wasser. Cambridge, UK: Cambridge University Press.

29. Ruvolo, M. 1994. Molecular evolutionary processes and conflicting gene trees: the hominoid case. *American Journal of Physical Anthropology* 94: 89–113.

30. Tutin, C.E.G., M. Fernandez, M. E. Rogers, E. A. Williamson, and W. C. McGrew. 1991. Foraging profiles of sympatric lowland gorillas and chimpanzees in the Lope Reserve, Gabon. *Philosophical Transactions of the Royal Society of London B*. 334: 179–186.

31. Crawley, J. S., and G. R. North. 1991. *Paleoclimatology*. Oxford: Oxford University Press.

32. Kortlandt, A. 1995. An ecosystem approach to hominid and ape evolution. In *Human Evolution in Its Ecological Context, vol. I: Evolution and Ecology of Homo erectus*, ed. J.R.F. Bower and S. Sartono, 87–96. Leiden University: Pithecanthropus Centennial Foundation.

33. Wrangham, W. R., and D. Peterson. 1996. *Demonic Males: Apes and the Origins of Human Violence*. Boston: Houghton Mifflin.

34. Kortlandt, A., and M. Kooij. 1963. Protohominid behaviour in primates. *Symposia of the Zoological Society of London* 10: 61–88.

35. Gregory, W. K. 1930. The origin of man from a brachiating anthropoid stock. *Science* 71: 645–650.

36. Grubb, P., O. Sandrock, O. Kullmer, T. M. Kaiser, and F. Schrenk. 1999. Relationships between eastern and southern African mammal faunas. In *African Biogeography, Climate Change, and Human Evolution*, ed. T. G. Bromage and F. Schrenk, 253–281. New York: Oxford University Press.

37. Moreau, R. E. 1963. Vicissitudes of the African biomes in the late Pleistocene. *Proceedings of the Zoological Society of London* 141.

38. Lovett, J. C. 1993a. Climatic history and forest distribution in eastern Africa. In *Biogeography and Ecology of the Rainforests of Eastern Africa*, ed. J. C. Lovett and S. K. Wasser, 23–29. Cambridge, UK: Cambridge University Press.

39. White, F. 1983. *The Vegetation of Africa*. Natural Resources Research. Paris: UNESCO.

40. Aubreville, A. 1949. Climats, forets et desertification de l'Afrique tropicale. Paris: Soc Ed. Geogr. Marit. Colon.

41. Peters, C. R., E. M. O'Brien, and R. B. Drummond. 1992. *Edible Wild Plants of Sub-saharan Africa*. Kew, UK: Royal Botanic Gardens.

42. Tutin, C.E.G., and M. Fernandez. 1985. Foods consumed by sympatric populations of *Gorilla gorilla gorilla* and *Pan troglodytes troglodytes* in Gabon: some preliminary data. *International Journal of Primatology* 6: 27–43.

43. Nishida, T., and S. Uehara. 1983. Natural diet of chimpanzees (*Pan troglodytes schweinfurthii*): long-term record from the Mahale Mountains, Tanzania. *African Studies Monographs* 3: 109–130.

44. Gautier-Hion, A., M. Colyn, and J.-P. Gautier. 2000. *Les Primates d'Afrique Centrale*. ECOFAC. Libreville.

CHAPTER 4

On Being a Ground Ape

Zanj

Forests on the east African coastal littoral likely habitat for emergent hominins some time between 7.8 and 6.2 mya. Possible anatomical changes in an eastern ape population centred on structural changes in the pelvis. Eastern African apes transformed into "eastern ground apes." Proposed "Evolution by River Basin" implies that separate hominin lineages might have originated in separate basins 5 to 6 mya.

The students were of several nationalities; but hunched over their task, they were, first and foremost, 20 young foraging primates. Each shuffled along a predetermined line, within a meter or two of their nearest neighbor, combing carefully through the leaf litter for any small animal they could find. Occasionally there was a grunt of satisfaction as some choice mollusk or invertebrate was uncovered and quickly dropped into a jam jar of formalin.

I could not but reflect that the quiet, sun-spotted forest floor over which this transect sampling was taking place had been the scene of just such grubbing, grabbing, and grunting by innumerable other primates; perhaps baboons last week, but much less familiar forms over the many millions of years that the Kiwengoma forest has existed.

Confirmation of the forest's very great age was one of the objectives of the students' activity. Many of the little animals they were so nonchalantly popping into pickle jars were new to science. Others were diagnostic of the age of the habitat because they belonged to ancient, conservative, and moisture-dependent lineages, well known for their inability to disperse or colonize new ground.

Kiwengoma, in the Matumbi Hills on southern Tanzania's Indian Ocean littoral, is a last vestige of the forests that once capped hilltops and filled valleys all along the east African coastline. The distinctness of this habitat has only begun to be documented and appreciated in very recent times (1–3). It is a discovery that has been masked by the fact that the coasts of eastern Africa have been prime habitat for modern humans for at least 40,000 years. Vast areas of cultivation or secondary vegetation suggest that human settlement, periodically dense, has degraded and impoverished the coastal forests over a very long period. The virtual absence of all large forest mammals (other than those, such as elephants and buffaloes, that can replenish their numbers by drifting down from drier hinterlands) also suggests long-term attrition (4). The extent of this degradation is perceptible in a narrow strip that runs (with only a few naturally dry interruptions) from southern Somalia to Natal (figure 4.1). In spite of this history of sustained human influence, studies of the flora and microfauna reveal that this has been one of the most important forest subregions in Africa (5, 6). The vestigial primate populations that have survived in forest enclaves not only provide confirmation of the former suitability of this region for forest primates, but their differences, at both the species and subspecies level, demonstrate that isolation from central Africa is of very long standing (7). It is not far fetched to suggest that before they were cut back and settled, the coastal forests were prime habitat for a diverse but isolated primate community that might once have included apes. But how long ago is "once," and what sort of apes?

Abundant evidence for the evolution of a specific coastal community of forest animals and plants convinces me that this is the region best positioned to provide the sustained isolation essential for speciation, during the early Pliocene, of an otherwise wide-ranging ape. The eastern coastal forests are my candidate for the birthplace of a unique but still hypothetical form of "ground ape" that was the immediate precursor for the earliest bipedal hominins, and hence ultimate ancestors for contemporary humanity. If this proves to be true, the students were not just conducting a biodiversity survey, they were, for a brief moment, replaying an activity of their remotest ancestors in the very spot where they, too, once searched for choice morsels.

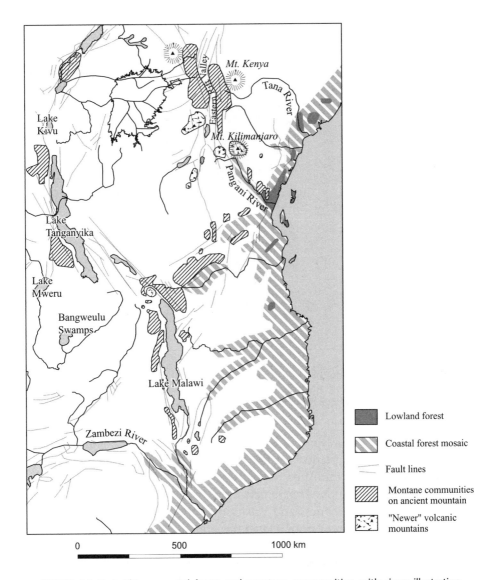

FIGURE 4.1 East African coastal forest and montane communities, with rivers illustrating pathways for forest animals into the interior.

Kiwengoma is but one vestige of a 4000-km strip of forests and wooded galleries that once ran down much of the length of Africa's Indian Ocean coast. The forests still shelter sufficient endemic species of animals and plants to have earned recognition among biogeographers as a "Center of Endemism" quite distinct from the main forests of central and western Africa (5).

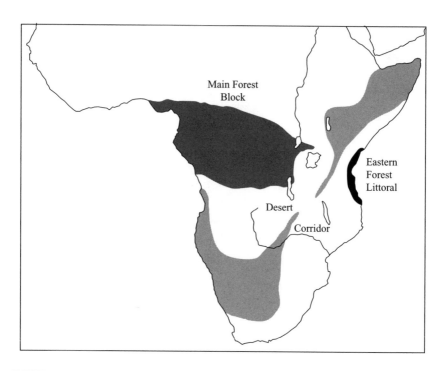

FIGURE 4.2 An arid corridor between Somalia and southwest Africa has periodically separated east coast forests from the main blocks of forest in western and central Africa. (After Kingdon, J. 1997. *The Kingdon Field Guide to African Mammals*. London: Academic Press.)

That distinctness lies not only in different climates, soils, and species. For the animals and plants that live there, the narrowness of the coastal strip implies both diminished population sizes and much greater susceptibility to fragmentation. Thus, smaller gene pools could have been as influential as the narrowness and isolation of the habitat in making east coast endemics different from related megapopulations living west of the Rift Valleys. Notwithstanding the narrowness of this habitat during very arid periods, forests and thickets of similar formation would have become much more extensive during humid periods. The distribution of many plants and animals testifies to the east coast being the core area for forest and thicket regimes that spread far inland, mostly along broad valley systems and up the moister slopes of inland hills and mountains (figure 4.1).

In chapter 3 I suggested that the split between chimp and human ancestors was precipitated by a change in climate that dried out a long tongue of land from northeast to southwest Africa. This change disconnected a small population of apes living along the eastern coastline from the much larger populations that inhabited the woodlands and forests of central and western Africa. What followed, I contend, was the emergence, in the earliest Pliocene, of a species that, for want of a name, I have labeled the "east African ground ape" or, less formally, "eastern ground ape." Beginning as no more than an intriguing idea in my mind during the early nineties, the concept of a squatting ape was reinforced by my belated reading of a seminal 1970 paper by Clifford Jolly (8). The ground ape name therefore originated as an artifact of my imagination; nonetheless, it described an anatomical and ecological projection that was informed by a lifelong habit of matching the peculiar anatomies of real animals to their unique but equally real lifestyles, as expressed in their ecology and distribution. Comparing fossil species with living ones has, of course, been the stock in trade of palaeontologists from the very beginnings of their discipline.

I was pleased therefore, when in 1994 the name "ground ape" became a scientific reality with the discovery, in Ethiopia, of fragmentary fossils belonging to a 4.4 to 5.8 million-year-old hominin species *Ardipithecus ramidus* (literally "stem ground ape") (9). In October 2000, the remains of a still older hominin species, *Orrorin tugenensis* ("original man" in the local Tugen dialect and nicknamed "Millennium Ancestor"), were retrieved from 6 million-year-old deposits in the Tugen Hills, above Lake Baringo, in Kenya (10). According to the model of evolution presented in these pages, both these fossil species are likely to represent direct derivatives of my "eastern ground ape." If this proves correct, their dating—up to 1.6 million years apart—would imply that ground apes and their more immediate descendants could have survived, perhaps in diverse forms, for at least a couple of million years.

Chapter 3 suggested that Late Miocene apes coming from Eurasia may well have spread, during clement times, from the Atlantic to the Indian Oceans. Consider then, the situation faced by the easternmost outliers of this population. Initially in touch, if tenuously, with their fellows in the west, these easterners could have first begun to be "stranded" by a general trend toward dry conditions (figure 4.2).

Following the extreme aridity that seems to have accompanied and followed the arrival of Eurasian immigrants about 10.5 mya, the next arid cycle peaked about 7.8 mya. This period may have marked the beginnings of isolation for eastern apes. If coastal origins are correct, the ground apes' definitive emergence from their enclave can now be linked to the end of the shorter but very much more severe arid cycle that climaxed about 6.2 mya. There are no fossils for the period between 7.8 and

6.2 mya to help us determine whether the evolution of a straight back was staggered over nearly 2 million years or something that developed later and much faster.

Consider now what might have been so odd about the east to trigger the development of such an aberrant ape. Even today, the physical and biological peculiarities of eastern African littoral forests are numerous, but the reconstruction of a realistic portrait of the eastern apes' habitat needs some less obvious factors to be taken into account. These include the greater extent of forests in prehuman times, fluctuating palaeoclimates, and the formerly pervasive influence of extinct animals as competitors, predators, and dynamic features and shapers of the landscape. If the distribution of some forest mammal species (monkeys, elephant-shrews, squirrels, antelopes) is anything to go by, the entire eastern seaboard could have been colonized, in a somewhat irregular fashion, by the ancestral protohominin ape. The initial spread might have depended on spells of particularly favorable climate. Later, some of the residual physiological limitations inherited from an ape ancestry might have declined under the influence of consistently drier, cooler, more open forest or woodland types. Such accommodations would have been most marked at the southern and, to a lesser extent, northern extremities of their range.

Setting aside, for the moment, all the later peculiarities of the hominin line, it is clear that the pelvis was the primary structure to be reorganized in the first hominins. That reorganization must have corresponded with some quite fundamental changes in the ecology and behavior of eastern apes. What might these changes have been?

Some 40 factors can be listed that might have differentiated the habitats of the hypothetical eastern apes from those of populations living further west. These peculiarities and some possible implications (or responses) for apes are shown in table 4.1.

These peculiarities bear closer inspection. Adaptation to the specifics of their finite and confining environment would follow the genetic separation of a small ape population in its eastern coastal enclave. During the most arid climatic phases, ape populations could have fragmented and their numbers dropped still further. Thus, periodic fluctuations in numbers and localized fragmentation within an already small gene pool could have speeded up both selection and genetic drift. By contrast, apes in extensive forests and woodlands to the west remained conservative because their much larger overall range embraced a much wider range of different types of wooded habitats. Less topographic relief in the western half of Africa should have made splits in population both rarer and less long-lasting, thereby inhibiting speciation (11).

TABLE 4.1. Evolutionary implications, for ground apes, of isolation in East Africa's littoral forest.

East African Littoral Forests (Compared to Central Africa)	Evolutionary/Behavioral Responses and Implications for "East African Ground Apes"
Isolation by Somali Arid Corridor	Genetic separation
Small overall extent	Small ape population
Long north/south extent	Extended "linear" populations
Gaps in north/south distribution	Fragmentation of populations facilitated
Lower canopies	Less volume of arboreal habitat
Shorter trunks	Easier climbing
Lower rainfall averages	Drier feeding conditions
Drier microclimate	Food decomposition slower
More seasonal	Fruit diet less continuous
More seasonally adapted plants	More underground storage organs
More erratic seasonality	Fruiting less reliable
Fewer superabundant fruiters	Fruit quantity diminished
Fewer fruit "staple" species	More reliance on other foods
More deciduous plants	Seasonal differences on forest floor
Some leaves shed year-round	Leaf-litter continuous
Marked seasonal leaf-fall	Seasonally well-lit forest floor
More ground-level plants & animals	More activity on forest floor
More resources at or near ground level	More ground feeding
More small animals at ground level	More incentive to hunt/handle prey
Great diversity of terrestrial animals	More diverse hunt/handling skills
Diversity of prey behavior	More flexible responses to obtaining food
Many small, scattered food items	More time spent foraging
Many food items small and diverse	Finer dexterity and coordination
Foods more diverse	More flexible foraging behavior
More nutritious and concentrated animal foods	Possible reduction in length of gut
Fewer seasonally superabundant foods	Foraging less clumped
Riverine habitats linear	Foraging more linear
Riverine resources more continuous	More continuous foraging
Nutrients concentrated near river banks	Resources mainly in the vicinity of banks
Forest/woodland galleries sometimes narrow	Home ranges sometimes sharp edged
Vegetation degrades or opens away from river banks	Sharper home range demarcation

TABLE 4.1. (*continued*)

East African Littoral Forests (Compared to Central Africa)	Evolutionary/Behavioral Responses and Implications for "East African Ground Apes"
Ecologically degraded borders to ranges	Restricted home ranges
Consistent resources in confined range	Home range well known
Undergrowth seasonally/locally dense	Vocal contacts episodic
More competition on floor (see next two entries)	More competitive encounters
Larger species (pigs, primates, etc.)	Group intimidation strategies?
Smaller species (rodents, reptiles, etc.)	Chase, catch, consume strategies
Predators more common	More alert; faster regrouping
Diverse large mammal encounters more frequent	More flexible group displays/threats
Unpredictable encounters more frequent	Frequent individual scanning and responses

The most important characteristics of the eastern forests, from the perspective of hominin evolution, would have been their long-term consistency as a habitat with rich and diverse but spatially restricted resources. These would have been restricted in the sense of being more dependent on groundwater, usually rivers, and less on continuous rain (although rain clouds and dew would have been important on hills and mountains close to the coast [6]). Resources would also have been restricted, in being distributed on, under, or, more generally, closer to the ground than they are in moist, dense, high-canopy rainforest.

One way of asking how apes might have responded to these limitations is to look at the feeding strategies of living species. For example, when contemporary chimps are under duress from a poor fruit season, they break up into smaller foraging units that scour the environment more thoroughly while trying to maintain their frugivorous dietary preferences for as long as possible (12). By contrast, the more terrestrial gorillas respond to the same pressure by maintaining their groupings but diversifying and enlarging the range of their foods to include previously ignored and less digestible plants (13) Another variant, better suited to eastern forests, would have been to diversify (by including more animal and underground foods) but also to spend more time and effort foraging for smaller (but still nutritionally rewarding) items. As observed in contemporary situations, these are stopgap routines for gorillas and chimpanzees. However, I am proposing that similar strategies could develop or be transposed into a sustained and systematic way of using a spatially restricted environment. It could also have been a strategy that allowed the eastern ground apes' range to be used much more intensively than has ever been observed among chimpanzees.

In the contemporary forests of eastern Africa, there are at least 50 species of large fruiting trees known to bear fruit edible for higher primates (14). Recent studies have suggested that any single group of chimpanzees exploits a similar number of species while their foraging range goes from as little as 5 km^2 in rich habitats to as much as 400 km^2 in poor ones (average 12.5 km^2 (15). There seems no reason to suppose that the eastern apes of the Late Miocene would have been any less adept at getting fruit, but the relative size of their ranges could have been influenced profoundly by changes or expansions in their diet. Use of the finer grained animal and plant potentials of the forest floor could have substantially increased the carrying capacity of eastern forests for apes. This food source could have created the potential for a wholesale transformation of their behavior and ecology, especially if the main effect was to contract the size of their yearlong range.

In terms of plant foods alone, there are 130 species of plants in the coastal and "eastern arc" forests that are already known to be edible for humans, chimpanzees, or baboons (14; see also appendix). Of these, 36 percent produce fruit in the canopy, whereas 40 percent offer food on or close to the ground (the balance comprises small trees of intermediate status) (figure 4.3). These figures refer to *known* foods, but there could be as many as 350 species of plants (16) that are theoretically possible, even probable, foods.

The main point in raising these hypothetical ratios is the contrast they suggest with the resources of central African rainforests and the unambiguous preference of living chimpanzees for fruit harvested in the canopy. By way of illustration, Wrangham et al. (17) estimated that, of their total feeding time, forest chimpanzees in Uganda spent 71.7 percent of their time eating tree fruits and only 17 percent eating herbaceous vegetation on the ground.

The micro animal resources of forest floors in the east are substantial. The leaf litter fauna is abundant, consisting most notably of termites, ants, snails, and land crabs but also many other small invertebrates (3). More than 100 species of reptiles and amphibians include 36 lizard and 34 frog species, while rodents, elephant shrews, mongooses and small antelopes are among the most common mammals. Forest floor birds, such as guinea fowls, francolins, ibis, crakes, pigeons, pittas, robins, and

FIGURE 4.3 Plan and section of transect A, in a lowland forest at Kiwengoma, Tanzania. (From Kingdon, J. 1990. *Frontier Tanzania Expedition 1989 Matumbi.* Interim Report. London & Dar-es-Salaam: Frontier Tanzania.)

thrushes, are particularly common (18) (their abundance is also a good indicator of terrestrial resources because they live mainly off the invertebrates).

Behavioral adjustments to living in smaller home ranges, to specific types, choices, and locations of foods, and to appropriate foraging techniques as well as more terrestrial use of their restricted habitat are sufficient, in my view, to have induced evolutionary changes. Furthermore, the nature of those changes implies a "precursor condition" well suited to becoming erect. So what were these evolutionary changes? Assuming that east African forests were viable habitat for Pliocene apes, what specialization or innovations might have been selected for?

If the eastern apes alternated between climbing fruiting trees and feeding on the forest floor, a vital statistic would be the relative amounts of time spent between these two activities. Another critical factor would have been how much daily movement was necessary. The spatial distribution, relative nutritional value, and size of their food items would have significantly influenced the speed, versatility, and energy demands of their daily movements. Foragers would have spent as much, or more, of

FIGURE 4.4 Sketches of imaginary squatting ground ape, illustrating altered balance between upper and lower trunk with an enlarged figure of the supposed proportions when standing.

their time searching leaf litter and excavating as they did in fruit-bearing trees or moving from one feeding site to another. Quadrupedalism would never have been abandoned if substantial distances had to be covered, especially if such journeys involved exposure to predators. Easy refuge in trees and a small home range were both essential prerequisites for habitual squat-foraging.

That apes needed a relatively secure and rich environment (rather than an insecure and demanding one) to become bipedal is, in my view, a crucial concept. Whether they are foraging or socializing, modern apes constantly interrupt their progress with pauses to manipulate a food item, handle a temporary tool, touch or hug a fellow, or make a gesture. All these manual operations require that they squat, lie down, stand on two legs, or become three-legged.

Although captive apes spend a lot of time squatting on their haunches or lolling about, wild ones are obliged to spend more time on the move. When on the ground, their quadrupedalism appears to be reinforced by the need to keep going or to be alert to danger. An ape's inability to free the arms and hands from supporting the body is therefore a frequent and obvious impediment to such activities. Seen in this context, carrying the body around on four struts was an obstacle to the full development of preexistent faculties (figures 4.4 and 4.5).

Dryopithecine apes must have suffered similar inhibitions in the Miocene. Any environment that was sufficiently secure and rich to per-

FIGURE 4.5 Sketches of squatting chimpanzees for comparison with ground apes.

mit foraging and socializing while seated on the ground could have effectively removed these inhibitions.

I think that what took place was less a case of bipedalism initiating new behaviors than the removal of frustrating constraints on many existing talents. These could be said to have been warped or at least hampered by the anatomy of a weight-bearing wrist and hand. Before becoming bipedal, the potential for more effective manual manipulations of foods or fellows must have been curtailed by the persistent intrusion of weight-bearing duties. A special dimension of this is amusingly illustrated by the "six-legged gait" of two apes walking side-by-side, each with one arm over their companions' shoulder. Humans in the same situation can simply walk hand-in-hand.

The main reason why a quadrupedal stance becomes inappropriate for systematic and sustained ground-searching is that only one hand is available at a time. As Jolly (8) was the first to point out, the solution is to drop onto the haunches and use long arms and agile hands to pick and choose. In this respect, his model derived from his studies of baboons and geladas in Uganda and Ethiopia. Many of his observations have been bulked out here with the emphasis shifted from grass seeds to a much more diverse diet, the ecological context being forest rather than grassy valleys and the geographic setting localized rather than generalized.

The hands, head, and shoulders of squatting apes would have had to swing easily from side to side as the forager investigated terrain within the arc of its arms' reach, and it was in this respect that apes would have had well-established assets. One aspect of "squat-feeding" that tempers the old joke of three legs coming between four and two is the likelihood that a feeding ground ape might, indeed, have propped itself on one arm from time to time. Nonetheless, the release of *both* forearms from weight bearing must have been the real evolutionary innovation that accompanied the acquisition of an erect back. Flexibility in the shoulders, elbows, and wrists were all advantages that may have been perceptible in Bishop's ape 14 million years earlier (19). However, if the orangutan is any guide, the maneuverability of limb joints was probably refined still further during the Oriental phase of ape evolution. In any event, these were attributes that would have been essential preadaptations for the developments that followed.

In contrast, other ape attributes, ones that also derive from brachiation, would have been decided impediments. The four highly versatile limbs of brachiating apes are bound to a trunk that is so compact that it can be levered through three-dimensional space (or over the ground) as a single, relatively inflexible unit. The neck and shoulders of a great ape embody great power and mobility. Heavy head and arms (especially in gorillas) depend on thick vertebrae, long nuchal spines, and well-developed muscles in the neck, shoulders, and arms. The forelimbs carry substantial weight whenever the ape walks. Such "over-development" at the front end of the animal is combined with a short, inflexible lower back. The hind legs are strong but short, and their anchorage is a huge, splayed pelvis that is closely integrated into a single trunk mass. Apes attempting to walk upright waddle on bent legs because massive overinvestment in arms and shoulders makes them "top-heavy" (20). The principal structures maintaining this homogeneous body mass are the short, rather straight vertebral column and close binding of a widely splayed, conical rib cage to the broad iliac wings of a very long pelvis. The "muscle-bound" nature of the upper thighs' attachment to this body mass ensures that the lower body lurches from side to side in concert with each stride, a style of locomotion that is very wasteful of energy (21).

The foraging behavior that I am proposing would have challenged the operational utility of this anatomical arrangement quite fundamentally. Evolving a functional separation of activities between shoulders and hips would have demanded that the operative unit of (frequently flexed) legs

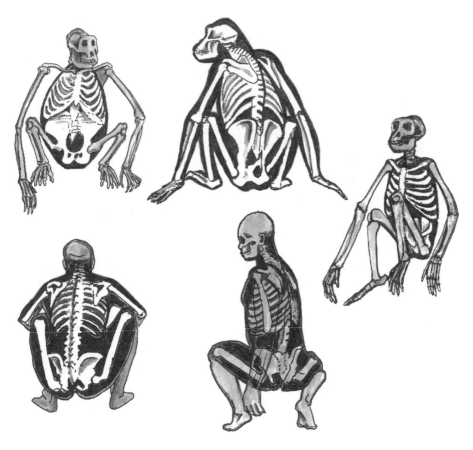

FIGURE 4.6 Ape proportions while squatting (top) compared with those of a ground ape (right) and a human (bottom).

and pelvis provided a base that was consistently firm on the ground (and equally stable in trees). Furthermore, that stability would have had to be relatively independent of activity in the upper thorax. To achieve this result, four major modifications of the skeleton would have been necessary:

1. As the main obstacle to flexibility at the waist, the apes' long iliac wings (22) fanning out to abut the rib cage, would have had to disengage from the upper trunk and retract to the point where they ceased to be one element in a unified thoracic mass to become basal structures within a more autonomous, specifically pelvic unit (figure 4.6). Because their broad and elongated pelvis makes up almost half the length of an ape trunk (23), any retraction had to be compensated for through a combined lengthening and strengthening of the lumbar region of the vertebral column.

FIGURE 4.7 A comparison between ape and human arm lengths, illustrating why apes and their immediate descendants would have benefited from long arms, and why I am too short-armed and long-legged to be a proficient ground forager. For comparison, sketch of imaginary reconstruction of a ground ape.

A less obvious effect of greater flexibility at the waist might have been a small but significant raising of shoulder height and vertebral erectness with each lumbar twist, because the pulling back of one shoulder brought the head and neck back, closer to the squatting body's center of gravity. A foraging style that persistently and habitually altered the head's balance on the neck might have selected for a more accommodating reorientation of the skull base in relation to the spine (24, 25). There could also have been selection for some reshaping of the individual components in the spine, the vertebrae, to accommodate to the increasing frequency with which they were aligned vertically rather than being cantilevered out at an angle (figure 4.7).

2. Integral to retraction of the ilia, the sacrum, as part of the same

pelvic complex, would have had to compress. This compression would have demanded a compensatory increase in breadth, if only to retain mechanical strength in the pelvic girdle. However, sacral modifications cannot be divorced from the bones' origins as vertebrae. The lumbar vertebrae of the lower spine also had to enlarge (26) and change their form to permit new demands for side-to-side rotary movement without a loss of mechanical strength. Compensation for such loss would have become important as the limbs on the upper and lower trunk diverged in function. As the lumbar region lengthened and the pelvis shortened, a newly flexible "waist" would have appeared (figure 4.8).

3. Maximum advantage for visual scanning (whether for food, companions, competitors, or predators) would have required that the back of a squatting ape be frequently straightened and as fully vertical as possible. The achievement of extra elevation through "twisting" has been mentioned, and scanning techniques could have included a tendency to turn not only the head but the entire head and shoulders. To sustain upright stance, it is not only adaptive to have the individual vertebrae become broader and stronger, the newly flexible body gains in strength and control over balance if the column and sacrum recurves within and immediately above the pelvis. This development would have initiated that well-known human characteristic, "lumbar lordosis" (27).

4. At the other end of the vertebral column, articulation with the skull would also have had to change. From their very first discovery, a different cranial balance in Australopithecines was cited as one of their decisively nonape features (28). In the context of this analysis, an oblique, cantilevered articulation for the skull would have been ill-suited to the sustained rotary panning of the head (or head and shoulders) that would have accompanied systematic litter-searching (or even the monitoring of surroundings from a branch-squatting position in the trees). To permit easy rotation of the head, the point of balance at the base of the skull would have had to be repositioned and the details of its articulation with a vertically aligned atlas and axis would also have had to be remodeled (29).

5. The development of seated feeding as an adaptation is made all the more plausible by partial convergences that have been reported in a fossil ape thought to have inhabited swamp forests in Italy and Sardinia 7 to 9 mya (figure 4.9). The teeth of this dryop-

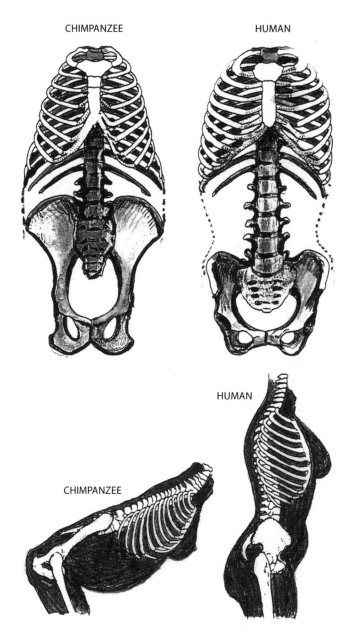

FIGURE 4.8 Remolding the oblique, platelike pelvis of a typical ape into an upward-facing basin disengages it from the thorax, while enlargement and multiplication of more flexible lumbar vertebrae creates a "waist."

FIGURE 4.9 *Oreopithecus bambolii*, a long-armed, tailless "swamp ape" from Italy, antici- pated early hominins in the modified form of its pelvis. (After Martin, R. D. 1990. *Primate Origins and Evolution. A Phylogenetic Reconstruction*. Princeton, NJ: Princeton University Press.)

ithecine swamp ape *Oreopithecus bambolii* are highly specialized and seem to be convergent with those of some monkey species. By contrast, the lower back and reduced hips converge with those of humans (30). Even more significant, in terms of corrob- orating my central proposal for the existence of an "east African ground ape," are structural convergences, short legs, and "plat- form-like" feet (well suited to terrestrial "shuffling" as well as to locking firmly onto branches). These adaptations broadly cor- respond with the theoretical morphology of my ground ape construct.

6. The anatomical modifications that I associate with squatting evoke data from some very unexpected quarters. For example, the talus, or heel bone, in a human fetus has minuscule facets associated with muscular attachments that are tensed during squatting. Where a juvenile or adult occupation (such as brick making in India) (31) involves prolonged squatting, these facets are retained or even become more developed (32). Otherwise they disappear in postnatal life. While one interpretation of this detail could be that the facets relate to the folded crouch of a fe- tus, another possible implication is that the facets are anatomical vestiges left over from a squatting phase in our evolution. An- other suggestive oddity is that tree kangaroos, who spend most of their lives squatting precariously on thin branches, have ex-

ceptionally well-developed facets and appropriately developed attachments on the talus (33).

7. Mobile shoulders, discussed in chapter 3, would have played a vital role in the foraging of east African ground apes. The brachiator's loose arms and shoulders, first developed to hoist a heavy body up, over, and under tree branches and trunks (34, 35), would have acquired a new utility by allowing easy swings and outreaches of the hands as the apes shuffled along, gleaning mostly small food items over the radius of their substantial reach.

One of the likely consequences of developing greater manipulative skills was a progressive compromising of two primary functions of the chimpanzee-type hand: namely, climbing and weight-bearing. Because a ground-foraging life in eastern forests supposedly made lighter demands on high-canopy climbing as well as on weight-bearing, the selective pressures maintaining or reinforcing these functions would have been the opposite to those operating in western and central African ape populations. It was not necessary for a strong climbing hand to be abandoned: on the contrary, excavation of roots and burrowing animals would have demanded considerable strength. (It should also be remembered that in the time since the chimp and hominin lineages diverged, it is possible, even probable, that chimp hands have become hyperspecialized for weight-bearing and heavy-duty climbing, both functions that might have involved some reduction or repositioning of their thumbs.)

So, to begin with, all that the eastern apes might have required was some minor "reversion" toward a more versatile multipurpose hand (36) including the large thumbs that characterized some early Miocene protoapes (37). The basal articulations of thumbs and lateral fingers in the "Lucy" skeleton suggest that the trend toward a broader, large-thumbed hand was perceptible in all early hominins (in spite of her retaining strongly curved phalanges) (38).

As for the hind legs, the more mobile species of Eurasian apes were presumably good medium- to long-distance quadrupedal walkers, as chimpanzees are today. If spatial mobility had given way to more stay-at-home squatting in eastern Africa, a likely initial trend would have been to retain broad, large feet but to reduce the muscle attachments, strength, and possibly the length of the long bones (39).

The differences between living chimps and gorillas show that more ground-living and less climbing in the latter leads to shorter toes and a blunter, less divergent big toe (36). This trait was taken still further in hominins.

A locomotory dimension of "squat-feeding" is that wherever resources were densely scattered, the squatting could have been punctuated by extremely frequent but very abbreviated "advances" in which the forager shuffled forward, without rising completely, or briefly stood up on its hind legs. Whatever the frequency of such standing, I contend that it was only after the vertebral column had become fully vertical that routine bipedal walking could evolve. The energy savings of a fully upright body posture are sufficient to have persuaded Ishida (40) that postural uprightness must have preceded bipedal locomotion. Using trained and untrained Japanese monkeys as models, he calculated a 30 percent saving in energy costs between fully upright and "bent" bipedalism.

It is extremely unlikely that apes could ever have become bipedal and savanna-dwelling in one step (41). The assumption that aboriginally forest-dwelling, quadrupedal apes moved out into savannas where their home ranges were larger than those of their forest predecessors is implausible on several counts. Instead, I contend that Pliocene ape populations in the eastern littoral forests had smaller, more constricted home ranges, and that they used smaller ranges more intensively.

Eastern forests were not only more riverine; their water and other resources also acted as nutrient traps or magnets, mainly transported in by animals but also through seasonal water flow and overspill. (Remember that fertile sediments gathered from a vast hinterland tend to deposit on or close to levees, as rivers meander over coastal plains.) In favorable localities, such an environment could have been consistently richer and more diverse, nutritionally, than the averages to which most western apes seem to be adapted today. An enlarged dietary base could have allowed smaller home ranges or, alternatively, permitted larger social groupings. The first option is the more likely because the finite boundaries of the range and its mainly linear orientation would have reduced the advantages of living in larger units (with their implication of expanded territories). Small, stable, well-integrated residential groups would know where to find food, water, shelter, or refuge in their often riverine home range. The risk of invasion or displacement by invading conspecifics would have been mitigated by range borders, with known neighbors being relatively short, predictable, and nonencircling. In living chimpanzees, increasing the frequency of intergroup encounters appears to promote tighter cohesion within each social unit (42). Would this imply that eastern ape groupings could have been less cohesive? In relation to "others" of their own species, possibly; but what about "others" of other species?

Ground dwelling would have exposed eastern apes to a greater diversity of predators, competitors, and other intrusive visitors, such as elephants, giant pigs, hippopotamuses, and other primates. These would have presented more of a challenge than in apes further west because such encounters could have significantly diminished access to vital resources. In these circumstances there could have been some selection for effective, group-coordinated, intimidatory displays and intolerance of competition from nonspecifics. For at least some of the eastern ground apes' descendants, the elaboration of such behavior could have been a major social correlation of becoming more terrestrial. Even so, the well-tried response of simply climbing a tree to get out of the way must have remained the easiest option. While defense of small, high-value feeding ranges could have promoted equally intensive group-based territorialism against other apes, the frequency of such confrontations could have been substantially diminished. A diversity of interactions with *other* species, many of them potential or partial competitors, and the incentives for group responses combine to imply a need for exceptional flexibility in behavior. Here, too, the utility of communication systems that allowed quick, socially coordinated but flexible counteraction to groups of large, often intelligent antagonists might have motivated subtleties of "intention movement," vocalization, and gesture (43) that could have been precursors to specifically hominin patterns of mind and speech. However, speculation on socioecological models is a well worked (if not over-worked) genre; the principal burden of this analysis is its ecological and biogeographic dimensions.

Could east coast origins have a bearing on subsequent speciation and dispersal of the "Australopithecines"?

Rich coastal forests would have remained the exclusive habitat of eastern "squatting apes" until two related developments were complete. The first was anatomical stabilization of verticality in the spinal column and repositioning of the head. Once the vertebral column had become a stable, upright pillar and weight had shifted downward, bipedal stance became not only feasible but, in terms of energy expenditure, the most efficient way of balancing. In the sequence of events that is suggested here, rising up on two legs must have been totally conditional on the anatomical, mechanical, and behavioral innovations that accompanied "squat-foraging" (8). The new balance of the upper body left no other option but to straighten the legs. In a real sense this "unbending of the legs" can be seen as subsidiary to the more revolutionary reorientation of the upper body. The development can be caricatured as "jack yourself up and do the same things faster and higher."

The next anatomical stabilization was for erect posture to become normal, so that standing and walking were structurally sustainable (44) instead of occurring as galvanic bursts of energetic activity (the costly form in which they would have begun). There are important consequences for the recognition that erect balancing—and even more so, walking—are distinct evolutionary increments from squat-feeding, because different populations of ground apes might have been slower or faster in their development of sustained walking. I suspect that minor regional variations in the exact ratios of arm lengths to leg lengths, in how independently mobile the upper body was, or in how well the thorax was balanced could have been influential in speeding up the selection for permanent bipedalism. Likewise, assuming that both hands became relatively independent foraging organs, small regional differences in dexterity, reach, handedness, and, perhaps, the development of specific tool-assisted foraging techniques could have become factors in the differentiation of regional populations. Even the adoption of habitual walking could have been slowed or speeded up by such factors.

Whatever their residual abilities at climbing trees and whatever resources were still up in the trees, eastern ground apes would have become progressively more terrestrial so that their main sphere of activity would no longer have been oriented toward the canopy but toward traveling and feeding on the ground. Nonetheless, tendencies toward more or less bipedal walking could have been strongly influenced by local environments. The rewards for foraging in a standing position would have tended to be greatest on the margins of forest galleries and glades (because fruiting shrubs and vines are commonest in well-lit situations). Manual foraging is common enough in chimpanzees but the acts of eating and moving have to be kept separate because the arms and hands are employed in both activities. For the ground apes, the continuous use of forelimbs and hands to collect and process food eventually replaced their use for support.

It can be predicted that the energy required to rise up on the hind legs was very great during the early stages of becoming a terrestrial forager. Probably it would have been too tiring to be used other than in sudden rushes, such as during escapes, during short food-transports, or in intimidation displays. Once the body had become poised vertically and body weight had shifted downward, mere straightening of the legs would have been sufficient for adequate support and balance. At this point, some reorganization of the pelvis and its associated musculature as well as elongation of the main levers (femur and tibia) would have been essential for a striding gait, but mere standing was scarcely problematic. The knee joints and ankles would also require some remodeling so that a newly rebalanced body weight could bear straight down through vertically aligned femurs and tibias. If this interpretation of hominin evolution is correct, reconstructions that depict hominids in bent postures are as misleading as the Piltdown jaw, because fully erect posture preceded bipedal gait!

Once the body is truly balanced above the pelvis, bipedal walking can become exceptionally efficient. What happens at the hip joints is critical because bipedal locomotion has some resemblance with the mechanics of an axle, as was pointed out by James Gray (45):

> A "propeller" of some kind or other exerting a backward force against the surroundings is an essential part of all self-propelling systems[;] . . . the "propeller" must act against the animal's surroundings—the outside world—and must move relatively to the rest of the animal. When we compare the propellers of animals with those of self-propelling machines made by man, we find one very striking difference. The inventions of mankind depend upon rotation. . . . Nature never uses a wheel: she uses rods or levers, and these can move up and down, or from side to side, but can never make complete revolutions about a stationary axis. . . . We must remember that a wheel on its axle is really only a series of levers, coming into action one after another. A six-spoke wheel rolling along on its rim can be regarded as six legs each ending in a foot; . . . only one foot is in use at a time but as soon as the toe of one foot leaves the ground, the heel of the next foot comes into touch with the ground; . . . As a propeller, however, it does not matter what the spoke (or leg) is doing after it leaves the ground so long as it is ready to take up its propeller duties again in the right place at the right time. . . . The limb is turning about its upper end, just as each spoke of a wheel turns about the axle—the difference between our legs and the spokes of the wheel is that our leg turns forwards about the hip joint when the foot is off the ground, and backwards when the foot is on the ground; but the spoke of a wheel moves continuously in one direction in a circle. Mechanically the propulsive effects are the same.

It may be a stretch to compare strolling with rolling around the landscape, but the mechanical simplicity of bipedal walking is a striking innovation. It is an innovation entirely dependent on perfectly upright balance in the spine.

A central question is, how long did such a major reorientation of the spine and limb-balance take to evolve? Assuming that the mechanism that separated chimp and human ancestries was climatic, I have looked

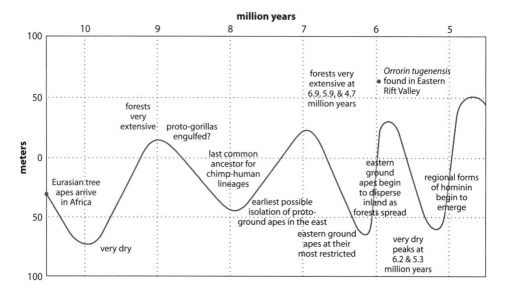

FIGURE 4.10 Sea level changes (due to freezing or melting ice-caps) as a guide to climate change. (From Moore, T. C., T. S. Loutit, and S. M. Greenlace. 1987. Estimating short-term changes in eustatic sea level. *Palaeoceanography* 2: 625–637.)

to records of global climatic change for a guide to the crucial events (figure 4.10). These, of course, are hostage to the relative accuracy of the climatological data, which increasingly suggest that rather short oscillations (typically about 4,100 years during the Ice Ages) may complicate the more spaced-out fluctuations of earlier models. If the suggestion is correct that a major period of aridity was responsible for the initial isolation, there are currently two possibilities. A prolonged but milder arid event is thought to have occurred between 8.4 and 7.5 mya. This would have allowed for a protracted, slow adaptation. A shorter, more severe period between about 6.5 and 6 mya implies a later, very rapid development. Molecular clocks would favor the later date, but the reconstruction of crude forkings in the family trees of living survivors may fail to register subtleties in extinct genomes. In any case, species are

> merely relatively stable biological entities occurring at particular times and places. Such "species" may have a relatively brief existence before being absorbed back into the parent population, or in effect becoming extinct. Species are less watertight biological concepts than populations that may have the evolutionary potential to develop unique (apomorphic) traits, but the fate of which will be dependent on external, environmental, competitive, and stochastic

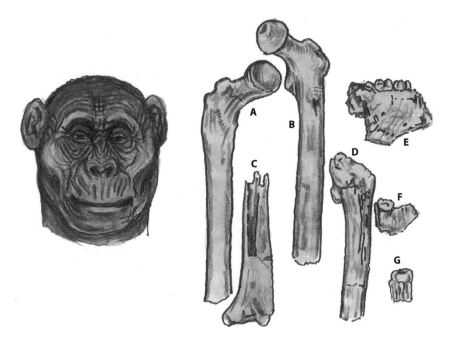

FIGURE 4.11 Left: Imaginary portrait of *Orrorin*. Right: Some diagnostic remains of *Orrorin tugenensis*. A. Left femur, posterior view. B. Left femur, anterior view. C. Right distal humerus, posterior view. D. Another left femur, anterior view. E. Left mandible (+M 2 and 3), lingual view. F. Third molar from right mandible. G. Upper third molar. (After Senut, B., M. Pickford, D. Gommery, P. Mein, K. Cheboi, and Y. Coppens. 2001. First hominid from the Miocene. *C. R. Acad. Sci. Paris, Sciences de la Terre et des planetes* 332: 137–144.)

(chance) factors. Some may simply die out, others may be swamped by other populations, while others may give rise to entirely new and distinctive taxa. (46)

It is not inconceivable that ground apes might have had periods of initial separation followed by some confluence or "absorption back into the parent population," only to diverge again in a final separation. Anyway, on current evidence the question is still open as to whether tree apes were transformed into ground apes in 500,000 or 2 million years (47).

At about 6 mya, after several hundred thousand years of drought, there was a very sudden and rapid warming in which warm, wet conditions became widespread and the first evidence for a fossil biped appears. *Orrorin tugenensis* or "Millennium Ancestor" is judged to be a 6 million-year-old chimp-sized hominin from Kapsomin in the Rift Valley in central Kenya (figure 4.11). It is, to date, the single most significant fossil species for understanding the beginnings of bipedalism (10). Although

fragmentary, the fossils represent some five individuals, males and females, from three collection sites. The hands and arms were essentially
those of climbers, so in this respect were still apelike. The three femurs,
instead, were quite unlike those of apes, and a well-preserved femoral
head proves that the upper leg had a relationship to the (still unknown)
pelvis that was more similar to that of a biped than to that of a quadrupedal ape. The vertebral column has been deduced to have been vertical. The relatively small, *thickly* enameled teeth are more human than
apelike in both shape and alignment. Collected in the Kenyan Rift Valley
in what has been extrapolated to have been a wooded savanna with
clumps of trees, both the location and the dating of these fossil hominins
are not inconsistent with the eastern ground ape model. The rapid retreat
of cold, dry temperatures and the spread of moist forests at about this
time would have helped eastern ground apes break out of their enclave
and move inland. The uplands of the Kenyan Rift Valley were a relatively
short distance away, linked by several major rivers. For example, the 300-
km Pangani river links the heart of the coastal forest complex directly to
the Rift uplands with their higher rainfall and fertile volcanic soils. This
river's association with hills and mountains starts with the Usambara
Mountains, close to its mouth, and continues past the Pare, Kilimanjaro,
and Meru mountains to the Rift rim itself. (Even today, the archetypically
coastal Suni antelope, *Neotragus moschatus*, ranges up this valley to the
Rift uplands.) Rivers may well have served as the major conduits for
ground apes, but the well-watered, raised rift floor might have been both
corridor and an attractive habitat (figure 4.12).

The establishment of this, the earliest known biped, in the interior of
eastern Africa, in a much cooler, higher, but still well-wooded environment has many implications for my model of "evolution by basin." In
the first place, it documents a very early head start for occupation of this
region. The fact that several individuals are represented in such a preliminary excavation suggests that these were already successful and, presumably, locally abundant animals. Such early success in occupation of the
Rift Valley could have helped provide the basis for further adaptation to
local conditions. Any later fossil from the same region becomes an obvious candidate for descendant status, but this anticipation must be tem-
pered with more than usual caution. A central tenet of this book is to repeatedly question the very natural assumption that the locality in which
a fossil is found is its region of origin (48). In later chapters I question the
local origins of many fossils (with some significant exceptions), but I do
identify *Kenyanthropus platyops* as one possible local descendant. Any
such suggestion must be extraordinarily tentative, given that a 2.5 mil-

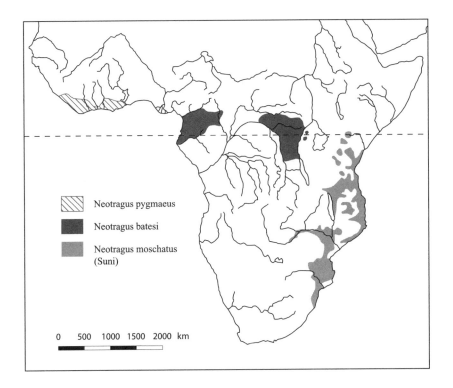

FIGURE 4.12 Map of Suni (*Neotragus moschatus*) distribution with the ranges of its nearest relatives in the forest. (After Kingdon, J. 1997. *The Kingdon Field Guide to African Mammals.* London: Academic Press.)

lion year gap exists between these two Rift Valley hominins. Nonetheless, the different anatomical permutations that characterize each species of hominin support my proposal that the hominin radiation began early and began diverse.

The Eastern Rift can be visualized as a north-south split that connects and runs through the centers of two gigantic uplifted domes (these, in turn, can be visualized as part of the buckling of Africa into a series of tectonic basins surrounded by plateaus and raised swells; figure 4.13). Before it split open to let in the Red Sea (thought to be in two phases, about 20 mya and then again about 5 mya), the northern dome embraced the whole of the Ethiopian highlands and southwest Arabia, all of 400 km across. The smaller but still impressive Eastern Rift Dome centers on southern Kenya and is outflanked by the giant volcanoes: Mt. Elgon and Mt. Kenya in the north, Mt. Kilimanjaro in the southeast, and further to the southwest, the now extinct crater of Ngoro-ngoro. Even more than a river valley, the Rift's floors and their surrounding slopes and mountains

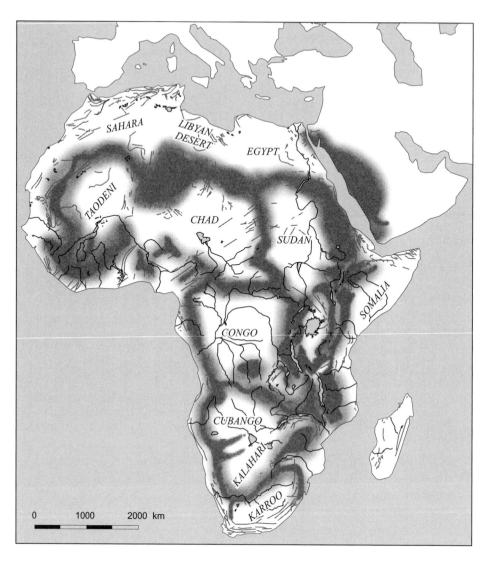

FIGURE 4.13 Schematic map of tectonic basins in Africa. The most extensive swells down the eastern side of Africa have split open to form the Rift Valleys and the Red Sea. (After Holmes, A. 1965. *Principles of Physical Geology*, revised edition. London: Thomas Nelson & Sons Ltd.)

constitute an environment of unparalleled diversity and richness within a relatively small ambit.

The other fossil hominin that could approximate to my postulated ground ape is *Ardipithecus*, dated from 5.8 to 4.4 mya (9) and also a Rift Valley fossil, but from the Ethiopian dome rather than the Kenyan Eastern Rift. Still very provisionally described, fragmentary fossil skeletons of a primitive hominin were collected between 1992 and 2001 close to the Awash river in the Ethiopian Rift Valley. The first fragment was found by Alamayehu Asfaw in December 1992. It has been named *Ardipithecus ramidus* (meaning "root-stock ground-ape"), and its skull seems decidedly chimpanzee-like in its jaws, teeth, and *thin* tooth enamel. However, the balance of its head on the spine (betrayed by a forward position for the foramen magnum, where the spinal cord enters the cranium) suggests that this animal was erect-backed, something approximating to my hypothetical eastern ground ape. Its hand bones, a relatively complete set, resemble those of a chimpanzee in the length and curvature of the fingers. The pelvis, legs, and feet have yet to be described, but its locomotion is thought to have been unlike any known primate, living or extinct (9).

This animal may be rather too distant (i.e., too far from the coast) to model the eastern ground ape, but the discovery of *Ardipithecus* in the then forested valley of a major river in northeastern Africa and its anatomy (such as has been reported, so far) accord well enough with an earlier east African forest origin. Its remains were accompanied by the bones of numerous monkeys and bush-loving kudu, so *Ardipithecus* was clearly a forest or thicket animal and not a savanna dweller. This hominin not only provides some support for my basic model of coastal ground apes colonizing inland basins; its difference from the much earlier *Orrorin tugenensis* also suggests the evolution of a particular local endemic form in Ethiopia—an animal that may have retained some of the features of its parent stock but also acquired new traits more suited to the very nonequatorial nature of northeast Africa. Once this animal had successfully adapted to Ethiopian conditions, its presence could be expected to have deterred any further hominin incursions until species arrived that had decisively different adaptations.

By 5.8 mya, the ice age peak of 6.2 mya was over, and the foothills of the Ethiopian uplands, if not their upper heights, would have become available to primates for invasion or recolonization, with east African lowlands the major source for such colonists. The Web Shebelle and the Awash share headwaters close to today's Ethiopian Rift Valley lakes (figure 4.14) such that a population of east coast origin dispersing up one river would have found no obstacle to its descent down another. The distance between the Awash Valley in the Ethiopian highlands and the east African littoral could suggest that this species had already acquired some mobility. However, the existence of appropriate habitats that were also contiguous could explain the initial spread of ground apes to Ethiopia.

It is therefore possible that both *Orrorin tugenensis* and *A. ramidus* rep-

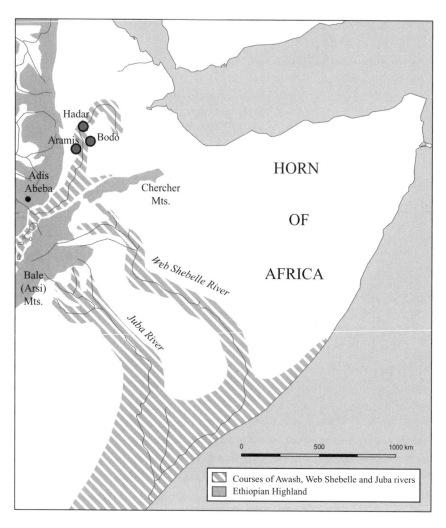

FIGURE 4.14 Map of Horn of Africa, illustrating courses of the Awash, Web Shebelle, and Juba Rivers flowing off the Ethiopian Highlands.

resent two samples of the postulated eastern ground ape, one relatively central, the other more peripheral. The former had rather small molars and thick enamel, like modern humans, while the latter retained the tooth form and thin enamel typical of apes. The contrasts raise many problematic questions. On the face of it, the Ethiopian species was around for sufficient time, perhaps, for enamel to thin (the thickness of enamel seems to be relatively labile) (49). But what about *O. tugenensis*? How near or far was each from its place of origin? Did they represent common widespread species or rarer, more localized endemics; and to what extent did their adaptations reflect solutions to special, local, or very general conditions? These are questions that cannot be answered until more is known about the anatomy and ecology of the two fossil species. In a biogeographic perspective, such questions should recur with the discovery of every fossil, but they tend to be downplayed or neglected in the understandably proprietorial contexts of fossil-finding and fossil interpretation. The very names of new fossils: "root," "stem," "original," "pre-man," "first" speak for their discoverers' hopes.

So what hints do the tattered fossils and my projection of the ground ape offer as to general appearance, lifestyle, and so forth? It is reasonable to infer that a prime limitation on prehominin apes would have been their dependence on tree shade. It must be uncertain at this remove whether Miocene apes were as physiologically constrained and ecologically limited to moist, well-watered, shady habitats as contemporary species are. However, because the climate, particularly at 5.5 to 6 mya, is thought to have been predominantly warm and humid over most of the tree apes' range, it seems likely. Greater tolerance of sunshine and desiccation need have developed only when dietary resources beyond the forest or woodland edge (and greatly enlarged home ranges) had become normal. That may well have been a separate and much later evolutionary development. If, as seems likely, ground apes gave rise to more than one set of descendants, they may well have differed in their physiological tolerances, especially if some populations were more exposed than others. In terms of their general appearance (i.e., the length, type, and distribution of hair and the range of skin colors), I think it unlikely that ground apes would have differed very greatly from living species, notably the chimpanzee.

All descendants of the ground apes would have been adept at scouring the floor, but their harvesting technique and adaptations might have been applied to "reaching out" and, perhaps, "up" in quite different ways, according to the environmental conditions in which they found their major foods.

Once the structures associated with erect posture, pelvis, backbone, and skull base were stabilized and the animals occupied a definable ground-combing niche, eastern ground apes could have expanded their range and responded to some of the ecological and climatic peculiarities of an increasingly far-flung range. The most obvious gradient concerns changes in temperature and seasonality between equatorial east Africa, temperate south Africa, and (to a lesser degree) upland Ethiopia. Additionally, populations could have diverged in their response to two dimensions of physical expansion.

The first dimension was upstream where forest galleries sometimes narrowed and, eventually, higher altitudes began. Looking at a map of Africa's rivers, it is obvious that the potentials for expansion along this axis are immense. (One should not, of course, restrict ones' awareness to the river systems that appear in an atlas; very modest tributaries could have provided habitat.) The lack of any physical barriers and gradual or staged ecological changes upstream suggests that this was an easy route to follow but one that involved very different distances from a coastal starting point. These differences, in turn, would have had far-reaching implications for the genetic isolation of expanding populations and for the relative mobility necessary for the exploitation of local resources.

The second dimension was lateral expansion so that animals could travel and forage further out and away from the riverine "spine" of their home range. This may appear an easy enough feat as even modest walking skills would have allowed short excursions, but the main hazards for a rather feeble, shade-loving primate were likely to have been greatly increased exposure to predators and physiological stress. However, the safety and shade offered by trees (even if these were more scattered) might have mitigated such hazards and offered some continuity for established and still partially arboreal behavior patterns. Of course, it is also possible that one or more populations could expand in both dimensions, but again, a sequence seems likely—first one, then the other.

An extensive north-south range has already been postulated. Between the Web Shebelle in Somalia and the Limpopo in South Africa, there are some 12 major rivers that run down to the east coast from deep in the interior. All of these could have offered arteries of expansion to coast-based populations of animals that were still intrinsically riverine. This expansion, when it took place, would have resulted in the colonization of separate and very different domains. Understanding biogeography in terms of basins is fundamental for aquatic biologists (50, 51). More recently, basin boundaries have been applied to hominin fossils from enclosed inland basins (52), but my proposal that ground ape ecology was focused on rivers invites the application of "basin evolution" to a very particular stage of early hominin evolution.

During their dispersal over a long coast and up numerous rivers, the degree of genetic isolation must have played a central role. For example, between Zanzibar and central Mozambique, a moist littoral links numerous relatively short river basins. In this region of ground ape distribution, gene flow between coastal populations and those in the immediate interior could be predicted to have been virtually continuous. Thus, populations east of Lake Malawi might be expected to have remained very similar to coastal ones. By contrast, early adaptation to decisively different habitats, such as the swampy flood plains of the future Lake Victoria basin or the fertile shores of Rift Valley lakes could have been much more conducive to genetic differentiation. Discovery of the Kenya Flat-face (*Kenyanthropus platyops*) could be taken to imply that some such development had taken place by 3.5 mya. If there was selection for different adaptations in basins that were physically and ecologically distinct, those basins deserve detailed scrutiny. If the basins and their inhabitants were different enough, more than one type of hominin might have emerged. The existing fossil record is consistent with just such a development.

Chapters 5 and 6 examine different basin regions as potential environments for two, three, or even more descendants of eastern ground apes. I think each probably started diverging as soon as it reached the headwaters of its respective river system. My analysis is rooted primarily in the two best-known fossil lineages ("Lucies" and South African man-apes) and the three newer, less well-documented lineages: the millennium ancestor (*Orrorin tugenensis*), the root-stock ground ape (*Ardipithecus ramidus*), and the Kenya Flat-face (*Kenyanthropus platyops*). I have tried to posit ecological contexts to explain some less-conspicuous differences. I go on to propose that what were initially quite subtle traits signify at least three divergent trends. Only one of these could culminate in humanity, and I pose the possibility that the hominin family tree might have started to branch at a much earlier stage than has so far been suggested. I plan to show that the fossil record is not inconsistent with this early basal branching, and much of the detail in these chapters is devoted to showing how such a history can be reconciled with the ecology of African environments and the present sparse scatter of fossils—fossils that are (like other species of living mammals) frequently found far from their region of origin. If the tree started to branch long before the time of most currently known and dated fossils, the possibility must be accepted that many of these hominins, so assertively stitched into the human genealogical chart by their discoverers, may belong to collateral, not directly ancestral branches.

REFERENCES

1. Moreau, R. E. 1963. Vicissitudes of the African biomes in the late Pleistocene. *Proceedings of the Zoological Society of London* 141: 395–421.
2. Kingdon, J. 1971–1982. *East African Mammals. An Atlas of Evolution in Africa* (3 vols., 7 parts). London: Academic Press.

3. Lovett, J. C., and S. K. Wasser. 1993. *Biogeography and Ecology of the Rainforests of Eastern Africa*. Cambridge: Cambridge University Press.

4. Kingdon, J. 1981. Where have the colonists come from? A zoogeographical examination of some mammalian isolates in eastern Africa. *African Journal of Ecology* 19: 115–124.

5. White, F. 1983. *The Vegetation of Africa*. Natural Resources Research: XX, 356 pp. Paris: UNESCO.

6. Kingdon, J. 1990a. *Frontier Tanzania Expedition 1989 Matumbi*. Interim Report. London & Dar-es-Salaam: Frontier Tanzania.

7. Kingdon, J. 1997b. *The Kingdon Field Guide to African Mammals*. London: Academic Press.

8. Jolly, C. J. 1970. The seed eaters: a new model of hominid differentiation based on baboon analogy. *Man* 5: 5–26.

9. White, T. D., G. Suwa, and B. Asfaw. 1994. *Australopithecus ramidus*, a new species of early hominid from Aramis, Ethiopia. *Nature* 371: 306–312.

10. Senut, B., M. Pickford, D. Gommery, P. Mein, K. Cheboi, and Y. Coppens. 2001. First hominid from the Miocene. *C.R. Acad. Sci. Paris, Sciences de la Terre et des planètes* 332: 137–144.

11. Kortlandt, A. 1981. Geological processes that might explain the divergence of the hominid, chimpanzee and gorilla lineages. Lecture delivered at the *11th Colloquium of African Geology*, Milton Keynes.

12. Tutin, C.E.G., and M. Fernandez. 1985. Foods consumed by sympatric populations of *Gorilla gorilla gorilla* and *Pan troglodytes troglodytes* in Gabon: some preliminary data. *International Journal of Primatology* 6: 27–43.

13. Tutin, C.E.G., M. Fernandez, M. E. Rogers, E. A. Williamson, and W. C. McGrew. 1991. Foraging profiles of sympatric lowland gorillas and chimpanzees in the Lope Reserve, Gabon. *Philosophical Transactions of the Royal Society of London B.* 334: 179–186.

14. Peters, C. R., E. M. O'Brien, and R. B. Drummond, 1992. *Edible Wild Plants of Sub-saharan Africa*. Kew, UK: Royal Botanic Gardens.

15. Reynolds, P. C. 1981. *On the Evolution of Human Behavior*. Berkeley: University of California Press.

16. Hawthore, W. D. 1993. East African coastal forest botany. In *Biogeography and Ecology of the Rainforests of Eastern Africa*, ed. J. C. Lovett and S. K. Wasser, 57–102. Cambridge, UK: Cambridge University Press.

17. Wrangham, R. W., M. E. Rogers, and G. Isabirye-Basuta. 1993. Ape food density in the ground layer in Kibale Forest, Uganda. *African Journal of Ecology* 31: 49–57.

18. Stuart, S. N., F. P. Jensen, S. Brogger-Jensen, R. I. Miller. 1993. The zoogeography of the montane forest avifauna of eastern Tanzania. In *Biogeography and Ecology of the Rainforests of Eastern Africa*, ed. J. C. Lovett and S. K. Wasser, 203–228. Cambridge, UK: Cambridge University Press.

19. Gebo, D. L. 1996. Climbing, brachiation and terrestrial quadrupedalism: Historical precursors of hominid bipedalism. *American Journal of Physical Anthropology* 101: 55–92.

20. Coppens, Y., et B. Senut, eds. 1991. *Origine(s) de la bipédie chez les Hominidae*, Cahiers de Paléoanthropologie. Paris: CNRS.

21. Shapiro, L. J., and W. L. Jungers. 1988. Back muscle function during bipedal

walking in chimpanzee and gibbon: implications for the evolution of human locomotion. *American Journal of Physical Anthropology* 77: 201–212.

22. Mednick, L. W. 1955. The evolution of the human ilium. *American Journal of Physical Anthropology* 13: 203–216.

23. McHenry, H. M. 1975. Fossils and the mosaic nature of human evolution. *Science* 190: 425–431.

24. Schultz, A. H. 1942. Conditions for balancing the head in primates. *American Journal of Physical Anthropology* 29: 483–497.

25. Moore, W. J., L. M. Adams, and C.L.B. Lavelle. 1973. Head posture in the Hominoidea. *Journal of the Zoological Society, London* 169: 409–416.

26. Benade, M. 1990. *Thoracic and lumbar vertebrae of African hominids ancient and recent: morphological and functional aspects with special reference to upright posture.* Ph.D. dissertation. University of the Witwatersrand, Johannesburg.

27. Aiello, L., and M. C. Dean. 1990. *An Introduction to Human Evolutionary Anatomy.* London: Academic Press.

28. Broom, R. 1950. The genera and species of the South African fossil ape-men. *American Journal of Physical Anthropology* 8: 1–13.

29. Dean, M. C., and B. A. Wood. 1981. Metrical analysis of the basicranium of extant hominoids and Australopithecus. *American Journal of Physical Anthropology* 59: 53–71.

30. Kohler, M., and S. J. Moya-Sola. 1997. Ape-like or hominid-like? The positional behavior of *Oreopithecus bambolii* reconsidered. *Proceedings of the National Academy of Sciences* 94: 11747–11750.

31. Singh, I. 1959. Squatting facets on the talus and tibia in Indians. *Journal of Anatomy* 93: 540–550.

32. Rao, P.D.P. 1966. Squatting facets on the talus and tibia in Australian Aborigines. *Archaeology and Physical Anthropology in Oceania* 1: 51–56.

33. Barnett, C. H. 1954. Squatting facets on the European talus. *Journal of Anatomy* 88: 509–513.

34. Fleagle, J. G., J. T. Stern, W. L. Jungers, R. L. Susman, A. K. Vangor, and J. P. Wells. 1981. Climbing: A biomechanical link with brachiation and with bipedalism. *Symposium of the Zoological Society of London* 48: 359–375.

35. Napier, J. R. 1963. Brachiation and brachiators. In *The Primates. Symposia of the Zoological Society of London* 10: 183–195.

36. Napier, J. 1962. Fossil hand bones from Olduvai Gorge. *Nature* 196: 409–411.

37. Marzke, M. W. 1986. Tool use and the evolution of hominid hands and bipedality. In *Primate Evolution*, ed. J. G. Else and P. C. Lee, 203–209. Cambridge, UK: Cambridge University Press.

38. Susman, R. L. 1994. Fossil evidence for early hominid tool use. *Science* 265: 1570–1573.

39. Sarmiento, E. E. 1994. Terrestrial traits in the hands and feet of gorillas. *American Museum Novitates* 3091: 56.

40. Ishida, H. 1991. A strategy for long-distance walking in the earliest hominids: effect of posture on energy expenditure during bipedal walking. In *Origine(s) de la Bipedie chez les Hominides*, ed. Y. Coppens and B. Senut, 9–18. Paris: CNRS.

41. Kingdon, J. 1997c. Ecological background to the possible origins of bipedalism in early hominins. Ms. and lecture for J. Sabater Pi Festschrift "Ecological Background to Human Evolution." September 1997, Barcelona University.

42. Goodall, J. 1986. *The Chimpanzees of Gombe: Patterns of Behavior*. Cambridge, MA: Harvard University Press.

43. Jablonski, N. G., and G. Chaplin. 1993. Origin of habitual terrestrial bipedalism in the ancestor of the Hominidae. *Journal of Human Evolution* 24: 259–280.

44. Napier, J. R. 1967. The antiquity of human walking. *Scientific American* 216: 55–66.

45. Gray, J. 1953. *How Animals Move*. Cambridge: Cambridge University Press.

46. Foley, R. 1999. Evolutionary geography of Pliocene African hominids. In *African Biogeography, Climate Change, and Human Evolution*, ed. T. G. Bromage and F. Schrenk, 276–281. New York: Oxford University Press.

47. Gore, R. 1997. The First Steps. *National Geographic* Feb: 72–99.

48. Croizat, L., G. Nelson, and D. E. Rosen. 1974. Centers of origin and related concepts. *Systematic Zoology* 23: 265–287.

49. Martin, L. B. 1993. Phylogenetic and adaptive significance of thick enamel in hominoid evolution. *American Journal of Physical Anthropology* (Suppl. 16): 139.

50. Beadle, L. C. 1981.. *The Inland Waters of Tropical Africa: An Introduction to Tropical Limnology*. 2nd Edition. New York: Longman.

51. Worthington, E. B., and S. Worthington. 1933. *Inland Waters of Africa*. MacMillan: London.

52. Feibel, C. S. 1999. Basin evolution, sedimentary dynamics, and hominid habitats in east Africa: An ecosystem approach. In *African Biogeography, Climate Change, and Human Evolution*, ed. T. G. Bromage and F. Schrenk, 276–281. New York: Oxford University Press.

CHAPTER 5

On Becoming a Biped

"Evolution By Basin": Domes, Rifts, and
Floodplains

Possibility of three or more bipedal lineages emerging from squatting ground apes. Explained by expansions out of coastal enclaves about 6.0 mya into separate basin complexes. Kenya Flat-face (*Kenyanthropus platyops*; 3.5 mya) possible precursor to large-brained Rudolf human ("*Homo*" *rudolfensis*). Were Lucies, *Praeanthropus* ("*Australopithecus afarensis*"; 3.9–2.8 mya) ancestral to the Robust *Paranthropus* species (2.3–1 mya) through ecological displacement by the later evolution of *Homo*? Displacement of ecologically specialized species a recurrent theme in hominin evolutionary history.

As every clown knows, losing balance is always good for a laugh. Playing around with exaggerated gaits and falling about seems to be entertaining for the great apes too, especially young chimpanzees. Yet the same animals can bring serious concentration to the task of crossing a swaying liana, tightrope style. In spite of ad-

FIGURE 5.1 Apes are top heavy.

mirable versatility in their arboreal acrobatics, it is clear that these heavy, relatively slow animals cannot begin to compete with monkeys. That the critical faculty of balance can be tested with daily life-and-death feats in the treetops yet be deliberately violated in play (1) testifies to the self-awareness that primates, especially apes, bring to their control over gait and body balance. Even so, their top-heavy body build (figure 5.1) still limits what they can do, and many feats, notably attempts at bipedal walking, are clumsy improvisations in essentially quadrupedal animals.

In chapter 4 I suggested that it was squat-feeding, not bipedalism, that induced changes in the upper body, backbone, and pelvis of ground apes. The special legacy of squat-feeding was to disengage a heavy, cantilevered upper body from an equally oblique pelvis, and to rebalance a more lightly built head and thorax vertically over a compact, basinlike pelvis. That these changes could improve balance on two legs would have been an almost accidental bonus, an anatomical by-product. It is on this premise that this chapter is concerned with the bipedal legacy as a separate increment to the acquisition of a vertical back. I also contend that the development of bipedalism was associated with inland expansions. At least three, perhaps more regions were involved in these ventures into the interior of eastern Africa, and a variety of fossils are now known from inland sites—all but one of them within reach of the east coast via forested rivers. At this point it may be useful to list all the presently known fossils together with their estimated dates and localities. The list has been lengthened substantially in recent years and will, no doubt, continue to be enlarged by new discoveries. As more is learned, it is likely that many of the names that are currently used will have to be revised. The first two species on this list were discussed in the previous chapter.

1. The "Millennium Ancestor," *Orrorin tugenensis*. 6 mya from the eastern Rift Valley in Kenya.
2. The "Root-stock Ground Ape," *Ardipithecus ramidus*. 4.4–5.8 mya from Afar, in northeastern Ethiopia.
3. The "Kenya Flat-face," *Kenyanthropus platyops*. 3.5 mya from the shores of Lake Turkana, Kenya.
4. The "Kanapoi hominid," *"Australopithecus" anamensis*. 3.9–4.2 mya(?), also from the western shores of Lake Turkana.
5. The "Lucies," *Praeanthropus ("Australopithecus") afarensis*. 2.8–3.9 mya from Tanzania and Ethiopia. (Some anthropologists think that the collection currently gathered under this name represent more than one species, but current specimens are too fragmentary to lend decisive support to such ideas.)
6. The "Chad hominin," *"Australopithecus" bahrelghazali*. 3–3.5 mya from southeastern Chad. (Named from a single fragment of a mandible, and until there is more material can be provisionally regarded as an outlying but close relative of the Lucies.)
7. The "South African Man-ape," *Australopithecus africanus*. 3.6–2.3 mya from South Africa.
8. The "Surprising Hominin," *"Australopithecus" garhi*. 2.5 mya from the middle reaches of the Awash River in Ethiopia.
9. The "Ethiopian Robust," *"Australopithecus" (Paranthropus) aethiopicus*. 2.5 mya from Lake Turkana, Kenya.
10. The "Nutcracker Ape-man" or *Zinjanthropus*, *"Australopithecus" (Paranthropus) boisei*. 2.7–1.4 mya from east and northeast Africa.
11. The "Swartkrans Ape-man," *"Australopithecus" (Paranthropus) robustus*. 2–1 mya from South Africa.

Such an abundance of species scattered down the whole eastern side of Africa and over a long span of time shows that once two-legged walking had been perfected, there were few physical impediments to the colonization of suitable habitats. In most of these species, the acquisition of two-leggedness was not accompanied by any enlargement of the brain, which remained essentially apelike in its size and general shape. Enlargement was set off by other and different selective forces that operated in but two lineages. This and the next chapter examine the possibilities for the emergence of three distinct bipedal legacies, but it is even conceivable that the earliest inlanders could have given rise to still further discrete populations—some, perhaps, still further afield than Chad.

To become a functional and habitual innovation, bipedal standing and walking needed decisive ecological and dietary incentives. Even today, it is context that determines whether an ape's balancing act is seriously functional or comically playful. If modern apes are any guide, the ground apes' intrinsic experimentalism would soon have led them to exploit all the locomotory possibilities of their own changing anatomy. It is in this sense that various forms of contagious imitation of random or playful idiosyncrasies may have come into play (what Richard Dawkins [in litt] has called the "Baldwin Effect"). Bipedalism (made "easy" by an erect back, a rebalanced head, and a downward shift in weight distribution) could have been propagated quite rapidly through "aping" that resembled fashion in that bipedal walking, frequently useful for various purposes, was also copied and admired. A primate predilection for imitation is the first ingredient in Dawkins's argument. The second is runaway sexual selection with the choosy ones (mainly females) preferring the best and sexiest walkers—a prehistoric precedent for the celebrated female taste for a "neat" male butt? Of course, any improvement in walking skills must have been predicated on genetic variation in the ability of individuals to walk on two legs. Here is a recurrent theme in human evolution: intraspecific selection for demonstrations of competence and skill.

Bipedalism also required subtle but significant anatomical adjustments to convert habitually bent legs into habitually straight ones. The most obvious of these changes would have concerned the creation of stable balance at the hip-upper leg joint through enlargement and reorientation of the buttock or *Gluteus* muscles. Formerly curved articular surfaces at the knee would have needed flattening and broadening to stabilize their newly vertical weight-bearing role. Feet and ankles would have had to be resculpted to absorb the new pressures of an upright body's weight bearing down on the many, closely bound bones of the foot (figure 5.2) (2). Fossils have begun to emerge that illustrate the exact progress of reshaping ape joints into hominin ones (3, 4), but all these innovations will eventually need to be located in much broader, yet detailed geographic, ecological, temporal, and genetic contexts than they have so far. The still

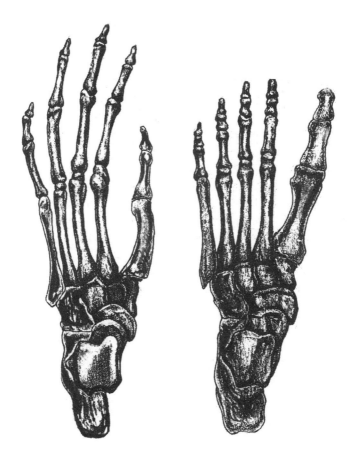

FIGURE 5.2 A "walking foot." Chimpanzee and human feet compared, showing changed proportions and alignment of toes, especially the first.

hypothetical existence of an east African ground ape provides a tentative but plausible framework for exploring the beginnings of bipedalism.

Because it might have been difficult to sustain genetic continuity over a 4000-km narrow coastal distribution (especially during climatic shifts that fragmented the forests), the existence of ground apes as a single gene pool could have been of relatively short duration. Indeed, the fragility of connections between different sections of coastline might have had some significant (but entirely unknowable) genetic consequences, such as differing resistance or susceptibility to diseases. Even so, the behavioral and anatomical transformations supposedly evolved during the period of initial separation from quadrupedal apes must have been inherited by all their successors. Although all would have remained partly arboreal forest animals as they began to expand inland, all eastern ground ape populations were probably capable of fully balanced and habitual standing and spasmodic walking. However, the rate at which both bipedal and manual skills developed from such rudiments might have differed substantially from one descendant group to another, especially if there were regional differences in the relative proportions of arms and legs. As it is, there is clear evidence that one line of hominins developed a "walking foot" and longer legs but retained a primitive apelike wrist (5, 6), while another line, in southern Africa, retained more apish feet but developed a flexible, more human-looking wrist (7, 8).

Decisive functional adaptation to living in drier habitats was less likely to have taken place in coastal forests than along the banks of distant inland rivers. During favorable climatic phases, the natural routes of population expansion would have been inland up broad forested valleys. The most likely times for early expansions away from the coast were during humid periods; just such a phase has been tentatively dated at 6 mya, and another set of moist conditions is thought to have climaxed about 4.8 mya. There is some evidence that oscillations were actually much more frequent, so the timetables suggested here may need extensive overhauling in the future. Nonetheless, it is the role of climates in the evolutionary process that is my main concern here, rather than the timing, which must of necessity remain tentative until palaeoclimatology becomes a more exact science.

It is conceivable that expansions inland took place at very different times and rates. We know, for example, that early hominins were already present in central Kenya at about 6 mya. Similar expansions might have been more delayed in areas with greater obstacles to spread (notably in the south). Given the extreme length of the coastal littoral forest, galleries that snaked inland could have funneled separate protohominin populations, at different times, into radically different regions with diverse ecologies. Each permutation could have offered different opportunities for subsequent speciation, region by region. In this and the following two chapters, I examine three such regions.

As with the ground apes, a prime limitation on actively evolving bipeds would have been a dependence on gallery forests, rivers, and streams. Part of this limitation might have lain in the physiology of a shade- and moisture-dependent animal, but the main restriction would have been the need for exceptionally rich and concentrated food sources. Improvements in walking could have been conditional on the ground ape's foraging technique: the manual sorting that so effectively scoured the floor could have begun to be applied to reaching out and up, to sources of food that were above and beyond the radial cone of a squatting ape's reach (9). Initially, a prime incentive might have been the many bushes that grow in glades, on forest edges, and in thickets and which produce great numbers of small berries; species such as jujubes (*Ziziphus*), wild raisins (*Grewia*), and caper bush (*Capparis*).

As with *Litocranius walleri* (an antelope of the acacia thickets that habitually stands on its hind legs to reach leaves and buds), the reward for this patient, painstaking plucking of such very small items would have been their high nutritional value. The protohominins' advantage over other animals that were feeding from the same resources would have been the ability to combine tree climbing (where necessary or possible) with the exceptional reach and maneuverability of their long arms, as well as the significant fact that their two hands could harvest simultaneously.

As was pointed out in the previous chapter, fully erect and truly dexterous animals could have expanded their ranges in two dimensions. The first dimension of physical expansion by protohominins was upstream, where forest galleries sometimes narrowed and higher altitudes began. The second dimension was simply a lateral expansion in which foragers could travel further out and away from the riverine "spine" of their home range. Leaving the riverine gallery forest is more problematic but need not have involved habitual movement out on grassy plains or open savannas, as is so often visualized (10). Although open land may have bordered ranges here and there, I contend that a "food gradient" might have interacted with increasing "costs" (mainly from predation) to limit the movements of eastern ground apes. Home range boundaries may have been drawn in nothing more conspicuous than a falloff in the relative ease and safety with which food could be obtained. Greater mobility in their descendants could have substantially enlarged lateral ranges without changing the essentially riverine character of their adaptive niche. I cannot envisage any great innovations in the physiology and general demeanor of apelike bipeds that were still tree and shade dependent. There is nothing in the anatomy or physical/ecological setting of fossils to suggest that the earlier hominins sustained habitual exposure to heat and sunshine. It is more likely that such adaptations came much later, especially for our own lineage. It is also very unlikely that bipedal walking induced an instantaneous incompetence for climbing. If modern humans living in forests and with totally inappropriate anatomies can become deft climbers, the protohominins, with their short legs, long arms, and strong hands and feet (and also forest dwellers), had no incentive or need to abandon such a useful skill.

The biogeographic expressions of these developments are quite simple, but if my suggestion is correct that the hominin radiation began with a coastal stock of ground-living apes, upriver dispersals from a 4000-km base provide a powerful conceptual model for understanding subsequent events in the evolution of hominins, especially the bushiness of their family tree.

Consider the extent of potential coastal forests. Between the Web Shebelle in Somalia and the Limpopo in South Africa, there are some 12 major rivers that run down to the east coast from deep in the interior. Some of these could have acted as temporary barriers to movement, but the seasonal nature of east African climates, especially during dry periods, makes it unlikely that eventual crossings were deterred for very long. Where rivers were at their broadest, some might have formed boundaries between discrete populations, but it is more likely, especially upriver, that basins would have provided the core areas of separate demes (genetically distinct subpopulations), with watersheds rather than river banks defining boundaries. Most Indian Ocean rivers could have offered arteries of expansion to coast-based populations of animals that were still intrinsically riverine. This expansion, when it took place, could have resulted in the colonization of separate and very different regions. With that colonization, "evolution by basin" could have effectively begun.

A discontinuous distribution over some 4000 km for the postulated ground apes may well have given rise to genetic differences in a purely coastal species. As soon as that parental population began to expand inland from its coastal base, the possibilities for genetic differentiation would have multiplied. To illustrate that this is not an imaginary scenario, consider the coastal squirrel, *Paraxerus palliatus*, that has at least five closely related sibling species or subspecies living around the headwaters of different rivers that flow off various mountain blocks inland (figure 5.3) (11).

The plethora of east coast rivers breaks down into some five major clusters; their position and differences can be summarized as follows, traveling from north to south.

1. The core of northeast Africa is the Ethiopian Dome; two large rivers, the Web Shebelle and the Juba, drain its eastern face and both exit together just *south* of the equator. During peak ice–ages, Ethiopia resembled a remote outpost of Eurasia: there were extensive glaciers and tundras at the highest altitudes, and narrow foothill forests were degraded, poor in species, cold, and dry. Large, frugivorous primates are not likely to have flourished under such conditions, and colonizations upriver from the equatorial coast would have been inhibited at such times. Colonization

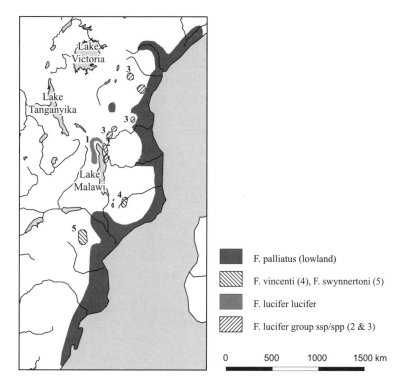

FIGURE 5.3 Distribution of some squirrels of the *Funisciurus* (*Paraxerus*) group. (After King-don, J. 1971–1982. *East African Mammals. An Atlas of Evolution in Africa* [3 vols., 7 parts]. London: Academic Press.)

from the west has always been inhibited by the Nile barrier, al-ways a major river that usually flowed over nonforested plains. Severe glacial periods that would have been sufficient to deter most primates are estimated to have peaked at 6.2 mya and again at 5.3 mya.

Warmer, wetter periods, supporting more extensive forests across a wider span of altitudes, might have held many more attractions for pri-mates; indeed, an abundance of hominin fossils in Ethiopia proves that the attractions were real. The more obviously favorable factors derived from the general biotic impoverishment; in a habitat with fewer pri-mates, there were probably fewer competitors, fewer predators, and fewer diseases. For a slow, frequently terrestrial ape, any lifting of these con-straints could have been of decisive importance, possibly compensating for a smaller choice of foods. Nonetheless, a more dispersed food supply might have demanded larger home ranges than at the coast, something that would have carried adaptive and evolutionary consequences that are discussed later.

2. The Tana, Galana, and Pangani rivers drain off the much smaller Eastern Rift Dome (figure 5.4). Here too, the uplands were high and relatively dry, but overall temperatures were ameliorated by the Rift's position on the equator. Although this region shares fauna and flora with Ethiopia, it is for the most part distinct; however, the main lowland connection between the two domes follows the shores of Lake Turkana (thanks to the Leakey family and the "Hominid Gang," by far the best sampled fossil fields in Africa). Here, fossils show that there were periods with few inhibitions to movement between these areas. Whether the Kenya highlands' affinities were with the Ethiopian dome to the north or with the other east African tropical highlands to the south may have depended on the vagaries of climate. Such ambiguities of affinity and periodically easy connections both to the north and to the south (not necessarily synchronous) make this an interesting region for the possible emergence of distinct forms of hominin. The Eastern Rift was a close destination for upriver dispersals from the central and richest portion of the coastal forest strip (opposite Pemba Island). Discovery of the "Millennium Ancestor" (*Orrorin tugenensis*) near Lake Baringo confirms the central position of the Eastern Rift and "Great Lakes" region in hominin evolutionary geography. Although drawing any sort of boundary around such a complex bioregion is fraught with difficulties, but its original core could be visualized where Lake Victoria now lies, with its northern limits Lake Turkana (former Lake Rudolf basin) and its southern limits the shores of Lake Tanganyika. Later, southward extensions would have been possible, even likely for some organisms. The many small lakes of the Eastern Rift Valley would have clearly marked the eastern margins of this "Great Lakes" bioregion. Two further hominins from the southwestern shores of Lake Turkana (one 4.1 million years old, the other 3.5 million years old) suggests implications that are discussed shortly.

3. The Rufiji River is the largest of a tightly packed and interdigitating succession of tropical river basins (notably the Wami, Ruvu, Rovuma, and some Mozambique rivers. These rivers drain the mountains north and east of Lake Malawi, and the whole complex of mountains and valleys is physically adjacent to the

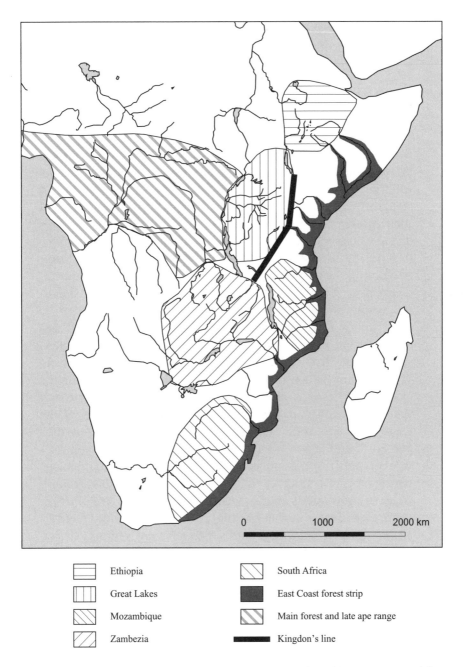

FIGURE 5.4 Map to illustrate basin evolution. Indian Ocean littoral forest mosaic and five basin regions described in text.

south-central portions of the coastal forest strip, presenting few barriers to movement by forest biota. Conditions inland, especially during moist periods, would have more closely approximated those at the coast than in any other region. Under these circumstances, really substantial genetic change in upriver populations might have been slow to appear. This is not to say that local, river-based subpopulations could not have existed, but the proximity of these basins and the likelihood that they were webbed together by similar ecologies during favorable climatic phases would have diluted tendencies toward isolation. The distribution patterns of some birds and mammals suggest ecological similarities between the more equatorial sections of the coast and their montane or riverine hinterlands (9, 10). It could be predicted, therefore, that this region might have been conservative in terms of hominin evolution, with inland populations closely resembling their coastal cousins because there were few impediments to gene flow. At present, there are no relevant fossil sites in the region to support or disprove such ideas.

4. The Zambezi River, one of the largest in Africa, drains a vast area of flat inland plateau, much of it no higher than 1000 meters, spanning the Tropic of Capricorn between 8.0° and 20.0° south. This position, which is influenced by both Indian and Atlantic Ocean climates, has probably helped to conserve riverine and shallow floodplain habitats throughout the vicissitudes of the Plio-Pleistocene. Today, temperatures range between about 10°C and 30°C, and rain falls for about 6 months during the year (November–April). Waters are conserved by slow, impeded drainage, which favorably modifies the local microclimate. Thus, groundwater sustains forests and woodlands along the margins of rivers, swamps, and lakes so that much of the Zambezi Basin is laced by an extensive web of water courses with relatively uniform vegetation. In spite of radical fluctuations in past climates, riverine habitats probably retained their viability as edaphic habitats. The headwaters of this river have long sustained one of the most important regions of edaphic grassland in Africa. Waterlogged soils inhibit tree growth and favor grasses. Long periods of flooding have caused unique ecological communities to evolve.

Upriver colonization of this region from the east coast would have been easy. However, once a population had become established in the interior, its isolation from the parental genotype could have been virtually total because the neck connecting coast and interior is relatively long and narrow. Furthermore, the parental gene pool on the coast might soon have been minuscule compared with the inland population (because of the much larger area of continuously interconnected habitat that the latter could occupy). Ecological differences between this and coastal riverside habitats might have reinforced and speeded up genetic changes between the two groups of inhabitants.

One of the most significant characteristics of floodplain habitats is that resources tend to be most concentrated close to the edges of flood waters. The annual cycle of these resources pulses out and in on either side of the riverine spine. Animals living off such resources therefore also tend to make lateral movements in and out on either side of their own stretch of water course. In the context of "evolution by basin," it is unfortunate that there are, as yet, no known early hominin fossil sites in the Zambezi Basin. However, it must be remembered that this vast area of Africa, certainly habitable for hominins, is interposed between the two main sources of fossils: east and south Africa. It is inconceivable that the Zambezi region played no part in the evolution of hominins; to ignore it for lack of fossils would be myopic.

5. The Limpopo Valley marks the northern border of a radically different complex of south African rivers all draining the 3000-meter-high Drakensberg mountain chain. The combination of more temperate latitudes and higher altitude means lower temperatures, more seasonality, and (on average) poorer resources for primates, confirmed by an impoverished primate fauna—both living and fossil. The ecological and climatic differences between a narrow Indian Ocean littoral and the interior are substantial, but they are not physically distant; there are also many small, short rivers to connect the two areas. When the seasonality of these steep south African rivers is compared with that of the Zambezi region, the pulsing of resources relates to altitude and temperature rather than to seasonally lateral movements of surface water. The contrast between seasonal movements going up and down river, rather than in and out from a home stretch of river, might have influenced a fundamental divergence between hominins in the two regions. There are many similarities between this area and mountainous parts of eastern Africa, but numerous complicating factors in the latter make the southeastern littoral and Drakensberg a much clearer model for discussing the environment's influence on hominid evolution. Crucially, there is also

an important endemic fossil record well within the drainage basin of the Limpopo and other rivers of the deep south.

Differentiation between south African coastal and inland populations would have been less significant than the overall genetic isolation of animals living in this continental extremity, far from their northern relatives. The isolation of this southern coastline has always been reinforced by a substantial ecological and climatic discontinuity between the mouths of the Zambezi and Limpopo Rivers. Here, dry "lowveld" grows on fast-draining, deep sands under a rainfall that is lower than adjacent areas to both the north and south (12). Furthermore, at all stages of evolution, incoming colonists of the far south, if successful, eventually acquire adaptive advantages over later invaders. The latter have less resistance to cold and less ability to survive on poor, tough diets, so southern Africa sometimes shelters species that combine conservative traits for their lineages with subtle adaptations to local conditions. A more detailed examination of this region is deferred to the next chapter.

"Last ape" to "first hominin" is only a conceptual transition (and a largely semantic one at that), but there could be a convenient geographic dimension whereby the former remained "squatters" that lived in moist, warm, nutritionally rich coastal lowlands, whereas the latter were (probably rather slow) "walkers" that took to a somewhat more exposed way of life beside rivers in the drier, cooler, and higher uplands. Because so much of the hinterland is an elevated plateau (much of it beginning to dome even before the Miocene), lowland forests in eastern Africa have long been mainly coastal, especially during periods of peak aridity. Even so, forests habitable to apes would have wound inland from the estuaries of the numerous rivers that drain the interior—then as now.

To explore the ecological dimensions of speciation in hominins, consider first the character of habitats bordering rivers within each basin. Combine these ecoprofiles with the boundaries of each biogeographic zone, and potential (if tentative) distribution patterns can be matched and allocated to known fossil hominids (13). Then correlate divergent ecosystems with divergent adaptive anatomy. In each case, compare what can be deduced from contemporary ecologies and the adaptive patterns of living species with those in the fossil record. In spite of a total absence of early hominin fossils from at least two of the basin regions listed earlier, the outline that follows is not contradicted by the still very sparse palaeontological facts.

A less accessible dimension is that of time. As suggested in the basin profiles, upriver expansions could have proceeded at different rates and with very different genetic outcomes. Thus, populations within some watersheds might have been conservative, slow, or incomplete colonists whereas other river systems invited more adaptive innovation or more thorough or faster colonization. Such differentials in the pace or rate of adaptation and diverging specializations may be crucial for understanding the hominin radiation as a whole. More specifically, I am suggesting here that such regional differences may have been part of the mechanism for the human lineage diverging from other hominins—a development that may have been earlier than the "australopithecine" stage that has been favored up to now (14). If changes in climate provided the impetus for eastern ground apes to expand inland, then the warm, wet periods that are thought to have followed the intense drought peaks of 6.2 and 5.3 mya were likely triggers for expansion. If the climatic calendar is right, many formerly arid areas would have become habitable after 6 and 4.8 mya, among them the uplands of northeast Africa.

One of the first demands to be made on a newly emergent hominid as it ascended its riverine highway would have been the challenge of lower temperatures. During the coldest periods, even lowland areas would have been cooler than at present, and widespread aridity could have been an absolute disincentive to eastern ground apes traveling very far upstream. However, a general adaptation to drier, cooler temperatures would have distinguished south African and Ethiopian hominins from those living along more low-lying, equatorial rivers. During later periods of global cooling, these upland populations may well have been forced to descend from the heights, but an acquired tolerance of cold and drought should have put south African and Ethiopian populations or groups at an advantage. Acquired tolerance for cold could favor their spread into newly cool areas (some perhaps vacated by less hardy natives) and some very far from the colonist's nuclei in the south or other cold highlands.

Other important criteria would have been rainfall and regional specifics of seasonality. Those entering the temperate woodlands of the south faced very different adaptive challenges to those in strongly seasonal, fast-draining, semiarid areas of northeast Africa. Speciation would have been encouraged by consistent differences in the distribution, abundance, reliability, and accessibility of terrestrial food sources as well as by temporal patterns of climate, seasonality, and ecology that would have influenced forage, breeding, and general survival.

If the ground-combing habits of eastern ground apes have been identified correctly, some continuity (if only partial or selective) should be expected in the foraging techniques of all their immediate descendants. The details of dominant soil types and relative fertility; the effects of

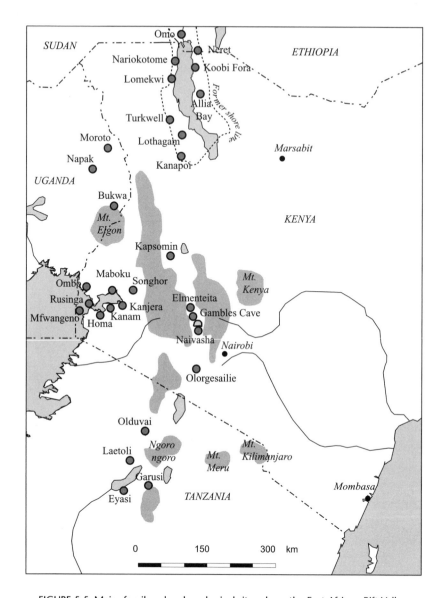

FIGURE 5.5 Major fossil and archaeological sites along the East African Rift Valley.

changing seasons on available foods, shade, tree cover, and distances to water; and the local mix of competitors and predators all would have played a role in diversifying the adaptive strategies of hominins living in separate basins (figure 5.5).

As more and more fossils are discovered, it will become increasingly possible to compare theoretical structures with raw fossils and reliable dates. So, allowing for still very sparse data, how well does this portrait of a basin-by-basin origin for hominins accord with the fossil record? The earliest hominins found so far, the "millenium ancestor" (*Orrorin tugenensis*, from 6 mya) and the "Root-stock Ground Ape" (*Ardipithecus ramidus*, from 5.8–4.4 mya) were described in the previous chapter. Fragments of other fossils, of similar but less certain age than those of *Ardipithecus*, have been found in central and northern Kenya, However, the next clear evidence for a biped comes from Kanapoi beside Lake Turkana, Kenya, dated between 4.0 and 4.2 mya. Here, in 1994, Peter Nzube Mutiwa discovered a nearly complete lower jaw that has become the type specimen for a species named the "Kanapoi hominid" (*"Australopithecus"(?) anamensis)*. Its chin is notable for resembling that of a chimp more than those of other hominins, but other body parts, notably knee, ankle, and elbow joints, indicate a fully bipedal hominin in which males were larger than females. The skull is apelike with parallel-sided rows of teeth, well-developed canines (especially in males), and a small brain. Apelike wrists, retaining the stiff locking mechanism and weight-bearing surfaces of a knuckle walker, are paradoxically combined with arms and legs that are more hominin in their proportions, the arms being less disproportionately long compared with the legs.

In the context of "coastal origins" and supposed diffusions up major rivers between about 6 and 5 mya, both the locality and the date are interesting because they suggest that during a prolonged moist period, newly mobile bipeds were crossing between separate watersheds. Kanapoi is in the middle of a lower lying saddle between the Ethiopian and Kenyan highland domes. According to the basin model, the immediate forebears of the Kanapoi hominid could have derived from south or north of Lake Turkana, they could have represented a Rift Valley species, or they might have come from still further afield. Only further study of existing fossils, or new ones, will determine whether this species could have descended from the "millennium ancestor" (4). As for the Ethiopian "root-stock ground ape" (*Ardipithecus ramidus*), this species seems to have had a different permutation of apelike traits in its skeleton than the hominins from later deposits, so it may belong to its own distinct, possibly dead-end line.

Even more puzzling is the 3.5 million-year-old old Kenya Flat-face, *Kenyanthropus platyops* (figure 5.6). This small-toothed fossil with a brain the size of a chimpanzee was discovered in 1999 on the eastern shore of Lake Turkana (15). *Kenyanthropus platyops* has so many facial and dental resemblances with the highly specialized but big-brained *"Homo" rudolfensis* that the latter is almost certainly its descendant and so might even

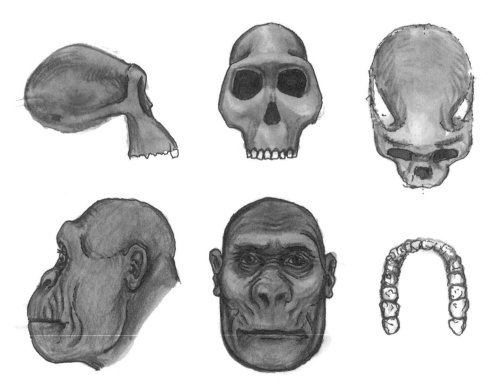

FIGURE 5.6 Bottom: Imaginary reconstruction of the Kenya Flat-face, with a sketch of the tooth arch. Top: Profile, frontal, and cranial views of *Kenyanthropus platyops*. (After Leakey, M. G., F. Spoor, F. H. Brown, P. N. Gathogo, C. Kiarie, L. N. Leakey, and I. McDougall. 2001. New hominin genus from eastern Africa shows diverse middle Pliocene lineages. *Nature* 410: 433–440.)

end up being called *Kenyanthropus rudolfensis*. These confusing changes and proliferations of name and lineage affinities effectively depose that most sacrosanct of attributes—brain enlargement—as the unique prerogative of humans. It is now clear that brain enlargement must have been shared by more than one lineage. The small, thickly enameled teeth of the Millennium Ancestor, *Orrorin tugenensis*, also suggest that this very early hominin might have occupied an approximately ancestral position to this particular line. Such an ancestry could, in turn, imply that the peculiar ecology of Rift Valley lakeshores might have invited the evolution of foraging and eating habits that demanded specialized teeth. It must be remembered that the intervening land between the Eastern and Western Rift Valleys was dominated at that time by the courses and surrounding flood plains of two large rivers (Nzoia-Katonga and Mara-Kagera) that are now mostly submerged under Lake Victoria. Then, as now, a large part of this territory ultimately drained into the Nile; however, the Rift Valley lakes, their associated rivers, and the then habitable Victoria Basin might have offered newly expansive eastern ground apes with an attractive, well-watered expanse that was open to colonization. The complex known today as the Great Lakes Region of central east Africa (a term that is not always consistent in its inclusion or exclusion of Lake Malawi) embraces a web of connections that can facilitate the movement of most waterside species. At its greatest extent, this region embraces several inland watershed basins but it is more simply treated, for the time being, as a single inland "Great Lakes" bloc.

By contrast with this apparently specialized lineage and in spite of its optimistically homocentric name, *Ardipithecus*, the "root-stock ground ape" poses a conundrum. Did this early hominin actually have descendants? let alone human ones? Or did it eventually decline and disappear as more advanced invaders from other basins encroached on its homeland? At present, there are no satisfactory answers to such questions: we do not know what sort of a population has been sampled, nor how localized or extensive its actual distribution might have been. We do not even know whether it is more useful or truthful to see *Ardipithecus* as the last of an older kind or the first of a newer kind, although students of its skull point out its close resemblance with that of a modern chimpanzee. If *Ardipithecus* is modeled as an autonomous species, already in possession of unique adaptations, its recovery from Aramis represents the discovery of an Ethiopian endemic close to (or even at) its place of origin. If modeled as a scarcely modified relative of a coastal Eastern ground ape, *Ardipithecus* becomes a very peripheral sample, perhaps even one that was without issue. Its possession of thinly enameled teeth is in striking contrast with members of the previously discussed lineage. Even allowing for relatively rapid enamel thinning or thickening, such differences still signify real adaptive trends, so it is difficult to relate this species to any descendant group. Around 4.4 mya, Aramis would have been a locality on the outer margins of what was habitable territory for ground apes. The date is late for anything like a ground ape, but it is not unusual for successful species to persist in appropriate habitats long after less conservative cousins have become common and widespread. The Awash provenance can be accommodated by my basin model because early and late, center and periphery, could have been linked by plausible events and a chain of continuous riverine galleries.

Given that our own lineage has been the only enduring survivor, could we ultimately have a northeast African derivation? The wealth of diverse fossils that has been found there suggests that Ethiopia has been close to, or a part of, the hominin highways of tropical east Africa. The Kanapoi hominid has been a puzzling fossil because the earliest fossils allocated to "*Homo*" had primitive limb proportions. Perhaps it exemplifies a particularly rangy type, just one of several newly bipedal hominins, one of many small twigs in a rapidly proliferating bush. If "basin evolution" led to rapid diversification, the very earliest hominins (unlike later, more specialized forms) might be expected to have retained a similar basic anatomy and adaptive niche. Discovery of the Kenya Flat-face contradicts this prediction. The increasing diversity of early hominin fossils raises critical issues as to how long local endemics could have survived before being replaced by wider ranging types. In a very fossil-rich area, the absence of an unambiguous fossil ancestor for modern humans makes an Ethiopian origin less likely.

The late appearance of fossil *Homo* could imply that our own lineage was a "late developer" that elaborated its peculiarities in a secure and rather distant refuge. One candidate for such a source is southern Africa. The situation of early hominins in temperate southern Africa would have been so different from their tropical cousins, and the implications of their differences so far reaching, that "basin evolution" in the far south merits a separate chapter, which follows shortly.

Supposing that the "eastern ground apes" were both highly localized and relatively poor colonists outside their narrow region of origin, we may well ask when, where, and how their descendants became more mobile. Such questions become especially relevant for the species "*Australopithecus afarensis*" (named in 1974 by Don Johanson (16) from a mandible found by Maundu Mulila at Laetoli in Tanzania; dated at 3.6 mya). This specimen very closely matches the jaw of the 3.2 million-year-old skeleton discovered later in the same year by Don Johanson at Hadar as well as a much earlier (1939) fossil maxilla, also from Laetoli and given the name *Praeanthropus africanus* in 1950 and 1955 (17, 18, 19). It is not generally appreciated that fossils of the species we now call *afarensis*, or "Lucy," were collected at Laetoli in Tanganyika by a German expedition just before the second World War. These were not allocated their own genus until the 1950s because the German collectors and authors, as well as the journal and its language, were unfamiliar to later collectors; hence the duplication of names. (There are, of course, those who would prefer to retain separate names for all three specimens.) For reasons that include but go beyond scientific priority (but may still be very provisional reasons), "*Australopithecus afarensis*" is, in the pages that follow, recognized by its less well-known but earlier name, *Praeanthropus* (19). Subsequent finds suggest that this was a rather stable and long-lasting species, the earliest possibly dating to 3.9 mya and surviving to at least 3 mya. The Hadar skeleton, popularly known as "Lucy" (and her species as "Lucies"), was at the time claimed to be on the direct line of descent for *Homo sapiens*, an assertion bolstered by the supposedly intermediate nature of her teeth and humanoid pelvis, backbone, arms, legs, and feet. In spite of the digits of her fingers and toes being long and curved (with a small thumb), the knuckles had slimmed down, and the middle finger was sharply tapered. Although these modifications imply the abandonment of knuckle–walking, there seems to have been some delay in modifying the knuckle walker's wrist, which was less flexible than one would expect from a fully bipedal animal. A similar delay in accommodating fully to upright walking can be seen in the wrist's lower limb equivalent, the ankle. By putting foot bones from different individuals together, researchers have determined that the ankle joints of *Praeanthropus* were still chimplike in their flexibility (20, 21) but that their big toes had become aligned with the other toes.

At the time of Johanson's discovery, there was nothing to compare with Lucy's skeleton as a source of information about early human evolution. However, the discovery of the Kanapoi hominid suggested the possibility of a still earlier hominin belonging to the same lineage. Meave Leakey considers *A. anamensis* a likely precursor for the Lucies. In the evolutionary model presented here, this supposition would deny a human ancestral position for *anamensis* (as is discussed shortly). Now that many more, often puzzling fossils have been found, it is feasible to examine Lucy in terms of her antecedents as well as her descendants. There is an increasing number of candidates for the latter, whereas the possibility that speciation began in separate basins more than a million years before the first appearance of "Lucies" suggests that her supposed status as universal hominin ancestor must come under closer and more skeptical scrutiny.

While their status as human ancestors has been questioned, Lucies are more generally acknowledged as having the right combination of features to be likely ancestors to some or all of the Robust Australopithecines or "Nutcrackers" (22, 23, 24). In the schedule presented here, Lucies are too late and too different to be direct human ancestors. Furthermore, the need to differentiate them from the original *Australopithecus africanus* lineage further justifies a return to their earliest, albeit somewhat misleading, name of *Praeanthropus* (19). From the perspective of basin origins, it is also relevant to ask in which basin this widely dispersed species might have originated? If separate basin beginnings were soon blurred by expansion into other regions, close attention needs to be paid to those anatomical features that link the early forms with particular adaptations—which, in turn, might have been passed on to later hominin lineages.

On present evidence, there seem to be two main possibilities (other than an origin independent of the basin model presented here). One is that *Praeanthropus* has a northeast African origin. The Kanapoi hominid is a recurrent candidate ancestor that provides some support for the region as source. Alternatively, considering their relatively late dates (vis-à-vis the Millennium Ancestor), both these early bipeds could represent already peripatetic members of the lineage that later gave rise to the Robusts. In this case, objections could go the other way because it has been suggested that the Robusts may have retained even more apelike limb proportions than Lucies; however, the skeletal material is poor and the differences are, on balance, outweighed by shared traits. If Robusts do derive from the Kanapoi hominid and Lucies, the whole group could owe their ultimate derivation to a region of Africa that, so far, has yielded no early hominin fossils at all: the Zambezi Basin.

The Zambezi Basin, or Zambezia, would have been one of four hinterlands (the others being eastern South Africa, Ethiopia, and the Great Lakes region), with the potential for inducing rapid evolutionary change during the first expansions of a coastal protohominin or eastern ground ape. The rest of this chapter centers on features of Zambezia that might have favored the emergence of the first truly mobile, open-country hominins. The next chapter discusses how the south African hinterland might have nurtured yet another man-ape with the potential to evolve into an altogether different sort of hominin, *Homo*.

The Zambezi and Luangwa Rivers, together with the southern headwaters of the Congo River, drain the mainly flat southern (Capricorn) savannas. This is a region that has long been dominated by very broad, grassy floodplains laced by tree-lined watercourses, bushy termitaries, and well-wooded levees. I have dubbed the core of this region "the Broken Hill Center of Endemism" for early hominins (25). Superficially, the environment is not promising for primates. Grasses and a moist grassland and swamp fauna and flora are predominant over extensive areas. Today, the prevalence of fire, overgrazing by cattle, and human settlement mask the fact that a primary control on vegetation in this region is edaphic (i.e., waterlogged soils inhibit the growth of most perennial woody plants but favor annual grasses). The distinction between edaphic and secondary grasslands is important because the great extent of floodplains around the headwaters of the Zambezi may represent a long-term center for the evolution and maintenance of grass communities in tropical Africa. Unlike fire-maintained grasslands, which are now much more extensive, edaphic types have probably declined under the influence of settlement and land drainage. Soils are often either soft or friable, and in terms of dietary potential, there is a yearround reliable resource in the rhizomes, roots, corms, and bulbs of many swamp-forest, marsh, or flood-adapted plants. The smaller vertebrate and invertebrate fauna is mainly terrestrial (and mostly only available while soils are wet) but is rich and varied. In spite of the diversity of other potential foods that are to be found on these flats, the underground storage organs of plants serves to buffer the diet of any animal that can get at them. Furthermore, research has suggested that, even today, this region has one of the highest counts in Africa for wild plants known to be edible to chimps, baboons, and humans (26, 27).

Exposure to the elements (e.g., floodwaters, fires, or the sun) and, less consistently, to predators must discourage primates straying very far from riverbanks. Substantial seasonal cycles, involving great fluctuations in the relative abundance of animals and plants, combine with (for much of the fauna) local or larger scale movements between dry and wet season ranges. Currently, baboons (*Papio* spp.), vervets (*Cercopithecus aethiops*), and (more marginally) gentle monkeys (*Cercopithecus mitis*) are the only simian inhabitants of these areas. If a primate could have exploited some of the dominant plants and animals and made some minor seasonal shifts in range, the situation could have been very different. With virtually no competition from other primates and less pressure from large carnivores (which are usually scarcer on floodplains), an extensive ecoregion could have become available.

The ground-combing ancestry of hominins and their powerful hands could have combined to ensure for them an effective niche within such floodplain ecosystems The most likely region for such a primary development to have taken place is the southern savannas, because it is the only biogeographic zone where the main fate of precipitation is to dribble itself away over vast, very flat, fertile plains. There are other big floodplains around Lake Chad and the Sudd, but they are much more distant from the east coast and have less well-marked geographic or ecological boundaries. Even so, the retrieval of a 3.2 million-year-old Lucy-like fossil in Chad (*"A." bahrelghazali*) hints at the mobility of this species or species complex (28). That mobility could have had its beginnings more than 4 mya or, if the Kanapoi hominid is any indication, even earlier.

The floodplains of Zambezia are significant because they occupy a large, recognizable chunk of Africa, and the great river itself provides a long, corridor-like link with the east coast. If hominin origins began with the expansions of semiterrestrial apes out of their coastal enclaves, it would be surprising in the extreme if at least one population had not made some sort of adaptive accommodation to this vast area of Africa.

Most palaeontologists agree in their characterization of the habitats in which both Lucies and the earlier Robusts lived. An association between these hominins and grassland fauna and flora does not imply that they were habitually out in the open, but rather that they were the inhabitants of the woody parts of a complex mosaic. That Lucies could range into upland valleys is implied by the proximity of some Ethiopian fossils to montane trees such as pencil cedars (*Juniperus*) and wild olives (*Olea*). Most fossils have been found associated with broad grassy valley bottoms in tropical and subtropical regions. Then, as now, these areas were typically lined by galleries of trees along the levees or by broader wooded fringes along the margins of seasonally flooded grassy "dambos" or on lakeshores. The evolutionary origin of such an association could depend on a degree of geographic isolation in an appropriate region, which the "evolution by basin" model satisfies. Once more, Zambezia seems a promising candidate for the region of origin for the Lucies.

When this association with wooded dambos is taken into account and their imagined status as human ancestors is set aside, what do Lucies look like?

At first glance, Lucies would have resembled bipedal chimps or upright versions of their immediate precursors, the little Dryopithecine tree apes of temperate Eurasia. The sexes differed substantially in size, with the males averaging 50 percent larger than females, standing 1.4 (1.1–1.7) meters high to the females 0.97 meters; estimated weights ranged from 30 to 80 kg.

Lucies were like modern apes in the relative size of their brain cases (the brain averaged about a pint, or 375–500 cc). They also resembled apes in having bony ridges over their eyes; flat noses; large incisors and canines (especially in males); long, curved fingers and toes; long arms; and heavy, dense bones (implying lots of forceful muscular exercise). Because they were closer to apes than humans in all these respects, as well as in evolutionary time, I would guess that their general appearance would have been more apelike than human. Hair would have been straight, black, or red-brown and long enough for infants to cling to with their sturdy digits. The skin may have resembled that of chimps in darkening or becoming polymorphically blotched with age. Unlike chimps, they had buttocks developed from short prominent gluteal "ties" between the pelvis and thighs. Condyls at the femur-tibia joints were flattened to transmit body weight down the legs (29).

The long-toed Lucies have been somewhat uncritically accepted as the makers of some world-famous foot tracks that are recorded in cementlike ash at Laetoli, the original site for the 1939 and 1978 Lucy fossils. The Laetoli fossil tracks suggest an apparently competent walker with rather short toes. Reliably dated to 3.7 mya, these tracks are remarkably early to fit into any of the present range of known hominin fossils. Lucy toes are, according to some scientists, difficult to fit to the tracks. One authority, P. Schmid (30) has portrayed Lucies as clumsy walkers in which the trunk swung from side to side with every step. Given how substantially the pelvis had been reorganized in Lucies and how energy-wasting such ape-like waddling must have been, this characterization may be exaggerated. Nonetheless, if walking was a relatively new faculty, recently derived from a squatting forerunner, their gait was, almost certainly, very much less elegant than our own (figure 5.7). Schmid has also raised doubts about the stamina of Lucies, pointing to the constraints on lungs and efficient oxygenation of the muscles because of their heavily built, relatively inflexible, apelike ribs over a conical thorax. Later hominins had lighter ribs with more lung space in barreling rib cages. Notwithstanding their limitations, Lucies were clearly numerous and widely distributed, so their broader ecological adaptations must have been quite successful for the conditions of their time. There is no direct evidence for social structures; however, extrapolating from living primate species, a big difference in the sizes of the sexes implies one-male groupings. Thus, overall group size might have resembled those of lowland gorillas (2–12 individuals, although group numbers probably varied with the local carrying capacity of their home ranges). In their sexual dimorphism, Lucies also anticipate Robusts rather than the better matched sexes of *Homo* species.

No less than contemporary chimps or baboons, lucies clearly made a living in a mosaic of habitat types that would have included glades, pans, even valleys, of more open grassland or herbaceous vegetation. However, their continued, apelike dependence on trees can be inferred from several lines of evidence and may have a fundamental bearing on the likelihood that they were off the main line of human ancestry.

The importance of trees was unlikely to have been due simply to the food that they supplied; as the descendants of ground apes, it can be assumed that much of their food would have been found by terrestrial foraging. Much more likely was Lucies' looking to trees as the obvious refuge from predators and for night roosts. That retention of trees as the ultimate refuge could have been one of the most fundamental distinctions separating the Lucy lineage from our own because, paradoxically, it could have inhibited the development of brainy children.

Long fingers, short legs, and chimpish heads not only signify some

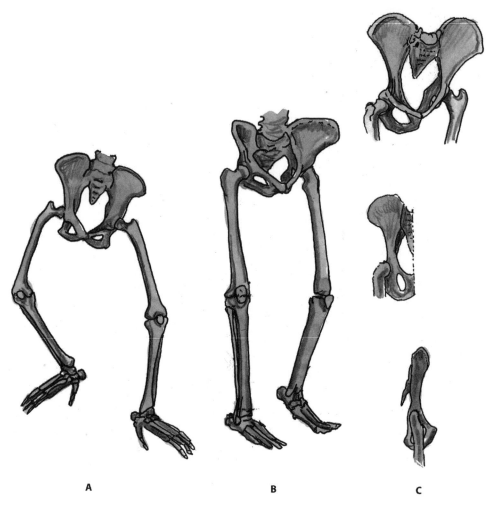

A   B   C

FIGURE 5.7 The proportions and stance of Lucy compared with those of a chimpanzee and human. A, C. Pan. B, D. *A. afarensis.* E, F. *H. sapiens.* (In part after Zihlman, A. L., and W. S.

continuity for arboreal or semiarboreal apelike habits; they also imply some continuity for the apish character of infants, who cannot get much more than clumsy, single-handed support while their mothers are clambering about in trees. Like chimps and gorillas, Lucy infants were likely to have had the ability to cling to their mothers' hair and body, an infantile precocity that would have been more apelike than human. The helplessness of human infants stands out when compared with those of other, mainly arboreal primates. It has long been recognized that a mother's

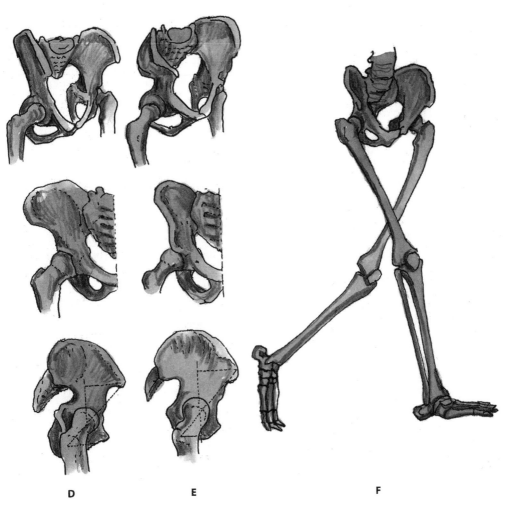

Hunter. 1972. A biomechanical interpretation of the pelvis of *Australopithecus*. *Folia Primato-logica* 18: 1–19.)

care of her helpless infant depends on terrestrial habits and the free hands of a more or less full-time biped (31). In the next chapter, I outline how, for one regional group of hominins, a strategy of predator deterrence rather than a rush to the nearest tree trunk might have severed their last and most enduring tie with trees.

To explain the ecological success of Lucies, there is no need to invoke special intellectual powers or even tool-using skills beyond those already demonstrated by modern chimps. Assuming that they were the only apelike primate over large stretches of territory, their success might best be understood as an extension of the diversification of immigrant Eurasian apes invading new African niches.

Lucies seem to me less like a generalised anticipation of *Homo* than an already versatile primate lineage adapting to an ecologically specialized role. This role was to exploit the plants and more easily harvested animals of a very particular habitat. The wide extent of dambos and valley bottoms in sub-Saharan Africa may help explain why Lucies became so widespread and common within 2 million years of "apes" becoming bipedal. This looks to me like success in a habitat that was relatively free of competition from other primates. Later, this lineage was challenged by coming into contact with a less habitat-specific hominin (with greater technical and social skills). I contend that its evolutionary response was to enhance its own habitat-specific specializations, perhaps becoming still more of a ground-oriented omnivore that was well-buffered by special skills in the excavation of subterranean plants and animals.

Instead of modeling Lucies as direct human ancestors, it might be more realistic to view them as an early trial model for bipedal ways of life. I think of them as a sort of evolutionary Hanuman (the monkey-headed hero of Hindu mythology), a hominin with a way of life that was still recognizably apelike, yet which shared some of our attributes, expressed within the rather specialized ecological setting of wood-lined dambos.

An important feature of floodplain ecology is the adaptation of shoreline and shallow water fauna to an annual cycle that follows the rise and fall of flood waters. This cycle ensures that a similar range of food types tends to be available throughout the year, and that the movements of animals mainly pulse in and out from the core of a permanent watercourse.

Valley-bottom forests and wooded levees similar to (but less extensive than) those of Zambezia occur in the complex ecological mosaic of east Africa, just to the north. The area now occupied by the very shallow Lake Victoria (then a plain with two large, meandering rivers) could have been prime habitat and a staging post for diffusion still further afield—eventually, no doubt, to the margins of the Nile Sudd, Lake Chad, and the Niger's inland floodplain (figure 5.8). *Praeanthropus* may have been one of the first hominins that was sufficiently versatile to migrate out of its home basin. The patchiness and wide separation of major floodplains and the Lucies' (or Kanapoi hominids') head start as colonists could be predicted to encourage their own subspeciation. Eventually, such mobility and ecological success can be supposed to have brought them into contact with other hominin lineages.

If the Lucy-Robust lineage originated in Zambezia, their northward

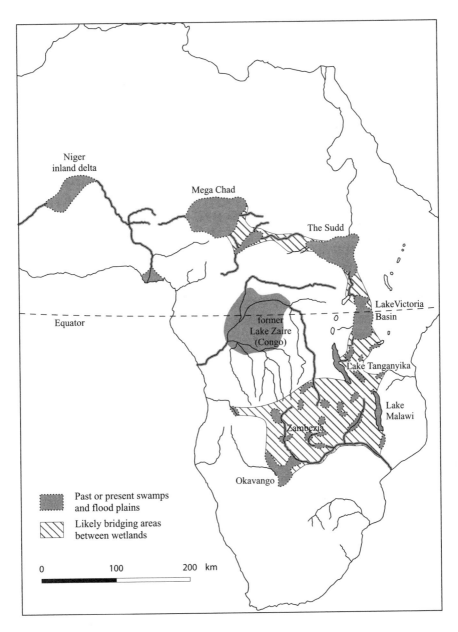

FIGURE 5.8 The web of tropical wetlands and floodplains linking Zambezia and Okavango with the Lake Victoria basin, the Sudd lake, and Lake Chad.

movement would have brought them into contact with conservative hominin populations that were still at the ground ape level of development (particularly in tropical eastern Africa). Further west, they would have encountered ancestral chimpanzees, something that might well have served to constrain their niche. Because these different but still closely related species could sometimes have been in competition for similar resources, an interesting mosaic of not-yet winners and not-yet losers might have developed. Such mosaics, in which conservative species (even ground apes, perhaps) could crop up as fossils long "past their time," will no doubt continue to make hominid prehistory a confusing field.

In the model presented here, colonists of separate basins would tend to look more and more alike the further back in time they go. Is it possible that Lucies (undoubtedly a very successful early pioneer species) were eventually displaced by contact with one or more competitors? They seem to disappear from the fossil record after about 3 mya, but did the Lucy lineage make its own adaptive accommodations to encounters with relatives from other hominin lineages? There are numerous precedents for an evolutionary story in which ecological "displacement" was followed by greater specialization in the displaced species. One example may be apt because it illustrates the nature of the process with a contemporary primate. It also exemplifies a different type of primate's likely route into more exposed habitats.

I remember being baffled by my first, childhood, view of Patas monkeys; one, a bounding streak of red and white, another stopping and standing up on two legs to peer back at us. The animals were traveling across open, sparsely treed plains in what was later to become the Serengeti National Park; viewing them as they ran, we momentarily took them to be some odd sort of gazelle. It was the rearing up on hindlegs and round, anthropoid faces that revealed we were seeing monkeys, not antelopes. Yet, seen close up and at leisure, Patas monkeys (*Cercopithecus patas*) are just long-legged, slender, and more colorful versions of their much commoner, very widespread relatives, the Savanna (or Vervet/Grivet) monkeys, *Cercopithecus aethiops*. These, in turn, are recognizably members of the great clan of African guenons, those archetypical forest monkeys.

Were we to classify animals as we do performers on a race track, the Patas would join greyhounds and thoroughbreds, not monkeys. Yet their ancestors were undoubtedly tree dwellers and must have passed through various arboreal or semiarboreal phases before becoming guenon greyhounds. Indeed, Savanna monkeys, from a common stock and now occupying just such a tree-to-ground niche, almost certainly displaced them

over the greater part of their former range (today, Savanna monkeys occur over most of sub-Saharan Africa outside forests and deserts [32]). The
most likely evolutionary sequence was that a few near-losers in the
proto–Patas/Savanna monkey contest escaped the competition, possibly
in some barren retreat on the edge of the Sahara. Eventually their descendants might have emerged as true terrestrial coursers (i.e., modern Patas
monkeys) and expanded into those steppelike zones where the Savanna
monkeys were decisively inferior.

This evolutionary jostle illustrates that the pioneers of a new niche can
be displaced by later arrivals from the same ancestral stock. The outcome
can be extinction for the pioneers or, in this instance, their displacement
into a still more extreme adaptive niche. The steppes and sparsely wooded
plains of the Sahel would hardly seem promising habitat for a guenon
monkey, yet the proto-Patas clearly brought some formidable advantages
to their survival: long legs, fast responses, acute eyesight, an eclectic diet,
and physiological toughness.

The sequence in which hominins appear and disappear from the fossil
record suggests that something similar might have taken place among
the very earliest species. The pioneering proto-Robusts or Lucies were the
first to enjoy the rich resources of an unoccupied niche only to cede part
(and eventually all) of that niche to late-coming *Homo* while their own
descendants survived by becoming more specialized.

Lucies would have been diurnal and, like Patas monkeys, vulnerable at
night; they must have relied on trees to sleep in. They might have for-
aged for long periods on the ground but, unlike Patas, only within easy
reach of trees where they could sit out encounters with big carnivores. It
may be worth remembering that modern lions were probably not yet part
of the African fauna but that other equally fearsome carnivore species
were abundant, although perhaps less so on seasonally flooded plains.

For any expanding ground ape population, the environment would
generally be moist and closed; but for the Zambezian pioneers, much of it
would have been flanked by drier, sometimes more open territory. So
long as riverine forests could meet all their needs and so long as they
were indefensibly vulnerable away from it, these pioneer species would
have remained forest animals; less concentrated foraging grounds, often
close at hand, would have been a foreign domain.

Between Somalia and Natal, the spacing out of rivers by latitude might
itself have reinforced genetic differences in hominin colonizers. Significant for the future could have been the potentials for later crossings of
watersheds deep in the interior. It is possible to subdivide Africa into ten
major biogeographic regions, each of which is more or less distinctive in

terms of vegetation, climate, seasonality, relative diversity, and the specialization of species and communities. Of these regions, five are in direct contact with east coast basins, and a further three are adjacent to these. Once one of these regions had been colonized by a biped and then adapted to, further divergences would have become possible. In various permutations, rift valleys, lakes, rivers, mountain chains, and ecological barriers define some of these inland biogeographic regions and provide barriers and bottlenecks. In some cases, regions or subregions broadly coincide with drainage basins.

Once the residents of a basin had broken their genetic connection with emigrants that had found their way out, the former might continue to adapt to peculiarities within the home basin, the latter to conditions outside it. Whereas long-term residents would have been under the least pressure to change, different mosaics of conservative and more derived traits could have developed as the two diverging populations pursued their separate fortunes.

In the context of the Savanna/Patas monkey simile, perhaps the arrival of a competitor that had evolved in another basin drove both "host" and invader to develop specializations that had their beginnings in adaptation to the particular ecology of their own "basin of origin." I am proposing here that Lucies (*Praeanthropus*) and "south African man-apes" (*Australopithecus* (*africanus*)) may have had different basin origins; if they or their sympatric descendants partitioned resources by diverging in feeding techniques, those divergences would have had to be built on established differences and strengths. These strengths, in turn, must have tended to correspond, closely, with adaptations that began in their places of origin. The present very sparse and patchy fossil record offers no confirmation for a supposedly northward spread of *Homo* (or *Homo*-like hominins) out of southern Africa, but there are hints of the supposed arrival there of a proto-Robust Lucy in southern Africa.

In terms of the "evolution by basin" model, it seems possible that substantial change in the very durable and successful Lucy lineage only followed direct contact with a sibling species that was able to withstand its competition. Such contact (and some divergence in both lineages) could have started in the far south. The eventual breakout of *A. africanus* descendants from their southern enclave probably set off a succession of competitive challenges and evolutionary change in both hominin lines. For the descendants of *A. africanus*, coexistence with Robusts might have served to sharpen the boundaries of their niche and shaped their ecology in ways that are difficult to second-guess. For the Robusts, this might have culminated in a diversity of nutcracker types and ecological shifts or a drift into more open habitats (33, 34). Nonetheless, it should be re-membered that in spite of Robust features being extreme, these were ef-fectively exaggerations of traits already diagnostic in their ancestral stock. The ancestral traits of Robusts must have included many subtle ecological and physiological adaptations as well as gross physical features, such as broad, crushing molars and big muscles and jaws to go with them. If that ancestor evolved in a particular basin, it is to be expected that the exag-gerations, as well as the subtleties, would include traits that had been adaptive for that basin. If so, what was that basin, and what sort of conti-nuity might be expected? More important, what external influence trans-formed Lucies (or Lucy-like "early" hominins) into the exaggerated, "late" Robusts?

I believe that it is significant, not accidental, that Lucies have never been found in southern African deposits. However, that absence must be qualified by the discovery of one important fossil. In 1984, a very shat-tered cranium was recovered from deep within the Sterkfontein Cave in South Africa. Still undated and known only by its university catalogue number (STW 252), this skull shares many features with the Lucies, no-tably large canines and incisors, a prognathous face, thin brow ridges, and modest temporal muscles. Its swollen molars and premolars are in-termediate between those of the Lucies and later Robusts. Here is the first intimation of continuity between *Praeanthropus* and *Paranthropus* (possi-bly meriting their eventual amalgamation under the latter name [35]). Its provenance also implies that STW 252 was already sufficiently distinct from the local southern African hominins to coexist with them—some-thing that true Lucies had apparently been unable to achieve. Could STW 252 signify the beginnings of ecological partitioning or specialization be-tween two formerly separated lineages that suddenly found themselves sharing territory? If so, could a late invasion of southern Africa by a Lucy-like population have triggered the evolution of Robusts? A date for this skull could be of special significance in answering these questions.

More reliable dating for a possible intermediate between Lucies and Robusts comes from a very distant source. In 1999, a small-brained fossil skull resembling that of a Lucy in several respects was described from Awash, Ethiopia, and dated to 2.5 mya. Named *"Australopithecus" garhi* (36) because its combination of inflated molars and protruding muzzle was a great surprise (Garhi means "surprise" in the Afar language), this hominin anticipates the Robusts in its molars but retains the prog-nathous profile of the Lucies. No tools were found embedded in the fos-sil-bearing deposit itself, but stone-gashed and hammered ungulate bones were found nearby, as were the bones of a long-shanked, short-armed hominin that might or might not have been of the same species. At Gona, nearly 100 km distant, the earliest stone tools ever found (dated at 2.6 mya) have been attributed to this species. But in the schedules and models suggested here, it seems more likely that these tools may have been the work of a species that was closer to our own human lineage but has eluded discovery at these sites (a not uncommon dislocation between bones and stones).

Schools of thought that can visualize the process of giant molars readily and rapidly reversing to much smaller proportions (and which are wedded to the belief that Lucies are direct ancestors) propose that this species sits in a bifurcation between Robusts and humans, and is ancestral to both (37). I think it more likely that *"A." garhi* illustrates, within a very speciose Lucy/Robust lineage, a general trend to evolving big molars.

The trend toward massive teeth, jaws, and skulls, with reinforced bony flanges to carry greatly enlarged muscles, is carried still further by a type of Robust that has turned up in deposits around Lake Turkana, 4500 km north of Sterkfontein. Although a very worn jaw fragment was first found (and named *Paraustralopithecus aethiopicus*) in 1967, it was only in 1985, with Alan Walker's discovery of the celebrated "Black Skull" (dated at 2.5 mya) (38), that the massive "nutcracker" hominins of about 2.3 to 1 mya found a direct link back to STW 252 (and by extension, Lucies). Alan Walker and his colleagues regarded the "Black Skull" (stained a glossy blue-black by manganese salts) as a fine specimen exemplifying an early form of Louis Leakey's so-called *"Zinjanthropus,"* or *Paranthropus boisei*. *Paranthropus aethiopicus* differs from its successors in features that are mostly recognizable in both STW 252 and *"A." garhi*: notably, its prognathous face, small brain, and vastly inflated cheekbones, which, in the Black Skull, provide attachments or enclose massive chewing muscles (figure 5.9). The temporal muscles in particular attach to a high crest erected over the small cranium of this male skull. The overall architecture of *P. aethiopicus* and its greatly enlarged molars and premolars confirm a relationship with the first-discovered and most famous of the "nutcracker" species, *P. boisei* (which tend to have dragged-down orbit margins) (39). Big molar teeth and deep jaws may also link *P. aethiopicus* with some exclusively South African fossils generally grouped under the name *Paranthropus robustus*, a less massively built nutcracker with distinctively squared-off, not slewed orbits.

Paranthropus boisei seems to have been the regional and variable representative of a more widespread group of advanced "nutcrackers" (40). As is to be expected from a wide scatter of (east African) sites with a timespan of 1.4 to 2.7 mya, there is great individual variation, even in a small

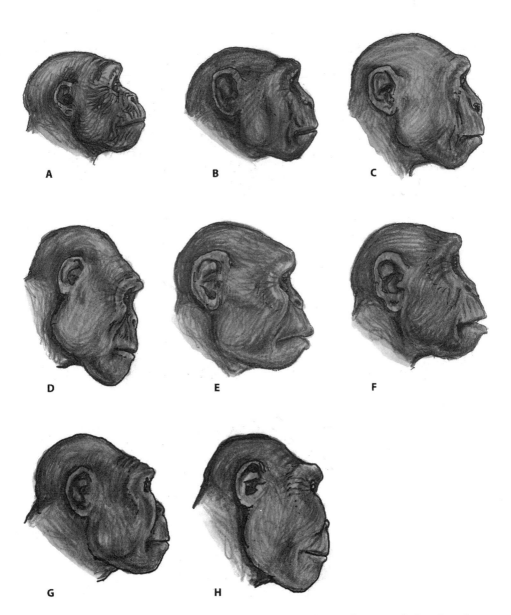

FIGURE 5.9 Imaginary profile reconstructions of the Lucy/Robust lineage with *Orronin* and *Ardipithecus*. A. *Orrorin tugenensis*. B. *Ardipithecus ramidus*. C. *"A." anamensis*. D. *"A." afarensis*. E. *P. aethiopicus*. F. *"A." garhi*. G. *P. robustus*. H. *P. boisei*.

FIGURE 5.10 Reconstructed portrait of the original "Nutcracker Man," *Zinjanthropus*, now known as *Paranthropus boisei*.

sample of fossils. There have been implausible suggestions that *P. robustus* derives from *A. africanus*, but a mosaic of resemblances between all the Robusts makes a common ancestry more likely (41, 33). Morphological peculiarities of *A. robustus* have been more convincingly associated with its southerly isolation from a tropical parent stock and some minor but significant "dwarfing" (42).

The incisors and canines of *P. boisei* were so diminished that the shortened, arched palate lay mostly beneath and behind the orbits. The enlarged jaw musculature pushed the cheekbones forward into the same plane as the front teeth, creating its characteristic flat face (figure 5.10). It has been suggested that a functional explanation for the curious shrinkage of the front teeth was the need to sort and sift gritty or coarse-coated foods with the lips, tongue, and fore teeth before drawing the residue back into the mouth for crushing and heavy chewing (43). Lip and tongue sorting could imply a progressively greater inclusion of soiled or rinded foods in their diet. Diminished front teeth also imply that hands, not incisors, must have extracted and processed foods in the first place. Given that these dental innovations come quite late in the evolution of the Robust lineage, we can envisage progressively improved skills in excavation or extraction; perhaps the use of digging sticks became more and more routine. With wood failing to fossilize, we may never know for certain; but very abraded ungulate long bones, clearly used as soil-breaking leverage tools, have been found in close association with Robust fossils, so tool-assisted excavation of foods seems to be more than likely (44, 45). In this respect, contemporary chimp technology shows the rudiments of spading or chiseling when apes from particular localities use hard palm ribs to open or deepen cavities in tree trunks that conceal food (46).

Paranthropus robustus (found by a schoolboy at a site called Kroomdraai and part of the same complex of calcite deposits as Sterkfontein and Swartkrans) was the first "nutcracker" to be described by Robert Broom in 1938; it has been dated to about 2 mya. In 1948, Broom discovered a different and younger fossil, a mandible, to which he gave the name "Swartkrans ape-man" (*Paranthropus crassidens*). Like variation in the *P. boisei* specimens in eastern Africa, the *P. robustus* specimens may be sampling a sufficiently broad time span (2–1 mya) to accommodate more than one temporal population, but coming from such a tight cluster of sites, no geographic element intrudes.

The Robust fossil record, if permitted to include Lucies as forerunners, suggests a single lineage that appears, in various impressively "staged" but highly variable forms, to span over 3 million years. The distribution of all the Robust and proto-Robust fossils ranges, very patchily, from a handful of sites in South Africa to a wider scatter in East Africa, Ethiopia, and as far afield as Chad, where the chin and foreteeth of a Lucy-like animal, *"A." bahrelghazali*, has been dated to 3 to 3.5 mya. How the distribution of the Robust lineage and its changes through space and time are interpreted must be fundamentally altered by reference to basin evolution and my suggestion of east coast origins. I think that Lucies may have owed much of their preeminence to a long head start in the colonization of the interior. While a more northern origin remains possible, I think the most plausible basin was that of the Zambezi.

Supposing that Lucies represent the earliest ancestors for the Robusts, the preeminence of *Praeanthropus* over a great stretch of Africa for nearly a million years suggests that they were singularly well adapted to diffuse along narrow forested or wooded watercourses. In this way, they could cross watershed boundaries with ease, a facility enhanced by their early "perfection" of bipedal walking. Identifying the precise ecological conditions that could provide a setting for such a successful adaptation as bipedalism as well as locating its geographic locus has exercised many minds and generated many models.

The route of enquiry taken here has been to jump forward to more specialized descendants, on the principle that the peculiar specializations of descendants may only be hinted at in earlier members of a lineage. The more specialized the adaptations, the more likely they are to originate in regions with appropriate ecologies.

One hint that the primary habitat of Robusts resembled the forests that line riversides and levees in the Zambezian Basin is that their fossils are often associated with the water-dependent, valley-loving reduncine, or waterbuck, lineage (47, 48). Where a resource evolves, consumers of that resource evolve too, so there is good reason to suppose, even in the absence of good fossil sites, that the principal valley-bottom antelopes, the waterbuck, kobs, or "Reduncines" may have evolved, for the most part, within "Zambezia." Not only are these typical fauna of the flat, grassy valley bottoms (going by the name of "dambo" in central Africa and "mbuga" in Kiswahili-speaking areas), but Zambezia is quite the most likely region of evolutionary origin for this group of valley-bottom grazing antelopes. It could also be the ultimate region of origin for Lucies and Robusts.

Robust Australopithecines have been shown to be consistent inhabitants of the wooded levees and river banks that meander through flood plains and "dambos" (ref). Lucies, too, are consistently retrieved from wooded riverine habitats, another point in which they seem to anticipate the preferences of Robusts (49). A common origin implies a common basin of origin; in terms of the "evolution by basin" model, Zambezia, the main center in Africa for broad flat floodplains, is the most realistic candidate.

If my proposal is correct, their appearances or eruptions outside this region may represent a succession of migrations in which the migrants adapted differently, every time, to the exigencies of time and place. Some, like the pioneering *Praeanthropus*, became relatively successful, stable, widely distributed, and long-lasting types, whereas others may have been unstable, patchily distributed, and more prone to local adaptation or variation. This tendency to local diversification might have been reinforced if Robusts developed tool-assisted techniques that differed from region to region.

I have already noted that floodplain omnivores may be better able to secure animal foods in the late flood season but rely more on plant foods in the late dry season. This strategy would have been an exactly opposite permutation to the strategy of an arid land omnivore for which the main seasonal dietary shift is from mainly vegetable foods in the wet season to mainly animal foods in the dry. Furthermore, it is precisely this sort of inversion that permits sympatric species in complex interdigitating habitats to coexist with minimal overlap in diet.

If Robusts relied on roots during the dry season, they would have moved in on a rootling niche (typically dominated by pigs and burrowing rodents) (50). This heralds a characteristic of all hominins—namely, to appropriate part, or all, of the niches of other species. Typically, the excavation techniques of hominids should have displayed a versatility of approach and a maneuverability of fingers, hands, and arms (even in the absence of tools) that rendered them decisively superior to pigs and rodents with their fixed digging styles. Even so, Robusts would have made dental and physiological adaptations that were typical of any other mammal entering a new niche.

A tough, soiled diet would have required a specialized digestion and dentition. Also, valley bottoms could have favored the development of some physiological tolerance of heat stress. These are the sort of developments that should be looked for if the evolution of massive jaws and teeth in robust Australopithecus are to be explained (51). The overdevelopment of their facial muscles and tooth buttressing is thought to have given Robusts odd proportions, with big, broad heads on bodies that were short and small by modern human standards.

Finally, taking the Robusts' occupation of floodplains as a starting point, consider what might happen when polyps of populations moved; imagine that populations had become established in Ethiopia. Here, floodplains are an insignificant element; but massive teeth and a tolerance of solar radiation would have permitted Robusts to thrive in drier, cooler savannas because they could harvest and digest some superabundant resources, such as local grass and shrub roots. In Ethiopia, many animals and plants that are common in eastern and southern Africa are rare or absent. The combination of fewer predators, fewer competitors, and fewer diseases could have permitted the invading hominids to become more abundant in this part of their range than elsewhere. Eventually, because the environment was so unlike the one in which they had evolved, the Ethiopian form could have become a distinct race or species—and perhaps one that lived at a higher density than some of its relatives. Certainly, it could no longer be characterized as a floodplain species (33), even if that had been, unequivocally, a launching pad in its evolutionary career.

Before passing on to consider hominin fortunes in the south, there are some final points to be made before taking leave of the Robusts. The robust lineage consisted of "pure bipeds"—probably quite competent at many of the things that modern humans do, and very likely better at doing some of the things that our own ancestors were doing at that time. More significant is the fact that later Robusts were clearly earning a living by different and much more specialized means than the humans or proto-humans with which they shared their environment. To observe that Robusts adapted to the challenge of difficult foods does not explain what pressures drove their apparently more "generalised" ancestors down the specialist pathway.

I think it possible that Robusts, like Patas monkeys, may have derived from a lineage that was able to range over a wider span of environments and eat a broader choice of foods only so long as they were alone in their occupation of this newly evolved, tool-assisted hominin niche. When others challenged their resources, or when they encountered other hominins with more developed tool kits, their drift toward extinction might have been deflected by selection for "enhanced adaptation." Competition for resources might have favored precisely those characteristics that derived from specific adaptation to their own place of origin, the adaptations that most differentiated them from their competitor from another basin. In retaining basin-specific adaptations, former basin endemics (notably those that had become more widespread) might have retained details of anatomy or aptitude that the competition lacked. These are the materials on which selection operates and, in this instance, could have been the starting points for successful specializations in the finding, processing, and consumption of tough, frequently subterranean foods (52). Relying on current fossil evidence, my reconstructions illustrate, in summary form, how I visualize the appearance of these stocky hominins.

The confrontation between two early hominins proposed here may have been one of the earliest of many instances in which competition between two lines drove one toward greater ecological specialization while the other—bolstered, perhaps, by more generalized technology and behavior patterns—seized the central, broad-spectrum ground. Such divergences are common enough in the ecological radiations of other animals. The possibility that the long-term outcomes of competition between different hominin technologies might parallel such adaptive divergences was, so far as I am aware, first proposed in my book "Self-made man" (25). It is a theme that repeats itself, with several variations, in the pages that follow.

REFERENCES

1. Hayaki, H. 1985. Social play of juvenile and adolescent chimpanzees in the Mahale Mountains National Park, Tanzania. *Primates* 26: 343–360.
2. Aiello, L., and M. C. Dean. 1990. *An Introduction to Human Evolutionary Anatomy*. London: Academic Press.
3. Stern, J. T., and R. L. Susman. 1983. The locomotor anatomy of *Australopithecus afarensis*. *American Journal of Physical Anthropology* 60: 279–316.
4. Leakey, M. G., C. S. Feibel, I. McDougall, and A. Walker. 1995. New four-million-year-old hominid species from Kanapoi and Allia Bay, Kenya. *Nature* 376: 565–571.

5. Lovejoy, C. O. 1974. The gait of australopithecines. *Yearbook of Physical Anthropology* 17: 147–161.

6. Latimer, B. 1991. Locomotor adaptations in *Australopithecus afarensis*: The issue of arboreality. In *Origine(s) de la Bipédie chez les Hominidés*, ed. Y. Coppens and B. Senut, 169–176. Paris: CNRS.

7. Richmond, B. G., and D. S. Strait. 2000. Evidence that humans evolved from a knuckle-walking ancestor. *Nature* 404: 382.

8. McHenry, H. M. 1986. The first bipeds: a comparison of the *Australopithecus afarensis* and *Australopithecus africanus* postcranium and implications for the evolution of bipedalism. *Journal of Human Evolution* 15: 177–191.

9. Jolly, C. J. 1970. The seed eaters: a new model of hominid differentiation based on baboon analogy. *Man* 5: 5–26.

10. Sarmiento, E. E. 1998. Generalized quadrupeds, committed bipeds, and the shift to open habitats: an evolutionary model of hominid divergence. *American Museum Novitates* 3250: 78.

11. Kingdon, J. 1971–1982. *East African Mammals. An Atlas of Evolution in Africa* (3 vols., 7 parts). London: Academic Press.

12. Acocks, J.P.H. 1988. *Veld Types of South Africa. 3rd Edn. Memoirs of the Botanical Survey of South Africa* No. 57. Pretoria: Botanical Research Institute.

13. Peters, C. R., and R. J. Blumenschine. 1996. Landscape perspective on possible land use patterns for early Pleistocene hominids in the Olduvai Basin, Tanzania: Pt. II. Expanding the landscape models. *Kaupia* 6: 175–221.

14. Johanson, D. C., and B. Edgar. 1996. *From Lucy to Language*. New York: Simon and Schuster.

15. Leakey, M. G., F. Spoor, F. H. Brown, P. N. Gathogo, C. Kiarie, L. N. Leakey, and I. McDougall. 2001. New hominin genus from eastern Africa shows diverse middle Pliocene lineages. *Nature* 410: 433–440.

16. Johanson, D. C.. and M. A. Edey. 1981. *Lucy: The Beginnings of Humankind*. New York: Granada.

17. Weinert, H. 1950. Uber dei Neuen Vor- und Fruhmenschenfunde aus Afrika, Java, China und Frankreich. *Zeit. Morph. Anthropol.* 42: 113–148.

18. Senyurek, M. 1955. A note on the teeth of *Meganthropus africanus* Weinert from Tanganyika Territory. *Bulletin* 19: 1–57.

19. Strait, D. S., F. E. Grine, and M. A. Moniz. 1997. A reappraisal of early hominid phylogeny. *Journal of Human Evolution* 32: 17–82.

20. Susman, R. L., J. T. Stern, and W. L. Jungers. 1984. Arboreality and bipedality in the Hadar hominids. *Folia Primatologica* 43: 113–156.

21. Lovejoy, O. C. 1988. Evolution of human walking. *Scientific American* 259: 82–89, 118–126.

22. Falk, D. and G. C. Conroy. 1983. The cranial venous sinus system in *Australopithecus afarensis*. *Nature* 306: 779–781.

23. Falk, D. 1986b. Hominid evolution. *Science* 234: 11.

24. Falk, D. 1990. Brain evolution in *Homo*: the "radiator" theory. *Behavioral and Brain Sciences* 13: 333–381.

25. Kingdon, J. 1993a. *Self-made Man. Human Evolution from Eden to Extinction?* New York: John Wiley & Sons.

26. Peters, C.R., E. M. O'Brien, and R. B. Drummond. 1992. *Edible Wild Plants of Sub-saharan Africa*. Kew, UK: Royal Botanic Gardens.

27. O'Brien, E. M., and C. R. Peters. 1999. Landforms, climate, ecogeographic mosaics, and potential for hominid diversity in Pliocene Africa. In *African Biogeography, Climate Change, and Human Evolution*, ed. T. G. Bromage and F. Schrenk, 115–137. New York: Oxford University Press.

28. Brunet, M., A. Beauvilain, Y. Coppens, E. Heintz, A.H.E. Moutaye, and D. R. Pilbeam. 1996. *Australopithecus bahrelghazali* une nouvelle espece d'Hominide ancien de la region de Koro Toro (Tchad). *C.R. Acad. Sci. Ser.* IIa 322: 907–913.

29. Jungers, W. L. 1982. Lucy's limbs: Skeletal allometry and locomotion in *Australopithecus afarensis*. *Nature* 297: 676–678.

30. Schmid, P. 1991. The trunk of the Australopithecines. In *Origine(s) de la Bipédie chez les Hominidés*, ed. Y. Coppens and B. Senut, 225–234. Paris: CNRS.

31. Conroy, G. C. and K. Kuykendall. 1995. Paleopediatrics: or when did human infants really become human? *American Journal of Physical Anthropology* 98: 121–131.

32. Kingdon, J. 1971. *East African Mammals. An Atlas of Evolution in Africa.* (Vol 1: Primates). London: Academic Press.

33. Grine, F. E. 1985. Was interspecific competition a motive force in early hominid evolution? In *Species and Speciation*, ed. E. Vrba, 143–152. Pretoria: Transvaal Museum Monograph.

34. Grine, F. E. 1986. Dental evidence for dietary differences in *Australopithecus* and *Paranthropus*: a quantitative analysis of permanent molar microwear. *Journal of Human Evolution* 15: 783–822.

35. Clarke, R. J. 1996. The genus *Paranthropus*: what's in a name? In *Contemporary Issues in Human Evolution*, ed. W. E. Meikle, F. C. Howell, and N. G. Jablonski, 93–104. San Francisco: California Academy of Sciences.

36. White, T. D., G. Suwa, W. K. Hart, R. C. Walters, G. WoldeGabriel, J. de Heinzelin, J. D. Clark, B. Asfaw, and E. Vrba. 1993. New discoveries of *Australopithecus* at Maka in Ethiopia. *Nature* 366: 261–265.

37. Locke, R. 1999. The First Human? *Discovering Archaeology* Jul/Aug 1999.

38. Walker, A. C., R. E. Leakey, J. M. Harris, and F. H. Brown. 1986. 2.5-MYR *Australopithecus boisei* from west of Lake Turkana, Kenya. *Nature* 322: 517–522.

39. Walker, A. C., and R.E.F. Leakey. 1988. The evolution of *Australopithecus boisei*. In *Evolutionary History of the Robust Australopithecines*, ed. F. E. Grine, 247–258. New York: Aldine de Gruyter.

40. Grine, F. E., ed. 1988. *Evolutionary History of the "Robust" Australopithecines.* New York: Aldine de Gruyter.

41. Wood, B. A. 1988. Are "robust" australopithecines a monophyletic group? In *The Evolutionary History of the "Robust" Australopithecines*, ed. F. E. Grine, 269–284. New York: Aldine de Gruyter.

42. Suwa, G., T. D. White, and F. C. Howell. 1996. Mandibular postcanine dentition from the Shungura Formation, Ethiopia: crown morphology, taxonomic allocations, and Plio-Pleistocene hominid evolution. *American Journal of Physical Anthropology* 101: 247–282.

43. Kay, R. F., and F. E. Grine. 1989. Tooth morphology, wear and diet in *Australopithecus* and *Paranthropus* from southern Africa. In *The Evolutionary History of the "Robust" Australopithecines*, ed. F. E. Grine, 427–447. New York: Aldine de Gruyter.

44. Weaver, K. F. 1985. Stones, Bones, and Early Man: The Search for Our Ances-
    tors. *National Geographic* 168(5).

45. Brain, C. K., and P. Shipman. 1993. The Swartkrans bone tools. In *Swartkrans.
    A Case's Chronicle of Early Man*, ed. C. K. Brain, 195–218. *Transvaal Museum
    Monograph* No. 8. Pretoria.

46. Whiten, A., and C. Boesch. 2001. Chimpanzee Cultures. *Scientific American* Jan
    2001: 61–67

47. Shipman, P., and J. M. Harris. 1988. Habitat preference and paleoecology of
    Australopithecus boisei in Eastern Africa. In *Evolutionary History of the "Ro-
    bust" Australopithecines*, ed. F. E. Grine, 343–382. New York: Aldine de
    Gruyter.

48. Spencer, L. M. 1997. Dietary adaptations of Plio-Pleistocene Bovidae: implica-
    tions for hominid habitat use. *Journal of Human Evolution* 32: 201–228.

49. White, T. D., and J. M. Harris. 1977. Suid evolution and correlation of African
    hominid localities. *Science* 198: 13–21.

50. Hatley, T., and J. Kapelleman. 1980. Bears, pigs and Plio-Pleistocene hominids:
    A case for the exploitation of below ground food resources. *Human Ecology*
    8(4): 371–387.

51. Rak, Y. 1983. *The Australopithecine Face*. London: Academic Press.

52. Walker, A. C. 1981. Diets and Teeth. Dietary hypotheses and human evolu-
    tion. *Philosophical Transactions of the Royal Society, London* B 292: 57–64.

CHAPTER 6

On Being a Manipulative Man-ape

Isolation in the South

Unique ecology of the southeast African coast and its hinterland. South African Man-ape, *Australopithecuns africanus* lineage (3.6–2.3 mya) approximates human ancestors. Chimpanzee behavior as a guide to latent behaviors and manipulation of tools. "Scavenging" graduating into "niche-stealing." The *A. africanus* grade of hominins a useful model of transition between ground apes and early *Homo*.

I once stayed some months in a country guest house in Greytown, in the South African province of Natal. While I was there, the garden temporarily hosted a variety of birds (I particularly remember the wagtails). Winter in the Drakensberg Mountains, which border Natal, is an ordeal that most of the smaller passerine birds escape by flying down to the foothills and coastal plain only to fly back to the uplands for the summer; some pass through Greytown.

Who could have guessed that this suburban birdwatcher's delight was actually the last vestige of a very ancient seasonal movement, by many more organisms than birds, up and down the hills of southeastern Africa. Still less could I have envisaged then that this passage might not only mimic similar shifts by early hominins, but that responding to seasonal change might have favored traits essential for the emergence of humans.

In the previous chapters, I proposed that distinct hominin lineages, including our own, might have emerged directly from different regional populations of the coastal ground ape. KwaZulu-Natal, with Greytown near its center, embraces a large part of the most southerly of these regions or "basins" (actually south Africa is an amalgam of numerous mini-basins with only a few largish rivers, the most significant being the Tugela). I claimed that this might be a special case because the entire area south of the Limpopo constitutes a distinct geographic and ecological entity in its own right.

Although all basins would have been open to colonization from the east coast, the prospects for speciation by would-be colonists was shown to differ considerably from basin to basin. So, if divergence between the human and nutcracker lineages can be explained by "basin evolution," consider again, in that light, what could be relevant differences between basins.

The point has already been made that the Southern Highlands (in today's Tanzania) were more continuously and intimately connected with the central region of ground ape distribution on the Zanj coast. This should have eased gene flow between coastal populations and those in the immediate interior. Theoretically, this should have inhibited interior regions close to the coast and more centrally positioned from becoming sources of substantial genetic innovation, but the possibilities should not be entirely dismissed. With Zambezia supposedly homeland to the Lucy/Robust lineages, the choice of theoretical homelands for other emergent species narrows to the two extremities of the ground ape's supposed range: south or northeast Africa and to the peculiar lakeshore habitats of the Rift Valleys and Great Lakes.

While Ethiopia is plausible, I have already indicated that the recurrence of Lucies in this area, from an early date and over a protracted period, diminishes the chances that two very early hominins coexisted there. If there was space for two species of early hominin to share the environment, it is difficult, on present evidence, to explain why Lucies so effectively masked the existence of other hominins in a region that is particularly rich in fossil sites. As for the Eastern Rift and Great Lakes region, the possibilities are discussed in the next chapter.

There are a number of environmental conditions that might have predisposed South Africa to be a region of choice for the birth of a *Homo*-like hominin from a ground ape ancestry similar to that of the Lucies. Because fossils have to be the main guide to events, there are the South

African Man-apes, conveniently lumped as members of the *Australopithecus africanus* lineage, which not only differ from Lucies (1) and Robusts (2) but, on the strength of their shorter faces, larger brain-cases, and less prominent eye ridges, have long been considered by numerous scholars to belong to (or lie very close to) the *Homo* lineage (3, 4).

Contrast Zambezia and South Africa as potential homes for newly bipedal ground apes. To exploit seasonal changes in their very flat environment, Zambezian hominins would not have had to move very far from the core of their riverine ranges or territories. The choice of food types available to them would have been more restricted than in coastal forest mosaics, a mix of distinctive plants and sessile animals. The overall distribution of social groups would have tended to be linear, with territories or home ranges strung out along water courses like beads on a string. Such populations could have retained something of the social and spatial stability that has been proposed for their coastal ancestors because their environment, although different from the oceanic littoral, would have had some general resemblances in its floral composition and structure.

South of the Limpopo, things were very different. Lying between 21° and 34° south, a long coastal strip narrows to as little as 20 km wide, with mountains rising abruptly out of the plain (their steep pitch interrupted, here and there, by stepped terraces) to rise as high as 3000 m before falling away over much shallower slopes beyond the watershed. Climatic contrasts, especially between the warmer, moister Indian Ocean and drier, cooler Atlantic drainages are dramatic today, and seasonal differences could have been even greater in the past. Southern winters may be short (from May to August) but night temperatures drop below freezing inland, and there are heavy frosts and even snow in the highest localities. Summers, instead, are long and hot with rainfall ranging from more than 1600 mm on the escarpments to 350 mm or less in the west and northeast. The interaction between closely adjacent environments—low to high, wet to dry, hot to cold—is so complex that some 30 vegetation types have been described, just for the eastern portion of South Africa (5, 6). There are more than 70 perennial rivers and more than 100 smaller streams flowing out of the mountains that flank this 1000-km stretch of coast. Within this complex, there are larger river basins, such as that of the Tugela. Flowing through more than 10 vegetation zones, this major river has margins up to 3000 m high, and many of its sources are close to north-flowing tributaries of the Orange and Limpopo Rivers, providing easy corridors for the further expansion of montane riverine fauna and flora (figure 6.1).

As was pointed out earlier, the well-watered eastern littoral is separated

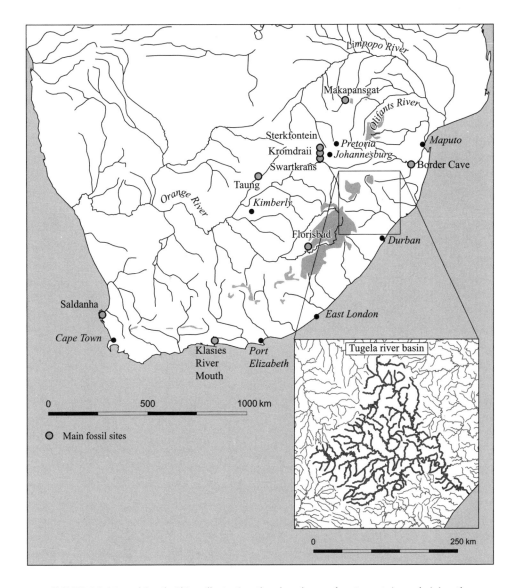

FIGURE 6.1 Map of South Africa, illustrating the abundance of east coast rivers draining the Drakensberg massif.

from the mouth of the Zambezi by a vast shelf of flat, deep sands devoid of surface water (7). This riverless coastal lowland is not an absolute barrier but has inhibited the spread of some forest and water-dependent species and must have acted as somewhat of a barrier to gene flow between established populations on either side of it. Other dry stretches of coast (or rivers) break the continuity of coastal flora and fauna between the Horn of Africa and Cape of Good Hope (8, 9), but the length and the southerly latitude of the Zambezi-Limpopo arid shelf ensures that South African coastal gene pools are by far the smallest. They are also the most decisively isolated and live in the most environmentally different of all east coast communities.

The far south would have represented a much greater challenge than Zambezia for the earliest hominins, yet *Australopithecus africanus* fossils attest to South Africa being home to a rather peculiar group of hominins. The estimates for their first appearance are almost as early as those for the Lucies: 3.55 mya to the latter's 3.7 mya (10). Unlike Lucies, which disappear about 2.9 mya, the South African Man-apes are still recognized at about 2 mya.

They also differ from Lucies and Robusts in showing less sexual dimorphism (implying significant differences in social structure) and no sign of the anatomical or dental adaptation to specialized diets that typify the Robusts. These differences could be interpreted in terms of technological and behavioral solutions to obtaining food. The complexity of ecotypes in southeast Africa could have invited an adaptive shift that diversified food and ways of acquiring it. Equally important, an initially tropically adapted ground ape's yearround preference for stable home ranges in coastal forests would have had to change. Slowly, as seasonal shifts up and down the mountains became a routine annual movement, the narrow strip of coastal lowland forests would have become a winter retreat. Eventually, as the hominins' overall range enlarged, the narrow coastal shelf might have become a marginal habitat for a much larger population of now upland or veldt-adapted animals. Assuming that the process of accommodating to the south African interior was a rather slow, drawn-out process, it can be supposed that the known fossils from Highveldt sites such as Sterkfontein and Swartkrans have, so far, sampled local hominins that were already long established in the area. In other words, they come from relatively late dates in the geographic spread and ecological adaptative history of a unique South African endemic hominin.

South African fossils give faces and figures to a peculiar lineage of hominins living in rather surprising environments. The reconstructions that follow try to relate the real fossils to a region with equally real and peculiar ecologies and a biogeography that is even more distinctive. South Africa is portrayed here as the most promising locale for early human evolution, but my broader preoccupation is with the process that led to the evolution of *Homo*. To my mind, the South African fossils illustrate this process very well, but it remains remotely possible that something very similar took place in some other African upland. The absence of appropriate fossils elsewhere makes South Africa more likely to be a real setting rather than an illustrative approximation. Although there must be elements of a theoretical construct in the account that follows, there are sound ecological reasons for firmly tying that reconstruction to southern Africa and its rich trove of existent fossils. In the context of my hypothetical east African ground ape and basin evolution, the roots of a southern African hominin radiation must also be tied to coastal origins in the deep south.

As it extended its range into the uplands, a prime challenge to the *A. africanus* lineage would have been to get access to resources that were contested or preempted by a variety of other species, mostly mammals. It is possible that one significant divergence between this species and the Lucies living further north could have centered on more coordinated social responses to predators and aggressive competitors rather than pell-mell scampers into the trees. Thus, achieving access to contested resources could have called for the development of complex strategic behavior. That contest would have involved fewer primates (other than baboons) than in more tropical habitats, but pigs, terrestrial and subterranean rodents, carnivores, and hyraxes are all known to have been abundant or diverse in southern Africa. Nonetheless, overall productivity has probably always fallen well short of the better tropical environments. Experimental enlargement of an omnivore's menu could, perhaps, have stimulated further elaboration of a range of tools, initially of tool types that can still be observed in rudimentary use by chimpanzees today. It could have been in the responses of a rather small, slow mover to these challenges that a slightly more terrestrial *A. africanus* might have differed most from the Lucies and later Robusts. In the process, these southern Man-apes might have begun to relinquish some of the innate responses that once helped them secure and process foods. For example, "instinctive feeders" tend to avoid any plant or animal that is protected by an impervious shell, a chemical screen, or off-putting behavior. From an early age, instinctive feeders ignore such organisms in preference for a well-established menu that is recognizably specific to the species. I am suggesting that these hominins might have become "niche-thieves" because they pursued technological and behavioral solutions to get access to foods that were previously only accessible to *other* species (ones that had evolved the techniques or physiologies to bypass or outwit their prey's defenses). In a sense, all adaptive advances involve moving in on the niches of precursors, but I am proposing a much faster type of takeover in which tools, techniques, and strategic intelligence all played pivotal roles. This is a concept that is developed further in subsequent pages and chapters.

Before an ever increasing list of fossils proved that the hominin tree was abundantly branched, *Australopithecus africanus* was allocated its place within a linear scheme of human evolution, largely because it was one of the earliest "ape-men" to be discovered. First described in 1925 from the complete face, jaw, and brain cast of a juvenile excavated from a lime quarry at Taung, in today's Botswana, its discoverer, Raymond Dart, appropriately called his find "South African Man-Ape" (11). The academic establishment of the day, as usual, gave his discovery a skeptical or openly hostile reception. Later excavation of a splendid array of fossils in South Africa has suggested that the category *A. africanus* may embrace a variable, continuously changing, highly localized, but relatively long-lived lineage.

The earliest hominin fossil in South Africa, dated by one analysis to 3.3 mya (but possibly as early as 3.55 mya) is "Little Foot," a nearly complete (but still only partially described) skeleton standing about 1.3 m high and retrieved from the bottom of what must have been a 13-m sink-hole at Sterkfontein (12). Argument surrounds its exact identity, but it is currently supposed to be a relatively early *A. africanus*. The majority of other specimens (some of them relatively complete skulls and a vertebral column, complete with pelvis and thigh bone) are considerably younger, dated to about 2.5 mya, while the youngest, among them Dart's original Taung child, have been estimated at 2.05 mya (10).

The existence of a distinct lineage of hominins restricted to south Africa is consistent with an "evolution by basin" model (although the "basin," in this case, consists of a great number of short rivers and supposed hominin habitats that spill over to include west-flowing tributaries of the Orange River, a substantial territory of Highvelt). Indeed, the fossils come from rather dry areas far inland, and some from well within the Atlantic Ocean watershed, which implies a long preceding period of adaptation to southern African veltlands.

According to my model, virtually all of these fossils should, by 3 million years ago, be late, possibly relictual, populations that are being sampled. This conclusion is irreconcilable with "linear" schools of thought that would have *A. africanus* become a descendant of *Praeanthropus ("A. afarensis")*. Because the two species were broadly contemporaneous and differed in key anatomical details, this is no longer plausible. In addition, in spite of having an extensive range in east and northeast Africa, typical *Praeanthropus* does not turn up in South African deposits (nor *A. africanus* in east African ones). If both species derived from a common coastal ancestor, early forms might still have been sufficiently closely related to exclude each other competitively from their respective basins or ranges. Later, the two lineages might have coexisted for a while, but only after their specializations had become sufficiently different for distinct niches to emerge and for their distinctness to have favored the spread out and away from their original, more limited range. In chapter 5, I pointed to STW 252, a late, Lucy-like fossil from south Africa, as possible evidence for just such a brief coexistence.

There are other reasons to suppose that the South African Man-ape belonged to a separate lineage from the Lucies and the Robusts. Both the thickness and the conformation of tooth enamel differs between the two groups (13, 14), as does the pattern of tooth rooting in the premolars of the mandible (15, 2). Although both of these features can vary among individuals, the South African Man-apes are consistently closer to humans in both of these features. Furthermore wear patterns on the surfaces of teeth suggest that members of the Robust lineage chewed many hard, scratch-inducing items in their diet, whereas *A. africanus* took softer, less abrasive foods (16).

Before they were shown to be broadly contemporaneous, Dart's "South African Man-ape" was both appropriately "late" and sufficiently alike to a Lucy to have persuaded some scientists that it was a direct descendant. The differences may be relatively subtle but, from an anthropocentric point of view, are significant. Distinguishing features of the South African Man-ape, when compared with Lucy, include a relatively larger brain; a smaller "muzzle" or face (17); much less difference between the sexes (18); and molars of a similar type, arrangement, and rooting pattern to *Homo*. There are also clear hints of *Homo*-like ancestry in the bones of the hands, which are decisively less like those of Lucies than those of humans (19–21). Likewise, the thumb is very unlike those of apes (and Lucies) in being well developed and constructed to be opposed to the fingers in various subtle permutations of grip and clasp. As for the wrist, a humanlike flexibility contrasts dramatically with the "tied-up," well-locked bones of Lucies (22). While the finger bones of the 3.3 million-year-old "Little Foot" skeleton are apishly curved, they are of similar proportion to those of later humans. By contrast, the limb proportions and feet, instead, are less like those of modern humans than those of apes, with the big toe still capable of deflection from the other toes, as in apes (23, 24).

In terms of external appearance, the generally cooler habitats of *A. africanus* might have required fur that was denser or of a different texture

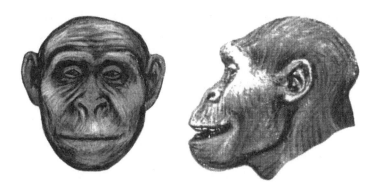

FIGURE 6.2 Reconstruction of *A. africanus* face, portrait and profile.

from that of contemporary lowland apes. A high premium on effective communication might have required particularly expressive faces with strategically placed tufts of hair on a facial disc with naked cheeks, nose, mouth, and brow. In terms of general demeanor, I envisage them as being vocal and continuously alert to one another's presence; possibly confident and ostentatious when in a group, they may have been much more retiring when isolated as individuals. Compared to the Lucies' more athletic, sexually differentiated, apish troopers, moving rapidly from point to point, I see them as somewhat more conformist, possibly more clumsy but prone to explore their surroundings manually and thoroughly (figure 6.2).

Now obsolete proposals for humans deriving from the same lineage as Robusts (25) were fueled by reference to the very flat faces of the later forms. Although the main mechanisms for this flat facial profile were *expansion* of the molars and jaw muscles and a radical (and apparently rapid) *reduction* of the incisors and canines, the end effect was superficially similar to the trend in *Homo*. Such resemblances easily distort supposed relationships when genealogical trees are built on a limited numbers of characters, taking little account of the fact that some similarities can actually be brought about by parallel but quite different forces. It could be held that the prognathic faces of *A. africanus* disqualifies them from the human lineage, yet when South African Man-apes are compared with their Lucy near-contemporaries, they have *less* prognathic faces and smaller canines, suggesting that they led the field in this very humanoid development (4).

South African anthropologist John Robinson was so impressed by these and other differences between *A. africanus* and the rest of the Australopithecines that he lumped the former into *Homo*, thereby asserting that they were direct ancestors to modern humans (26). Furthermore, when *"Homo habilis"* was first proposed as an early form of human (27), the name and its implications were rejected by some scholars on the argument that the anatomy was too similar to that of *A. africanus*! However, the resemblances between all early hominins are so great, and the precedent has become so well established, that *Australopithecus* has remained the most widely used name for all the early hominins. If *Australopithecus africanus* and *"afarensis"* are incontrovertibly confirmed as separate progenitors of two lineages, one solution could be to distinguish all the Lucies and robusts from *africanus* and its kin with the generic name *Paranthropus* (which some authorities currently reserve exclusively for the Robusts). In the meantime, I distinguish Lucies from South African Man-apes by using the earliest name for a Lucy fossil, *Praeanthropus* (28).

There is another newly discovered difference between the very earliest southern Man-apes and the Lucies that not only emphasises their distinctness, but also hints at a possible factor in the latter's earlier dispersal and the former's initial confinement to south Africa. When several sets of 3.6 million-year-old hominin footprints were discovered at Laetoli, Tanzania, they were soon matched up with *Praeanthropus* foot bones from slightly later deposits at the same site. In both tracks and bones, the big toe or hallux was found to be aligned with the other toes (although still clearly differentiated from them in both), but the known specimens of Lucy toes seemed rather long to fit the tracks. The easy strides marked out in the cementlike volcanic pavement imply that whatever species made them was an accomplished walker on feet that were nonetheless different from those of modern humans (29, 30).

Now, the articulating foot bones of the South African "Little-Foot" (an early *A. africanus* from Sterkfontein) have been assembled, it seems that the big toe in this hominin was not so firmly aligned with the other toes as in Lucies and thus might have been able to swing outward when a grip was needed. Furthermore, the orientation of articular planes in the knees of *A. africanus* fossils indicates a bow-legged gait that was emphatically less suited to fast movement than the straighter legged gait of the Lucies. There is, therefore, the strong implication that South African man-apes might have been slower or less accomplished walkers. Nonetheless, they *were* walkers, and it is relevant to ask how they might have compensated for their inferior abilities. To further complicate the issue, foot bones that are reliably allocated to *"Homo habilis"* from a much later deposit in Olduvai (specimen O.H.8., dated to 1.8 mya) also suggest a poorly developed arch and a less complete transfer of weight onto the big toe than in modern humans (31) (although it was aligned with the other toes in this specimen). There is also evidence that the semicircular canals of the inner ear in *A. africanus* are more apelike than in *Praeanthropus*, which has been interpreted in terms of the latter being a better bipedal balancer (32).

The delayed development of a compact, weight-bearing foot in the *A. africanus* lineage might seem to have a bearing on the confinement of this species to South Africa. In contradiction to this is the apparent anomaly of a similarly less evolved foot in the Ethiopian and east African "*Homo habilis*" long after Lucies had demonstrated their ability to disperse on legs and feet that (in spite of limitations) anticipated our own in many respects. This anomaly affirms a genealogical connection between southern *A. africanus* and eastern *H. habilis,* yet both species were unambiguously bipedal.

Understanding the anomaly could hinge on the presence of different selective forces on the two ancestral populations. It can be assumed that both descendants of ground apes were decisively terrestrial foragers. However, Lucies, from the start more numerous and with a much larger overall range, effectively pioneered the colonization of areas outside their region of origin, thereby gaining a substantial head start over the *Homo* lineage. Five factors could have contributed to this head start. First, with the clear exception of their lower limbs, the Lucies' adaptive modification of the preceding ground ape anatomy involved fewer and less radical reorganizations than in the *Homo/Australopithecus* lineage. Accommodations to living in more exposed environments might, therefore, have been quicker to appear because fewer components of change were involved. Second, there may have been a more apelike emphasis on plant parts (both above and underground), and, marginally, their food supply might have been more dispersed. Third, the relatively low-lying Zambezi region is more centrally placed for a wide spread across Africa. Fourth, well-wooded drainage lines along valley bottoms provide easy corridors of expansion from one basin to another. Finally, if the big toe was aligned earlier than in the *Homo* lineage Lucies may have become faster walkers and therefore better colonists. This difference need not contradict the probability that Lucies remained competent climbers and stayed close to trees as refuges. Likewise, less developed feet in the *Homo* lineage need not automatically mean more arboreal habits: if slower movements on the ground were offset by well coordinated and effective social responses to predators, there could have been less immediate selection for rapid change in the alignment of toes. For species of animals that defend themselves by using toxins or noxious smells, by mimicking poisonous models, or by group-mobbing potential predators (with painfully loud shrieks), it is common to be slower or to adopt ostentatiously confident behavior. The idea that mobbing and bluffing were an effective counter to the patent vulnerability of early humans is well supported by the frequency of such strategies in nature, and by many hints of the effectiveness of such behavior in later (and even very recent) human history.

The suggestion that *A. africanus* may be ancestral to *Homo* is as hotly disputed today as it was in 1925. Now, such prejudice can only be reinforced by the proposal that our own pedigree was slower off the mark in developing efficient walking than the Lucy-Robust dynasty! Another source of disbelief is the implication that the South African man-apes may also have been slower to relinquish arboreal carryovers than Lucies (a possibility at least not contradicted by the woodland biota with which the South African fossils are associated) (33). Again, the retention of some apelike limb proportions in early *Homo* is confirmed by the skeleton of a 1-m-high *habilis* from Olduvai that has been dated to 1.8 mya (34) Calculations suggest relatively short legs and long arms—proportions that indicate this species still climbed trees.

The tiny size of the Handy-man could be taken as evidence that divergent weights and sizes were key factors supporting the coexistence of early Robust and human lineages, but the earliest so-called *Homo* also included a larger and smaller form. The fossils we have are so few and already so loaded with theoretical significance that it can be difficult to concede that some, possibly all, of the currently known fossils of *Australopithecus africanus* and early *Homo* could represent populations that lived at too late a date to be direct ancestors. That probability actually reflects our desperate dependence on so few fossils, so fortuitously located in time and space. Yet, even if the presently known *A. africanus* fossils turn out eventually to be sideline buddings off our own lineage, their significance is scarcely diminished as guides to the time, place, and nature of *Homo*'s beginnings.

Before pursuing the contentious idea that both *A. africanus* and "early *Homo*" derived from a south African population of ground apes, let's return to the situation faced by the putative ancestors of these primates in this, their southernmost, outpost more than 5 mya. For an isolated population of ground apes, movement inland must have been severely inhibited by the combination of low temperatures and sparse resources during the southern winter. However, warm rains would have transformed these uplands every summer, and these montane habitats were so close to the narrow subtropical coastal plain that, in some instances, they were only a few kilometers upstream, albeit *steeply* upstream. These contrasts would have been exaggerated during periods when the overall climate was milder or more severe, with the uplands most attractive during the former, less so during the latter. Between about 5 and 4.5 mya, when conditions were supposedly at their warmest and wettest, inland penetration would have been at its easiest. In these circumstances, ground apes in this constricted southern enclave could have been tempted not only to expand their range into the uplands but also to become less strictly sedentary.

Local movement, driven by seasonal changes along short stretches of river, is a modest innovation; indeed, in its very earliest manifestations, it need have involved little more than a localized shift between summer range upstream and winter range downstream. Nonetheless, the longer term implications of seasonal movements could have been immense. To begin with, travel would have been eased by a majority of the rivers being short, falling along direct courses straight down to the sea. Furthermore, the spacing apart of streams is and always has been rather even (averaging about 5 km) so that single groups might, in many instances, have laid claim to single valleys. This arrangement would have been a subtle departure from the ranging behavior of river dwellers in flatter terrain, with more uniform environments such as in Zambezia. Today, as in the past, the southern coast and its riverine hinterlands differ physically and ecologically from equivalent habitats a thousand kilometers or more further up the coast. From a partly frugivorous ape's perspective, overall productivity in South Africa's eastern seaboard forests was a lot lower than in the tropics, with peaks and troughs corresponding to seasonal highs and lows, especially at higher elevations. In these circumstances, more reliance on animal foods could be expected. Expanding ranges upriver into the Drakensberg mountain chain would have faced no physical impediment; but the ecological barriers, formidable in the dry, cold winter, would have been trivial for much of the summer.

The Greytown garden wagtails hinted at a well-known annual movement of birds up and down the mountains along much of the southeast African littoral. In Natal, at least 76 species are altitudinal migrants, flying down to lower elevations in the winter (35). The majority of these are typified by insectivorous species such as wagtails, robins/thrushes, chats, and fruit or seed eaters, such as red-winged starlings, mousebirds, finches, and a hornbill. They shift between the coastal lowlands (sealevel up to 900 m) and medium elevations (900–1500 m), a distance that may be no more than a few km to 100 or more. Others range higher still, up to 3000 m high and more than 200 km away, and some make seasonal descents on both eastern and western sides of the mountain chain (36).

During dry centuries or millennia, cold-sensitive species might even have vacated the southernmost stretches of coast such that already constricted geographic ranges would have been squeezed still further and small gene pools would have become smaller still. It can be supposed that medium-sized, partly frugivorous primates would have been particularly sensitive to such constraints. If my model is correct, the south African hominin lineage would have passed through a particularly narrow genetic bottleneck at around 5.3 mya (currently thought to represent a global trough in low temperatures. See figure 4.10, p. 138).

The ground ape ancestry that I have proposed implies that an enlarged menu, in which the humblest of foods became acceptable, was linked to confined ranges within which seasonal change could be accommodated without movement. How would a descendant, offshoot population respond to the opposites of an expanding home range with physical separation of strongly seasonal resources?

If the choice of foods was not to shrink as animals moved inland from the coastal plain, one mitigation could have been to intensify the ground ape strategy with more thorough probing of the habitat and the development of new techniques for finding, gathering, and processing food. An indirect hint that this might have been the case comes from differences in the structure of hands between *A. africanus* and *Praeanthropus*: the latter retained more apelike elements, whereas the former, by 3.3 mya, had developed very large thumbs and flexible wrists. Sessile, mainly plant foods, particularly if available in bulk, make few demands on manipulation; by contrast, many small and diverse items, including active animals (uncovered by sustained grubbing and turnover of detritus) require much more manual agility and greater ingenuity of handling. While I am proposing that both species must have descended from the same coastal stock, *A. africanus* would seem to have intensified the eastern ground apes' supposedly omnivorous, small-item foraging, with long-term consequences for the anatomy of their hands. Its northerly siblings, instead, seem to have suffered less pressure to abandon apelike hand skills.

During their earliest isolation in the deep south, grubbing about in forested river courses was probably the most productive strategy. What about the lateral expansion of home ranges, away from the immediate vicinity of rivers and streams? This was an emancipation from a purely riverine existence that must have been well advanced by the time of the earliest south African fossils. The incentives to enlarge the reach of their foraging would have been dietary, with pressure to do so perhaps greatest at higher altitudes where food sources would have been more dispersed and certainly more seasonal. For any predominantly terrestrial animal, improvements in either mobility or security could have helped such a dangerous expansion in range. I have already suggested that Lucies, faced

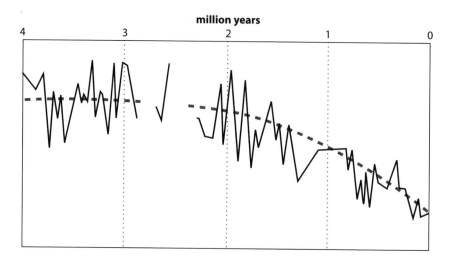

FIGURE 6.3 Fluctuations in carbon isotope records suggesting environmental changes in Africa. Data indicate a general trend toward cooler, drier, and more grassy habitats, especially during the last 1 million years. (Derived from De Menocal, P. B. 1995. Plio-Pleistocene African climate. *Science* 270: 53–59.)

with similar opportunities and costs, probably became more mobile. In the deep south, security might have been improved (even if marginally) by some facultative advance over ground apes in the use of hands; techniques might have included an increased use of tools, notably such old ape favorites as the breaking and flourishing of boughs and branches to chase off enemies (37).

Leaving these possibilities aside for the moment (and remembering that these short-legged, long-armed animals must have been slower and more vulnerable than both quadrupedal monkeys and their more mobile, sexually dimorphic Lucy cousins), their "lateral expansion" poses real problems. Fossils and their sitings prove (38, 39) that by about 3.3 mya (thought to be a cool period), *A. africanus* were living in landscapes that were dominated by dry woodlands and relatively open veltlands (where natural fires played a larger role in maintaining grass and shrublands than was the case in more tropical latitudes) (figure 6.3).

Perhaps more important was their successful coexistence with a wide spectrum of large savanna mammals, including predators. To have developed an effective strategy for that coexistence is more problematic than a mere physical emergence out of the forest. I am convinced that no reconstruction of prehistoric existence can be realistic without taking account of daily physical encounters among numerous and diverse species of

large animals, many implying serious disruption or competition for resources for a slow, small, and far-from-agile hominin.

As the southernmost population of ground apes expanded still further
inland, beyond the forested heights, the frequency of confrontations between species likely increased. This change would have been particularly
acute wherever and whenever water was scarce and caused numerous
large animals to come in from the surrounding catchment areas. On the
face of it, numerous predators and large mammals should have discouraged hominins from venturing away from their riverine bases, yet fossil
sites for *A. africanus* prove their presence very far inland by 3.5 mya and
imply an ability to live away from riversides (40).

To reconstruct that expansion, consider first the physical structure of
veltland habitats. The setting is South Africa, but many of the observations that follow could well apply to other regions. I do not need to disguise my hunch that this was a likely setting for the emergence of certain
human traits deriving from a coastal ground ape stock. Nonetheless, my
main intention is to illustrate some generalized challenges that were
faced by human ancestors, wherever they evolved. Some of my observations might apply as readily to an east African upland (Rift Valley) region
of origin, but the fossil and ecological evidence for this is still far too slender to attempt any sort of alternative scenario.

When rivers flow through drier hinterlands, banks and purlieus are often densely thicketed and webbed by the well-trampled and (in the dry
season) mostly smooth pathways of large herbivores, from hippos and
rhinoceroses to herds of elephants and antelopes, all coming and going
to water. For all but the smallest terrestrial animal, there is little choice
but to use and reinforce these paths; they are as much components of the
ecosystem as are tree trunks, river courses, and termitaries. Non-path
makers that regularly use such paths intrude on the domains of much
larger, group-living animals, and encounters become sufficiently routine
to demand equally routine responses. If hominins, in their search for
food, traveled any distance away from the riverine core of their range,
they too would have had to develop routine responses toward such encounters (figure 6.4).

Before pursuing the implications of other animal species as challenges
to ecological expansion, return to the details of foraging behavior. Such
behavior could have begun with the mix of ground-quartering, tree-shin-
ning, and stand-up berry picking suggested earlier for eastern ground
apes. Initially, a steep, forested habitat would hardly have called for much
in the way of bipedal skill; but as flatter plateau lands were reached and
home ranges enlarged, there could have been selection for a somewhat

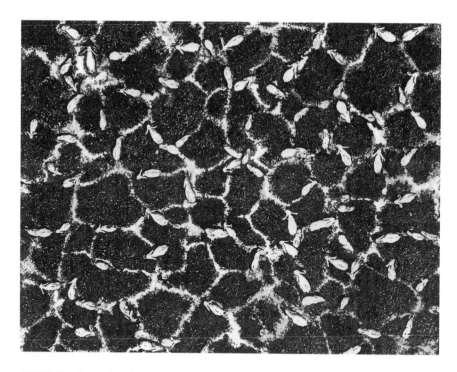

FIGURE 6.4 The earliest hominin habitat? Thicket with numerous paths made by browsing elephants. (Photo courtesy of A. Root [pilot, I. Parker].)

sturdier, better balanced type of walking. However, improved gaits on the ground need not have excluded a continued facility for climbing trees.

Larger ranges and more changeable environments (which were therefore less predictable and well-known) must have imposed changes on the pattern of foraging. If eastern ground apes, knowing their restricted home ranges intimately, formed small cohesive social units, it can be assumed that members would have been in very frequent, if not continuous contact.

As with all apes, females would have been more mobile between groups than males. Overall, however, I suggest that both ground apes and Lucies were likely to have remained permanent, strongly territorial residents of as long a stretch of river as served their needs. Instead, the South African isolates, moving in response to seasonal and altitudinal changes along physically well-defined valleys, frequently encountering other dwellers in the same valley system, could have been under strong pressure to modify their behavior. More specifically, foraging groups might have become less stable, and territorial defense might have been significantly modified. In many instances, fellow valley dwellers, numbering 100 or more, might have been kin, but the environment alone would have encouraged social accommodation between larger numbers of more loosely connected individuals. Under these circumstances, narrowly territorial behavior could have declined, and some sort of (clan?) system, embracing a variety of associations within a larger area, could have developed.

The distribution and nature of food would have determined what sort of accommodations were made between individuals. Seasonal changes in the abundance and distribution of food could have encouraged periodic dispersion of individuals and inhibited permanent close relationships both between and within sexes, favoring greater social flexibility. If something of the earlier ground ape habits of intensive, close-range, small-item foraging was relinquished only very slowly, competitive confrontations over food might have been relatively infrequent. Tolerance between weakly acquainted individuals belonging to loosely connected, "valley-by-valley" clans would have been encouraged by resources that were consistently well dispersed or by foods that were occasionally superabundant—the former promoting a safe dispersion, and the latter promoting temporary, noncompetitive convergences.

Shortfalls in food could have elicited two very different outcomes: intensified conflict or further dispersal. The former would have enhanced territorial responses, the latter reduced them. Strong defense of territory implies well-defined boundaries and, for a social species, their protection by the whole group. Here again, seasonality could have promoted cohesive territorial behavior during periods when individuals were together and relaxation when they were dispersed. Periodic confrontations could be expected between different "clans" on the borders of their ranges, but frequent dispersal of small units of fluctuating composition could have diminished the overall frequency and importance of intraspecific territorial behavior within the clan.

However, this is not to say that coordinated group behavior need be insignificant; to the contrary, larger numbers of relatively independent individuals might have had cause to rally, especially where seasonally concentrated sources of food were vulnerable to consumption by competitors, or where access was barred by other large species. In such circumstances, effective rallying of dispersed "clan" members and the displacement of competitors could have become essential for "lateral expansion" (away from river courses) and, quite possibly, for survival.

Extrapolating from chimps and other primates, the medium would have been loud, far-reaching calls; however, is it possible that the functional emphasis of calls shifted away from intraspecific group defense of small territories toward group defense of temporary resources against other species? Such a shift in the emphasis, objective, and targets of their behavior would have carried many implications for the process of humanization.

If changes in the distribution of food were best met with dispersal during one season, convergence in another, rallying calls could have had the social effect of promoting cohesion among otherwise scattered individuals. Loud vocalizations, under these circumstances, could have signaled opportunities for food that would remain inaccessible in the absence of a group response. Loud calls are merely the more assertive and obvious manifestations of a species' signaling system. In any animal, vocalizations have to be tailored to circumstances. At certain times of the day and night or during certain seasons, insects, birds, and other animals (including primates) put up a sustained barrage of sound. In the midst of all this noise, hominin vocalizations would have had to be both distinctive and capable of modulation for effective use at both loud and quiet levels. All signals are designed, either genetically or through learned intercourse, to be answered with a reciprocal signal or by an appropriate behavioral response. All communication, including speech, is built on this principle. Signals are also meant to be exclusive. Many naturalists have experienced the contradictions of trying and failing to interact with another species. Imitation of their signs, even if momentarily stimulating, must end in mutual incomprehension. This result should not surprise; natural selection has elaborated such a diversity of animal signals such that in the midst of all that oral, visual, and olfactory cacophony, somewhere, somehow, every species will have its own exclusive channel. We are meant to be excluded; a failure to share more than a mere gesture or two with another species speaks to the plain fact that we are excluded from 99.9 percent of the natural world's message systems. Notwithstanding this, most species have no difficulty deciphering a threat when they meet one.

If South African Man-apes had even the weakest resemblance with modern chimps and gorillas, they can be assumed to have had ample skills in driving off competitors, predators, even mere "disturbers," by noisily throwing or wielding sticks, stones, or debris (figure 6.5). Such mobbing displays range from the rearguard actions of single gorilla males to joint confrontations between all-male chimp rivals to noisy advances by mixed-sex, mixed-age chimpanzee parties toward other species in their path. Whatever the details of these displays, their effectiveness is observably enhanced by being social; from the viewpoint of competitors, massed ape advances effectively simulate a single multiheaded, multivoiced threat that is best answered with retreat. Early hominins might

FIGURE 6.5 Noisy display. Chimpanzees can make intimidating bipedal displays and may even wave branches or sticks while charging.

have followed the typical primate option of retreat up trees often enough, but their newly emancipated forelimbs would have been well suited to scratch up and toss debris or snatch and wave branches (41). Such displays need not have differed greatly from those used by gorillas and chimpanzees today to deter disturbances during intensive feeding. Anyone who has been confronted by the uproar of wild apes in face-to-face encounters on the ground, as I have, knows that they can be as intimidating as a street gang out for a rumble and are best responded to with a rapid retreat.

One of the advantages of mounting a noisy display whenever other animals obstructed access to resources would have been that most large herbivores soon learn to sidestep unnecessary disturbance. Where large numbers of medium-sized monkeys competed for a similar range of foods, any ability to deter them from consuming a common food supply could have been crucial. It is not only great apes that employ startling displays; porcupines, zorillas, ostriches, and plovers are but a few of the relatively helpless, slow, or small species that brazen things out with ostentatious displays of offensive or distracting noise, movement, or smell. For example, by flashing their black-and-white wings and holding them open in an apparently immovable posture, plovers can force large herbivores to deflect from stepping on eggs or young. Plovers, like ostriches and most antelopes, can readily discriminate between potentially harmful and harmless situations simply by observing gaits or trajectories. They then behave accordingly: fleeing, hiding, distracting, or deflecting with their various species-specific displays. In those localities where both displayers and the displayed-to meet on a regular basis, the longer term outcome is that both learn to save energy through mutual avoidance.

In spite of a menu of behavioral options and some potential for learning refinements, the displays of most species (including "bluffing" or "mobbing" routines) are stereotyped and genetically fixed. Apes, instead, can elaborate or adapt both individual and group actions directed at their own or other species (42). These elaborations are frequently specific to a region or group and can justifiably be described as cultural, learned behaviors (43). For example, an adult male chimpanzee in the Mahale Mountains of Tanzania deliberately and repeatedly chose to display in a particular ravine where he could dislodge large rocks and set them rolling to thunderous effect as they bounced downhill from rock to rock (44). Each arrival in the same steep valley provoked a similar display from the same individual. In other localities, tree buttresses (with acoustic properties well known to the local chimps) provoke violent heel-hammerings and hooting screams from more than one of the adult males. Rollable rocks or drummable buttresses elicit actions that combine attention-getting with the self-stimulation of strenuous exertion. In these cases, the trigger is strictly localized and has properties that have to be learned.

The stimulus need not be tied to a place. Chimps often make very loud vocal choruses when their routines of daily life are disturbed or their senses or bodies are jolted by the physical impact of thunder, heavy rain, a leopard's rasping cough, or the rumbling shudders of an earthquake. A common response is to scream, hoot, run, bounce, or engage in violent bouts of slapping, tearing, thumping, or dragging of vegetation. An "unusual" event or an "unusual" place typically provokes such self-advertisements, particularly in males. A super-signaling male often intimidates or coerces other members of the group, who may make appeasing gestures, draw closer, or otherwise interrupt or discontinue their previous activity.

The screaming and churning of leaves and branches can also intimidate predators. While evidently satisfying for the signaler or signalers, such displays also affect other group members and change what they are doing. An apparently arbitrary and individual action can therefore have social effects that are directly traceable to idiosyncratic behavior in an individual signaler.

Much quieter messages have also been developed by particular chimp populations. For example, Mahale chimps have the custom of rapping their knuckles on a dead branch or ripping leaves through their front teeth in a fast, open-lipped gesture (45). Again, males seem to be the main signalers, females the targets; but these subtler messages (commonly interpreted as courtship gestures) share features with the louder ones. Typically, they coerce or awaken desired social or sexual activity in a formerly neutral or routine setting. Chimp routines commonly demand quiet alertness from all members of the group (notably when human hunters, leopards, or even elephants are around). Sudden eruptions of attention-seeking self-advertisement indicate an abandonment of habitual caution and a dramatic change of activity for both signaler and respondees.

It would be easy to interpret these shifts from discretion to bold self-assertion as mere innate responses to external events and see their function as an advertisement of group territories, alarm calls against a predator, or courtship rituals. Parallels with the behavior of birds or mongooses support the supposition that much of the initial action is, indeed, innate; but there is also evidence that many chimp behaviors are learned, not instinctive. Knuckle knocking, leaf stripping, rock–bouncing, and even "rain dancing" all seem to be culturally learned quirks or local traditions rather than expressions of genetically inherited chimp instincts. Fellow group members must therefore learn to interpret their "meaning" through experience. Arbitrary signals can elicit quite specific responses. Suppose that the frequency of such behavior, even on a seasonal basis, was increased to the point at which the displayers (whether envisaged as "alarmists," "mobbers," or "bluffers") could repeatedly observe diverse reactions in other species. Like the Mahale male chimp, such displayers might begin to "shape" their performances to make them more effective for the situation or the species in question. This modification may seem very modest, but it carries with it the implication of some control over emotion and its manipulation toward ends that are "judged" on the basis of experience and memory. In these circumstances, the voice's carrying power, if differentiated in response to different contexts, has the potential for long-distance communication to other members of the group about "events" as well as "emotive states." In African apes, vocalizations are the longer range component, not only in aggressive displays but also as part of a larger repertoire of mostly close-up signals that mediate group activity. Using sounds, movements, and gestures, individuals signal their "intentions" to others and thus "persuade" them to join in an activity. Male chimpanzees trying to build alliances before testing an opponent commonly recruit partners with energetic gestural and vocal displays (46). If *A. africanus* commonly needed to make alliances against other species, and at a distance, such an invitation to join might well have had to encode information about the species and situation if it was to be persuasive. The observable behavior of chimpanzees can therefore provide some basis for suggesting how a communication system that depends on arbitrary, self-elaborated signals could develop and be applied to plausible resource-gaining situations (47).

Bluffing or mobbing displays may have also had some less direct effects that helped shape the unique ecological stategies of early humans. Frequent and elaborate vocalizations could have helped initiate a form of "hands-on" interaction with the environment and its other inhabitants that became an intrinsic part of an entirely new ecological strategy.

For example, an apparently minor elaboration in bluffing displays could have had significance for later, more explicitly offensive attacks on potential competitors and enemies and the development of tools and missiles. This would have begun with improvements in the accuracy of throws; captive chimpanzees can spontaneously develop an ability to throw stones underarm and sometimes with considerable force and accuracy (proving that some of these motor abilities must have been latent in all early hominins). Likewise, related motor skills in using hammer stones are common in both wild and captive chimps, as are powerful coordinations of arm and wrist while poking, clubbing, twisting, and swinging long branches in what are effectively extensions of the forelimb. Manipulating plant material is also intrinsic to nest-making. All the actions involved in hurling, hitting, hammering, and prodding are likely to have their origins in the two behaviors of competitive/aggressive display or active foraging because, in apes, they appear in either aggressive or exploratory contexts. (Poking or prodding actions include enlarging holes in tree trunks with a pounding pestle made from a hard palm rib and "termite-fishing," a much-publicized foraging technique of chimpanzees (figure 6.6). This behavior seems to be the miniature application of a much more general and probably innate propensity for exploring hollows—one that is even shared by thorn-wielding finches!) Thus, the beginnings of systematic tool use could have emerged from a combination of solitary, close-up foraging techniques; nest-building;

FIGURE 6.6 Foraging techniques in chimpanzees: probing with variously sized sticks, pounding with variously shaped stones.

and social, expansive, mobbing behavior—contrasts that will be taken up again later.

Diets and the elaboration or invention of novel techniques to obtain food are central to understanding how a forest ape of restricted distribution expanded both its range and its foraging repertoire to become *Homo*, living in a variety of habitats. Does narrowing the field of inquiry down to eastern South Africa and identifying the fossil species *A. africanus* as the most likely ancestral lineage help bridge this gap?

Over the years, south African man-apes have been portrayed as everything from the helpless vegetarian victims of carnivores to "killer-apes" themselves. The earliest fossils of *Homo* (distinguished from *A. africanus* by somewhat larger brains and an association with crude stone tools) have suffered from similar swings in opinion. There has been a growing consensus, however, that early *Homo* included scavenging in its repertoire of food-finding techniques (48, 49). Objectors have probably been swayed as much by popular notions of disgust for scavengers as they are by the supposed slenderness of the evidence, which consists primarily of *Homo* remains apparently associated with tools and the bones of prey that were apparently too large to be easily overcome by such an insignificant primate. When some of the fossil bones of antelopes were examined more closely, cuts were found that resembled tool marks, and some of these appeared to be superimposed over the tooth marks of carnivores (50, 51).

This persuasive evidence for "scavenging" in early *Homo* hints at the nature of the processes that allowed forest-based animals to expand their ranges. South Africa provided a plausible setting for first steps in that progression, and *A. africanus* fossils appear to sample hominins in the midst of that adaptive shift.

Scavenging, in the popular sense, could have been the late manifestation (after *Homo* had become a full member of the open country fauna) of a much larger trait in our lineage. This trait has less to do with a taste for meat from dead animals than with particularly acute sensitivities to other animals as guides to hidden or potential food sources. This sensitivity could have been to the animals themselves, as prey, but progressively tended, more and more, toward appropriating some or all of the other animals' own subsistence. This technique is most explicit in the term *scavenging* but also embraces robbing honey, chasing off pigs from sources of roots or truffles, and displacing or discouraging any other species from food sources they might otherwise be able to use. Scavenging, as commonly understood, is typical of open-country species. I contend that its beginnings, in forest-based members of the *Homo* lineage, could have provided a major mechanism not only for expansion out of the forest, but for a diversification of diet that led away from species-specific diets and, even more significantly, a rapid elaboration of flexible technologies for obtaining food.

Today, we can observe the convergence of individuals, even accretions of species (such as vultures), in response to food-finding cues from a larger community of food finders. The cues may be the sight of actual predator or prey, scent, sound, or indirect signs of disturbance. We can also see that scavengers, through the weight of numbers or "unnerving" behavior, can actually displace primary predators. The coordination required to carry off such a dangerous maneuver must also be sufficiently sustained to provide insurance against the obvious dangers of counterattack (52). There are wider implications for human evolution when "scavenging" is seen as part of an enhanced sensitivity to food-finding signals as well as an ability to subvert resources identified or captured by others.

Sensitivity to food-finding clues is nothing very unusual in itself, and it commonly involves other species as well as conspecifics. Many mammals and birds rely on indirect clues to the current whereabouts of food: vultures, sea-birds, waterfowl, and wading birds watch or listen to one another; scavenging carnivores watch the vultures; and predators are specialized to track their prey via scent, sound, and even vibration. Even terrestrial frugivores are alert to the calls of trumpeter hornbills or turacos. In every instance, tracks, traces, or the behavior and activities of other animals are reliable guides to predictable foods in momentarily unpredictable or changing locations.

All such foragers quickly learn to respond to any stimulus that leads to their preferred foods, but every species is already adapted to a specific dietary range; their learning from others is in the service of appetite and appetites are, to a greater or lesser degree, inherited and species specific. Thus, the vulture pursues only those clues that will lead to such soft tissues of dead animals as are extractable by its short hooked beak. The whole edifice of ecological complexity is built on just such adaptive specializations, as when the clawless otter, with probing fingers, finds and extracts crabs from murky waters and quickly crushes them between uniquely designed, broad, flat molars.

Were the earliest members of the human lineage any different? Were they simply "opportunistic omnivores that retained a marked preference for fruit," exceptional only for the number and variety of foods in their diet? In terms of diet alone, maybe, but the processes involved in getting an ever-expanding range of foods, the dietary flexibility that went with increased choice, and the lack of any obvious modifications of claws, teeth, or senses were unprecedented. This flexibility could have been based on the enhanced ability to discover, interpret, and share with other group members indirect clues to potential foods, as described earlier. It was a faculty that could have built on two established skills—one shared with many other species (especially predators) and centered on the close observation of other species, the other based on the ape's predatory intolerance of competitors that was mentioned earlier. Combining the predator's alertness to other species' behavior with a periodic intolerance of actual and potential competitors (as food choices increased), this unprecedented foraging technique might better be termed "niche-stealing" than scavenging.

The main brakes on any species expanding the number of different things it can eat are its own inherited anatomy and physiology and the precedence of other, more highly specialized competitors. Plants and animals protect themselves with such a barrage of devices that they require equally specialized devices in those predators that can overcome them. We know that later humans responded to the need for food-extracting "devices" with tools and cooking procedures, not anatomical adaptations. But how are we to bridge the gap between some kind of Miocene ape and *Homo*? I think the larger context of that transition is likely to be hidden in the detailed behavior of a slow, ill-equipped little primate challenged by seasonal habitats and strong inducements to expand out and away from narrow gallery forests. As for a specific geographic context, the peculiarly seasonal rivers of eastern South Africa are particularly promising sites for a hominin response to such challenges (53, 54).

Here, as elsewhere, adapting to the environment involves food plants, landscape, and various seasonal changes, but it is the sheer number and variety of large animals that presents a greatly underrated challenge. Fossils of large mammals and carnivores recorded from south African deposits are, doubtless, an incomplete sample, but the list is still long. Some 24 genera of large mammals include elephants, rhinos, pigs, hippos, and buffalos, while some 10 genera of larger carnivores include many types of hyenas, a wolf, saber-toothed tigers, lions, and leopards (55). The main primate competitors were baboons. Such variety is the expression of dense and complex partitioning among the many species that are present. Any living space or eating place is already subject to intense competition and continuous interference from numerous species, each with a different life strategy. Navigating through such a hazardous environment would have called for an unremitting awareness of other species as well as an ability to make fine discriminations and learn to anticipate different classes of behavior, in different species, for different ends.

Awareness of "others" includes members of the same species. Enlarged home ranges imply less detailed (or at least less up-to-date) knowledge of the land and, at times, fellow occupants. Other home range inhabitants have the potential to be competitors in one situation, useful guides in another, so that every encounter with a conspecific should have called for a more refined calculation or evaluation of behavior; such a response could have been just one of several incentives to develop greater self-awareness (a step toward human consciousness that is discussed later). "Awareness" is entirely dependent on the senses of vision, hearing, touch, and taste.

Clearly, the most significant innovation here would have been a greatly enhanced potential for the hands and an increasing "awareness through manipulation." Although the capacity of early primate hands to convey sensory information through touch can hardly be in doubt, the process could scarcely have differed from olfaction, hearing, or taste in that such responses to sensory input were essentially "hard-wired." Such instinctive responses allow instantaneous reactions to stimuli coming from the environment. If my portrayal of ground ape ecology is correct, the conversion of hands into gathering and processing organs would have been accompanied by a greater diversification of food types, requiring refined handling techniques and the frequent challenge of judging novel properties of plants and animals (including their relative edibility). Whereas smell and taste clearly rule the feeding of primates that have their faces close up to their potential food, it is possible that visual-tactile assessment and learning might have become surer guides to edibility for hand foragers that were actively enlarging their dietary repertoire. Such a switch might have slowed judgment, but it could have introduced the critical element of interacting with the environment through a link-up of visual judgment, practiced manipulation, and continuous learning. In that combination lay the ultimate secession of "instinctive" feeding techniques and the emergence of a larger relationship with the environment that increasingly depended on learned visual judgments mediated through the hands (and ultimately through tools, which are discussed in the next chapter).

My contention that human success has been built on the appropriation of niches from an ever-increasing list of species will be clear enough by now. Made possible only through the application of tool-assisted techniques to otherwise inaccessible or difficult foods, "niche stealing" implies an ability to take over resources from other animals. Among those resources is space. On this argument, the start of humanity's technological career could be indicated by signs of a decisive hominin expansion. Many scholars have opted for the first signs of australopithecines at about 4 mya. Alternatively, did *Homo* arise from an early form of *A. africanus* (earlier, that is, than any of the presently known fossils)? If South African *Australopithecus* developed as a species of restricted range, separate from the Lucies and robusts, its eventual "breakout" from its enclave could reflect the start of our uniquely human relationship with other species and communities. "*Homo habilis*" has, for a long time, been envisaged as the first "*Homo*," but there are reasons for caution in accepting this assessment.

When Louis Leakey (27) named "*Homo habilis*," his justification for classing that animal as *Homo* included the apparent association of *habilis* bones with stone tools. Immediately, there were three classes of objection. One was that there was no proof that the owners of the bones were the owners of the tools. The second was that the stones were not tools. The third was that there was too little difference between "*H. habilis*" and *A. africanus* (56). The first two objections have been overcome in part by more bones and stones as well as a greater study of stones. The third objection was more solid and enduring and hinged on very real resemblances between Leakey's OH7 (a specimen called "Jonny's Child" after its discoverer, Jonathan Leakey) and Robert Broom's STS71 from Sterkfontein, originally named *Plesianthropus transvaalensis*. The latter has long been viewed as an "advanced *A. africanus*," yet the "*Homo habilis*" resemblances are so close that they must challenge the wisdom of drawing a generic line between fossils with so much in common. These questions aside, the resemblances demonstrate that evolution is a continuum; they also underline the fact that our categories are artifacts no less primitive than Oldowan choppers or scrapers.

The application of stone tools to survival (as an adaptive trait) has long been taken as an appropriate criterion to justify the generic name *Homo*. But it is as well to remember that the aptitude for developing and using tools, not just stone tools, may have been several million years in the making. If I am correct in seeing the *africanus* lineage as a product of isolation in the south as well as being a likely progenitor of *Homo*, it is worth considering, once again, how Man-apes might have entered the "Big Game" country that surrounded their point of departure: the riverine forests of southern Africa. Here may have been the testing ground for a new way of life— a testing that may have gone on in this isolated southern enclave for well over a million years.

For a smallish, slow, and defenseless primate to outwit or outcompete large, intelligent mammals would have required more than tools alone. For their rudimentary technology to be effective, it would have had to be at the service of a stategically minded animal and, given their modest size, one that could share coordinated responses to frequent challenges from larger animals. With very many more unpredictabilities to contend with, a capacious memory and ingeniously flexible thought must have overtaken preprogrammed instinct as their primary asset. One anatomical expression of this ability is an enlarged brain. Although the enlargement is very modest in *A. africanus*, there is a measurable difference when compared with Lucies. One social outcome of a strategic approach to subsistence would have been a strong tendency for all members to draw into more compact, often larger groups whenever moving out into more open habitats.

Translating the overall trends into an evolutionary sequence, it can be predicted that *A. africanus* formed loose, flexible units while in secure, resource-variable surroundings but aggregated closely when more exposed. This strategy could have been a transitional state of affairs. I suggest that it was preceded by the consistently small, compact, and stable territorial parties of their immediate ground ape precursors. Successors such as "*Homo habilis*" more consistently exposed, would have tended to maintain larger, better coordinated groups on a more permanent basis.

It has been proposed that the species currently called "*Homo habilis*" should be downgraded to *Australopithecus* (31); this course has been followed here as a strictly provisional arrangement. Both "*Homo habilis*" and the cluster of fossils now described as *Australopithecus africanus* may or may not be eventually classed as *Homo*; but until an incontrovertible line of descent can be proven, it is probably wiser not to make too strong a commitment to one branch on an increasingly bushy tree of candidate ancestors. As for the Lucies and Robusts, if they prove to be a single lineage that is separate from the south African man-apes (an effectively certain conclusion), both must relinquish the name *Australopithecus* because this name was first applied to the South African Man-ape *A. africanus*. In these circumstances, Lucies either keep their first name of *Praeanthropus*, or the entire lineage becomes *Paranthropus*, the earliest name for a Robust (57).

All this emphasises the extreme subjectivity of current classifications—reminding us, again, that names are inadequate artifacts for such an incompletely understood yet dynamic field as the study of human evolution. Until a year or so ago, a line was drawn between "*Homo habilis*" and *Australopithecus africanus*, their possible immediate forebears. This line was justified by marginally larger brains and initially uncertain tool associations. However arbitrary or provisional, that boundary takes us away from an exclusively southern African focus and serves to mark the end of this chapter.

REFERENCES

1. Tobias, P. V. 1980. *Australopithecus afarensis* and *A. africanus*: critique and an alternative hypothesis. *Palaeontologica Africana* 23: 1–17.
2. Olson, T. R. 1985. Cranial morphology and systematics of the Hadar formation hominids and "*Australopithecus*" *africanus*. In *Ancestors: The Hard Evidence*, ed. E. Delson, 102–119. New York: Alan R. Liss.
3. Robinson, J. T. 1963. Adaptive radiation of the australopithecines and the origin of man. In *African Ecology and Human Evolution*, ed. F. C. Howell and F. Bourliére, 385–416. Chicago: Aldine.
4. McHenry, H. M., and R. L. Skelton. 1985. Is *Australopithecus africanus* ancestral to *Homo*? In P. V. Tobias, ed., *Hominid Evolution: Past, Present, and Future*. New York: Alan R. Liss.
5. White, F. 1983. *The Vegetation of Africa*. Natural Resources Research XX, 356 pp. Paris: UNESCO.
6. Acocks, J.P.H. 1988. *Veld Types of South Africa*, 3rd Edn. *Memoirs of the Botanical Survey of South Africa* No. 57. Pretoria: Botanical Research Institute.
7. Cooke, H.B.S. 1964. The Pleistocene Environment in Southern Africa. Ecological Studies in Southern Africa. *Monogr. Biol.* 14: 1–23.
8. Coe, M. J., and J. D. Skinner. 1993. Connections, Disjunctions and Endemism in the eastern and southern African mammal faunas. *Transactions of the Royal Society of South Africa* 482: 233–255.

9. Grubb, P., O. Sandrock, O. Kullmer, T. M. Kaiser, and F. Schrenk. 1999. Relationships between eastern and southern African mammal faunas. In *African Biogeography, Climate Change and Human Evolution*, ed. T. G. Bromage and F. Schrenk, 253–281. New York: Oxford University Press.

10. Foley, R. A. 1999. Evolutionary geography of Pliocene African. Hominids. In *African Biogeography, Climate Change, and Human Evolution*, ed. T. G. Bromage and F. Schrenk, 328–348. New York: Oxford University Press.

11. Dart, R. 1925. "*Australopithecus africanus:* The Man-Ape of South Africa" *Nature* 115: 195–199.

12. Brain, C. K. 1981. *The Hunters or the Hunted? An Introduction to African Cave Taphonomy*. Chicago: University of Chicago Press.

13. Beynon, A. D., M. C. Dean, and D. J. Reid. 1991. On thick and thin enamel in hominoids. *American Journal of Physical Anthropology* 86(4): 521–536.

14. Grine, F. E. 1986. Dental evidence for dietary differences in *Australopithecus* and *Paranthropus*: a quantitative analysis of permanent molar microwear. *Journal of Human Evolution* 15: 783–822.

15. Walker, A. C. 1981. Diets and teeth. Dietary hypotheses and human evolution. *Philosophical Transactions of the Royal Society, London* B 292: 57–64.

16. Kay, R. F., and F. E. Grine. 1989. Tooth morphology, wear and diet in *Australopithecus* and *Paranthropus* from southern Africa. In *The Evolutionary History of the "Robust" Australopithecines,* ed. F. E. Grine, 427–447. New York: Aldine de Gruyter.

17. Lockwood, C. A. 1997. Variation in the face of *Australopithecus africanus* and other African hominoids. Ph.D. dissertation. University of the Witwatersrand, Johannesburg.

18. Sigmon, B. A. 1991. Evolutionary changes in the reproductive system and mating strategies after hominoids became upright bipeds. In *Origine(s) de la Bipédie chez les Hominidés*, ed. Y. Coppens and B. Senut, 267–274. Paris: CNRS.

19. Richmond, B. G., and D. S. Strait. 2000. Evidence that humans evolved from a knuckle-walking ancestor. *Nature* 404: 382.

20. Clarke, R. J., and P. V. Tobias. 1995. Sterkfontein member 2 foot bones of the oldest South African hominid. *Science* 269: 521–524.

21. Gore, R. 1997. The First Steps. *National Geographic* February 72–99.

22. Ricklan, D. E. 1990. The precision grip in *Australopithecus africanus*: anatomical and behavioral correlates. In *From Apes to Angels: Essays in Anthropology in Honor of Phillip V. Tobias*, ed. G. H. Sperber, 171–183. New York: Wiley-Liss.

23. McHenry, H. M. 1986. The first bipeds: a comparison of the *Australopithecus afarensis* and *Australopithecus africanus* postcranium and implications for the evolution of bipedalism. *Journal of Human Evolution* 15: 177–191.

24. McHenry, H. M., and L. R. Berger. 1998. Body proportions in *Australopithecus afarensis* and *A. africanus* and the origin of the genus *Homo*. *Journal of Human Evolution* 35: 1–22.

25. Grine, F. E., ed. 1988. *Evolutionary History of the "Robust" Australopithecines*. New York: Aldine de Gruyter.

26. Robinson, J. T. 1956. The dentition of the Australopithecinae. *Transvaal Museum Memoir* 9: 1–179.

27. Leakey, L.S.B., P. V. Tobias, and J. R. Napier. 1964. A new species of the genus *Homo* from Olduvai Gorge. *Nature* 202: 308–312.
28. Strait, D. S., F. E. Grine, and M. A. Moniz. 1997. A reappraisal of early hominid phylogeny. *Journal of Human Evolution* 32: 17–82.
29. Leakey, M. D., and R. L. Hay. 1979. Pliocene footprints in the Laetolil beds at Laetoli, north Tanzania. *Nature* 278: 317–323.
30. Tuttle, R. H. 1981. Evolution of hominid bipedalism and prehensile capabilities. *Philosophical Transactions of the Royal Society, London* B 292: 89–94.
31. Wood, B. A., and M. Collard. 1999. The human genus. *Science* 284: 65–71.
32. Spoor, C. F., F. Zonneveld, and B. Wood. 1994. Early hominid labyrinthine morphology and its possible implications for the evolution of human bipedal locomotion. *Nature* 369: 645–648.
33. Sikes, N. E. 1999. Plio-Pleistocene floral context and habitat preferences of sympatric hominid species in east Africa. In *African Biogeography, Climate Change, and Human Evolution*, ed. T. G. Bromage and F. Schrenk, 301–315. New York: Oxford University Press.
34. Johanson, D. C., and B. Edgar. 1996. *From Lucy to Language*. New York: Simon and Schuster.
35. Johnson, D., and G. L. Maclean. 1994. Altitudinal migration in Natal. *Ostrich* 65: 86–94.
36. Berruti, A., J. A. Harrison, and R. A. Navarro. 1994. Seasonal migration of terrestrial birds along the southern and eastern coasts of southern Africa. *Ostrich* 65: 54–65.
37. Kortlandt, A. 1980. How might early hominids have defended themselves against large predators and food competitors. *Journal of Human Evolution* 9: 79–112.
38. Butzer, K. W. 1980. Palaeoecology of the South African australopithecines. In *Proceedings of the 8th Pan-African Congress on Prehistory, Quaternary Studies, Nairobi, 1977*, ed. R. E. Leakey and B. B. Ogot, 131–132. Nairobi: International Louis Leakey Museum Institute of African Prehistory.
39. Rayner, R. J., B. P. Moon, and J. C. Masters. 1993. The Makapansgat australopithecine enviroment. *Journal of Human Evolution* 24: 219–231.
40. McKee, J. K. 1999. The autocatalytic nature of hominid evolution in African Plio-Pleistocene environments. In *African Biogeography, Climate Change, and Early Human Evolution*, ed. T. G. Bromage and F. Schrenk, 57–67. New York: Oxford University Press.
41. Wescott, R. W. 1967. Hominid uprightness and primate display. *Amer. Anthropol.* 69: 78.
42. McGrew, M. C. 1992. *Chimpanzee Material Culture: Implications for Human Evolution*. Cambridge, UK: Cambridge University Press.
43. Whiten, A., J. Goodall, W. C. McGrew, T. Nishida, V. Reynolds, Y. Sugiyama, C.E.G. Tutin, R. W. Wrangham, and C. Boesch. 1999. Cultures in Chimpanzees. *Nature* 399: 682–685.
44. Nishida, T., ed. 1990. *The Chimpanzees of the Mahale Mountains: Sexual and Life History Strategies*. Tokyo: University of Tokyo Press.
45. Whiten, A., and C. Boesch. 2001. Chimpanzee cultures. *Scientific American* January 2001: 61–67.

46. Whiten, A. 2000. Primate Culture and Social Learning. *Cognitive Science*. 24: 477–508.

47. Boesch, C., and H. Boesch-Aschermann. 2000. *Chimpanzees of the Tai Forest: Behavioural Ecology and Evolution*. Oxford: Oxford University Press.

48. Shipman, P. 1986. Scavenging or hunting in early hominids—theoretical framework and tests. *Amer. Anthropol*. 88(1): 27–43.

49. Blumenschine, R., J. A. Cavallo, and S. D. Capaldo. 1994. Competition for carcasses and early hominid behavioural ecology: a case study and a conceptual framework. *Journal of Human Evolution* 27: 197–214.

50. Potts, R. 1988b. On an early hominid scavenging niche. *Current Anthropology* 29:153–155.

51. Weaver, K. F. 1985. Stones, Bones, and Early Man: The Search for Our Ancestors. *National Geographic* 168(5).

52. Kruuk, H. 1972. *The Spotted Hyena. A Study of Predation and Social Behavior*. Chicago: University of Chicago Press.

53. Maguire, B. 1980. The potential vegetable dietary of Plio-Pleistocene hominids at Makapansgat. *Palaeont. Afr*. 23: 69.

54. Peters, C. R., and B. Maguire. 1981. Wild plant foods of the Makapansgat area: a modern ecosystems analogue for *Australopithecus africanus* adaptations. *Journal of Human Evolution* 10: 565–583.

55. Turner, A., L. C. Bishop, C. Denys, and J. K. McKee. 1999. Appendix: A locality-based listing of African Plio-Pleistocene mammals. In *African Biogeography, Climate Change, and Human Evolution*, ed. T. G. Bromage and F. Schrenk, 369–399. New York: Oxford University Press.

56. Robinson, J. T. 1967. Variation and the taxonomy of the early hominids. In *Evolutionary Biology*, Vol. 1., ed. T. Dobzhansky, G. Hecht, and J. Steere, New York: Appleton-Century-Crofts.

57. Broom, R. 1938. The Pleistocene anthropoid apes of South Africa. *Nature* 142: 377–379.

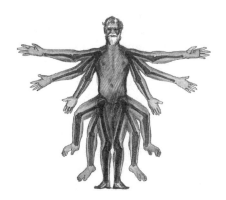

CHAPTER 7

On the Uncertainties of Becoming Human

Main-line, Side-line, or Parallel Humans?

P ossibility of several *Homo*-like lineages raises new questions about modern human ancestry. Where does the category *Homo* begin in the fossil record? The problem emphasised by coexistence of *Homo* (?) *rudolfensis* (2.4–1.85 mya) and *Homo* (?) *habilis* (2.3–1.7 mya). Was their allocation to *Homo* premature? Did the rudolfensis lineage emerge before 3 mya? the habilis lineage after 3 mya? Arguments exist for each as potential Modern human ancestors. The importance of accurate identification of differences in early humans and hominins to build an incremental view of evolution.

I n 1758, when Carl von Linne (Linnaeus) borrowed the Latin word for "man" or "human" to name the scientific genus "*Homo*" and added "*sapiens*" as our specific designation, he was naming what he believed was a single unique creation. Never could he have envisioned that his nomen, *Homo*, might one day accommodate as many species as his "*Equus*" with its horses, asses, onagers, and zebras.

Up to the 1960s, even anthropologists, persuaded by a handful of fossils that seemed appropriately laid out in space and time. and perhaps influenced by pre-Darwinian concepts of a "Chain of Being," sustained the belief that modern humans were the culmination of a single line of descent. It was the concept of a "chain" that gave rise to the popular analogy of "missing links"—a status that the media have conferred on almost every new fossil find over the last century. At the root of this concept of a single, linear ancestry was the assumption that bipedalism was the unique legacy of modern humans and their immediate forebears. Today, any examination of hominin evolution has to countenance proliferations, not only of early "australopithecines" but of more advanced bipeds right up to the emergence of truly modern humans. It must even be open to the possibility that separate but parallel lineages might have shared some of the features we have always envisaged as the most exclusively human: notably, bipedalism and enlarged brains.

The accumulation of generically "human" fossils is now immense (and the pile is still growing), but near the bottom of the heap, excavated from deposits in eastern Africa that are in the region of 2.3 million years old, are two forms until very recently judged the earliest worthy of the name *Homo*. Instead of a single-stranded chain (albeit one that was mutilated by "missing links") to connect us with our primitive forebear, the discovery of new fossils demands the more organic analogy of a thicket or a tree. So, in a genealogy that begins with several branches, it is not a few chain links that are missing but, it would seem, an entire tree trunk! So when and where are we likely to find that trunk? And is it really missing? Or must we choose between different trunks?

A final resolution of such questions is still illusive, but recent discoveries have multiplied the possibilities and greatly increased the interest and complexity of human prehistory. In the previous chapter, I suggested that *A. africanus* could be a close offshoot of the *Homo* trunk and is, in any case, an informative guide to the regional, ecological, and behavioral origins of our remotest beginnings as bipedal hominins.

At issue in this chapter is the starting point for calling a fossil "*Homo*." Should the name be awarded only to possessors of the major attributes of a modern human? If so, what criteria are to be used? Should our separation from the living chimpanzee be indicated by a change of name at that point of divergence, thereby making all fossil hominins *Homo*? Alternatively, if the "australopithecines," or earliest hominins, represent more than one lineage, as is suggested here, should the one that led to us (by my reckoning, an open choice between something resembling an early form of *A. africanus* and the brand new *Kenyanthropus* lineage) be subsumed in *Homo*? Should *Homo* begin with the "Erects," an assortment of African and Asian hominins that certainly includes fossils close to our direct ancestors? Alternatively, *Homo* could be reserved for Moderns alone, which would amount to a ceremony of divestiture in which the Erects would be stripped of their entitlement and presumably would have to revert to their earliest name, *"Pithecanthropus."* In that event, there would still be difficulties in drawing lines between these "ape-men" and other grades of possible predecessors, including *Australopithecus africanus*, *"Homo habilis,"* and *"Homo rudolfensis."*

Of several major sticking points, the one that is particularly germane to the theme of this book is the widespread assumption that the acquisition of bipedalism was a single event in our evolutionary history. If my model of basin evolution is correct, we will need to know much more precisely, increment by increment, the progress of bipedalism in at least two, possibly more lineages. If the *Australopithecus* lineage developed bipedalism from a starting point similar to that of the Lucy/Robusts and *Kenyanthropus* lineages but along a different evolutionary trajectory, how did the possibility of a "late start" influence the efficiency and functional anatomy of being a truly competitive biped? It is possible that the precise sequence of changes, notably in the proportions of individual limbs, facial structures, and tooth proportions, might have been different in three or more separately evolving lineages. Proportions could have diverged in, say, a squat little semiarboreal upland form and a lanky, effectively terrestrial lakeside type (1). Such differences could, in turn, have been influential in determining the long-term futures of the two lineages. One form of bipedalism could have been best suited to a specific ecosystem, the other could have been more readily remodeled to suit a wider range of adaptive traits. All emergent hominin species were probably sufficiently mobile as bipeds for other factors, such as technological aptitude, to have overridden limb proportions (and even residual tree-climbing skills) in deciding the fate of one species over another. Even so, much more material, especially limb bones, will be needed before such questions can lead to anything much more than intriguing speculation.

Up to the present, almost all discussion of speciation in early *Homo* has revolved around skull fragments. Indeed, calculations of brain size and an enlarging brain has come to signify the true marker for *Homo*. Thus, it was an estimate of 674 cc that prompted L.S.B. Leakey and his colleagues to designate his first specimen of *"H. habilis"* (Olduvai hominid no. 7, described in 1964) to *Homo*. Yet subsequent specimens of the same species have brain sizes estimated as small as 510 cc (2), while the largest (hotly contested) estimate for *A. africanus* has been claimed to approach 600 cc

(3, 4). So, even allowing for under- and overestimates, there are overlaps or a continuity in brain sizes between these two rather similar species, one of which was so confidently allocated to our own genus by L.S.B. Leakey and his colleagues. Now the picture has been further complicated by the work of L.S.B.'s daughter-in-law and granddaughter: their discovery and description of *Kenyanthropus* seems to represent a small-brained precursor for the large-brained "*Homo rudolfensis.*" Even an enlarging brain now seems to be the attribute of quite separate lineages. (The principal external expression of this enlarging organ is a higher, more rounded profile to an unridged cranial vault.)

Another well-established criterion for *Homo* has been small teeth and jaws, (a clear enough difference when the comparison is with the nutcrackers or apes but rather subtle when the comparison is with *A. africanus*, which do have relatively large molars). Likewise, the more rectangular outline of true *Homo* cheek teeth versus round-edged contours in australopithecine teeth is less clear-cut with *A. africanus*. A less formal objection has been the "requirement" for flaked stone tools to be associated with fossils, something that has become somewhat of a de facto criterion for *Homo*. (For example, grudging acceptance of Leakey's "*Homo habilis*" was probably eased by his Olduvai specimen sharing its bed with crude "Oldowan" stone tools.) Here again, this could be a flimsy basis for judging where to put the boundary between humans and protohumans, because chimpanzees use tools, even stone ones. Nonetheless, it remains a practical criterion, and *A. africanus* fossils have yet to be associated with stone tools.

Were all the evidence before us, which it is not, there are two directions from which we could approach this sort of taxonomic boundary drawing. One popular route is to travel back through the acknowledged lineage and keep the name *Homo* until there is a conspicuous break or branch in the supposed ancestral line. Another approach is to work forward from a broad primate base and identify the parting of the ways between the *Homo* lineage and all apes or, say, *Australopithecus*. There are precedents for the last two approaches: Jared Diamond, tongue in cheek, called humans the third chimpanzee (5); Ernst Mayr, instead, distinguished all hominins from apes by casting all the australopithicines in with *Homo* (6). If "evolution by basin" proves to be a correct model of hominin speciation and were South Africa or another "basin" to prove the source for *Homo*, then a natural break would fall between the particular basin dwellers that evolved into humans and all the others. If South Africa came up as the ultimate source, only the australopithecines now known as *A. africanus* would be eligible to become *Homo*. Alternatively,

Homo could be judged to begin with the first fossils that were unequivocally directly ancestral to Modern humans, a criterion that might include African Erects but would disqualify Asian Erects and Neanderthals and would certainly exclude *"Homo" habilis* and *"Homo" rudolfensis* (7), the subjects of this chapter. On the present (admittedly very inadaquate) evidence, the precursors of these two species are more certain than their descendants. The first is most closely anticipated by members of the *Australopithecus africanus* complex, whereas the second has similarities with *Kenyanthropus platyops*. I will, very provisionally, treat *"Homo" habilis* as *Australopithecus habilis* and *"Homo" rudolfensis* as *Kenyanthropus rudolfensis* (figure 7.1). (In this taxonomic jungle it will already be apparent that I have followed David Strait and others in returning Lucies to their earliest generic name of *Praeanthropus* [8].)

While *"Homo africanus"* could provide us with a "missing trunk" growing in southern African soil, it would not explain why humans speciated into two and, very soon thereafter, into still more species. Which of the branches led to us? Which were dead ends, and why the branching anyway? The recurrence of dead ends at later stages of human evolution emphasises the importance of finding a general explanation for both "branching" and extinction in *Homo* species. That, in turn, may help us understand the very new discovery of our own lineage winding its way through these thickets of human speciation and extinction. In any event, the metaphor of an increasingly bushy tree of human evolution has decisively displaced the older one of a chain.

Many still reject the evidence, claiming that large, big-brained types and smaller, small-brained types can be explained by sexual differences or else accommodated within the natural variation of single, highly variable species. Yet there are two hominin fossils, both with enlarged brains, both from East African sites, and their metric differences are accompanied by structural peculiarities that prove they are two distinct species (9, 7).

Both species may yet be shown to derive from as yet unknown fossils living in as yet unknown localities, effectively displacing *A. africanus* as a model of the *Homo* ancestral stock. In the meantime, the "evolution by basin" model plausibly envisages at least one of the big-brained hominins expanding out of the geographically confining enclave of southeastern Africa. Until very recently, all three species (*africanus, habilis,* and *rudolfensis*) were judged to be sufficiently alike to need no hypothetical intermediates to bridge their differences. That theory has changed with the discovery of *Kenyanthropus platyops*, which is a very suitable candidate ancestor for *"Homo" rudolfensis*.

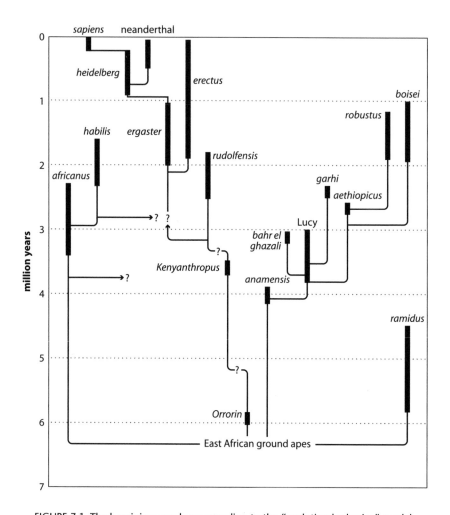

FIGURE 7.1 The hominin genealogy according to the "evolution by basins" model.

Of the two, the best-known and earliest to be discovered is *"Homo" habilis* (10) (reputedly nick-named "Handy-man" in 1964 by Raymond Dart). This was a variable but generally diminutive animal, with a small brain (ranging from 510 to about 680 cc). The cranium was rounded, as in *A. africanus*, but was broader, particularly at the base and with reduced areas of attachment for neck and jaw muscles.

The body was generally rather small. Indeed, one individual has been estimated to have weighed 24 kg and stood no more than 1 m high (11). Among the many arguments that the Handy-man was nothing but a Handy-ape have been persistent references to the closeness of its resemblance to *A. africanus*. This hypothesis is reasonable enough, and one Olduvai specimen in particular (OH 24) has more resemblance than most. Before capitulating to the majority view, Richard Leakey was for a while unwilling to rate his own 1.9 million-year-old Koobi Fora *habilis* skull (ER 1813) as *Homo*, mainly because of her very small size and brain (510 cc) (12). His caution may have been well founded.

The Handy-man has a short "dished" face, rather like *A. africanus* but with slightly less protruding front teeth. The broadly arched tooth row juts out from wide orbits behind a low, sloping forehead. The cheek teeth, or molars, are generally (but not always) rather narrower than those of *A. africanus*, while the incisors are, by comparison, relatively large. The proportions of arms to legs (measured as a ratio of humerus to femur) is calculated at 85%. This is exactly halfway between those of a chimpanzee (in which these bones have the same length, i.e., a 100% ratio) and a modern human (in which the ratio is 70%).

The alignment and fusion of the big toe with the other toes is thought to have been incomplete in *H. habilis* (13), in contrast with the feet of Lucies, which, in spite of very long curved toes, are judged to have had more resemblance with modern human feet.

To some scholars, this feature suggests partial arboreality; to others, it signifies a mere carryover from the preexistent apelike structure (explanations that are not mutually exclusive). Considering the animals' size, they were indeed quite likely to have climbed up into trees to sleep at night. A broader dependence on treetops as sources of food or refuge must be less certain, particularly if a broader strategy of facing off predators while foraging on the ground was already developed.

Envisage "Handy men," then, as rather diminutive animals scampering, rather clumsily, in well-bonded gangs, along pathways made by other animals through a predominantly thicketed but patchy environment. They were often noisy and boisterous: many of the animals that they encountered in their path would have been harassed with shouts, snatched-up missiles, or the shaking or clapping of broken boughs. Differences in size between males and females were measurable but inconspicuous, and all mobile children, of various ages, would have been as active as the adults in most of their activities. I would guess that they probably retained a rather apelike distribution and length of black or brown-red hair. Skin color, too, probably encompassed the variation seen in modern chimpanzees, but naked faces were probably made more expressive by strategically placed brows, mustaches, and other expressively sculpted tufts. At present, the most reliable early dates for *"Homo" habilis* (2.3 mya) come from Hadar, Ethiopia, and Baringo, Kenya. Still earlier, from a

2.6 million-year-old Ethiopian site called Gona, come thousands of flaked stones that have been attributed to *"Homo."* This is probably a safe attribution, because similar, if later, tools have been found with a partially toothed palate and maxilla of *"Homo" habilis.* Well-identified and dated specimens of this species come from Olduvai, Tanzania, and Koobi Fora on Lake Turkana from a time span of 1.7 to 1.9 mya. Other specimens are less securely identified and dated, and the possibility that some South African specimens (notably, former *Plesianthropus,* STS7, that are currently classified as *Australopithecus africanus* may prove to be *"Homo" habilis* could push the known time span for this species back still earlier. The actual beginnings of this, supposedly the most primitive type of *"Homo"* yet discovered, might be substantially earlier than the current fossil record suggests.

A second species (for more than a decade, known only by the catalogue number of a single specimen, ER1470) comprises a skull (without a mandible) that was found by Bernard Ngeneo on the shores of Lake Turkana (formerly Lake Rudolf). It has, until very recently, been known as *"Homo" rudolfensis* (figure 7.2). A lower jaw of the same species (ER1802) and some other fragments also derive from the shores of Lake Turkana. Another fossil that seems to belong to this species or lineage is a single mandible from Uraha, on the shores of Lake Malawi (14). Provisionally dated at 2.4 mya, this specimen is as early or earlier than the oldest *habilis.* In spite of the narrowness of this difference in date and in spite of the rarity of *rudolfensis* specimens, the possibility has to be kept in mind that Rudolfs or their immediate ancestors might have dispersed at an earlier date than the Handy-man. In the time frame that is suggested here, the Rudolf specimen "1470" is particularly late (from around 1.85 mya, although early reports erroneously put 1470 at 2.9 mya). So far, nothing is known about its skeleton except that the body, judging from the size of the skull, would have been substantially heavier and taller than that of *H. habilis* and well within the normal range of variation in modern humans.

Unlike *H. habilis,* the cheekbones of the Rudolf specimen were as wide as those of *A. africanus,* but all three species had similarly shaped, thin-walled crania. *Homo rudolfensis,* larger, with a higher forehead and less prominent brow–ridges, had a very long, vertical, slablike face with broad cheeks and large front teeth. A shortened palate from front to back had the principal effect of aligning the thick-bodied incisors and canines in a broad, clipper-like blade across the front of the flat face. Could this mean that the jaws of 1470 were peculiarly well suited to some form of forceful biting? If so, it was a permutation opposite to the one found in the Robust or "nutcracker" lineage, in which the action of chewing with giant

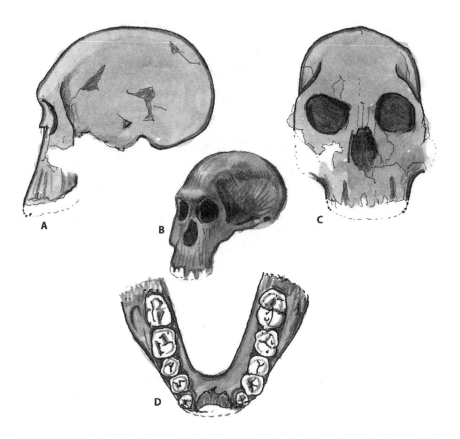

FIGURE 7.2 The fossil evidence: A–C. The "1470" skull. D. *"Homo" rudolfensis* mandible from Uraha. (After Bromage, T. G., F. Schrenk, and F. W. Zonneveld. 1995. Paleoanthropology of the Malawi Rift: an early hominid mandible from the Chiwondo Beds, northern Malawi. *Journal of Human Evolution* 28: 71–108.)

molars must have taken precedence over nibbling by the much smaller front teeth. Deepening of the face lengthens the leverage and strengthens the force of bites and chews, which suggests that the first crushing bites into food by *H. rudolfensis* might have signified some special staple foods, particular ways of dealing with them, or both. So unusual is this combination that it not only raises questions as to what sort of specialized feeding ecology was involved, but it also implies a narrowing of dietary options. Lakeside habitats might have yielded various foods that were best dispatched or preprocessed with broad, clipper-like jaws, but a mere handful of specimens makes the ecology of Rudolfs pure guesswork. There are some overall resemblances between the skull and jaws of 1470 and the Robusts that have led dental experts to suggest that both species made

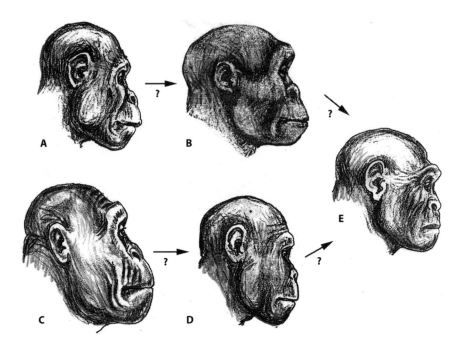

FIGURE 7.3 Which route to Modern humans? A. South African man-ape (*Australopithecus africanus*). B. "Handy-man" ("*Homo*" *habilis*). C. Kenya Flat-face (*Kenyanthropus platyops*). D. Rudolph human ("*Homo*" *rudolfensis*). E. *Homo ergaster* (erect). Not to scale.

parallel adaptations to more roughage in their diets (15). Coarse foods could imply that both species had responded to overall climatic change (16) and generally drier habitats (the rather amorphous conclusion of several scholars). Alternatively, both species could have derived from a specific region or regions, as is suggested here, where the exigencies of a dominant ecosystem slowly forced them to adopt specialized diets (figure 7.3).

These species, then, used to be the candidates for earliest *Homo*—both living at much the same time, both characterized by some significant and shared innovations in the proportions of their skulls, and both presumed to be the fabricators of crude stone tools.

Before elaborating further on what they might have had in common, consider what overall trends might be signified by smaller teeth, reduced chewing muscles, and bulging foreheads. Such traits have been summarized as "neotenous," a shorthand and overspecific term for one of the most important mechanisms controlling evolutionary change. Neoteny or "paedomorphism" describes the carryover, into adult life, of juvenile structures (including behavior). It is essentially a consistent alteration in the relative timing of various developmental processes (17). In this instance, selection would seem to have favored delays or slowings in development and even some minor suppression of selected adult features (such as brow ridges).

It is equally possible for selection to extend growth or speed up development in what appears to be the opposite of neoteny—as in birds that can fly immediately after hatching from the egg, or bats that are apparently independent of their mothers a few weeks after birth. In the example of megapode birds, speeding up the development of wing feathers is essentially "spending more time in the egg" and can be shown to be adaptive for a parentless hatchling against its predators. Likewise, selection may favor speedier development in one feature and slowing in another (with different ratios of the two, even in closely related populations). In this way, differences can emerge from a common matrix that quickly define an adaptive divergence. Thus, from an early stage in their branching, gorilla-faced, ape-brained Robusts contrasted with the more child-faced, bigger-brained so-called "*Homo*" (18).

What could be the overall adaptive advantages and costs for early hominins in delaying or suppressing adulthood? If a major characteristic of childhood is dependence on parents, one might suppose that it was mothers and other, mainly female, adults that carried the main cost of slower development. It has already been remarked that prolonging a fetal pattern of development so that babies that become brain-heavy and helpless for months after birth requires fully terrestrial, arm-free mothers. Another implication of prolonging childhood is that mothers must not only devote a great part of their life and energies to the care of offspring but, given their joint vulnerability, must live within a society that enhances security for all. If both sexes and all ages are to some degree "less adult" than, say, an ape or a nonpaedomorphic hominin, then all members of a social group have a shared interest in finding security together. Neoteny could have enhanced sociality by juvenilizing all classes, making them all more "dependent" on one another (particularly during perilous encounters with predators). The more frequent a group's exposure to such dangers, the more interdependent its members. Universal vulnerability could therefore become an asset if, as I outlined in the previous chapter, slow and defenseless little primates faced off the competition by rallying close together. The principal expression of this group effort might have consisted of putting on active and noisy displays and generally sharing well-coordinated "strategic plans" that would consistently put the whole group at more of an advantage than would any set of individual reactions or arbitrary initiatives.

In many species of animals, not just higher primates, one of the incentives to forming a social group is to find safety in numbers and benefit from more eyes on the lookout or ears on the alert (19). In all cases, however, the animals remain hostage to events, rushing to join others in defense of a territorial boundary, or fleeing, en masse, from a predator. What may have distinguished *Homo*-like lineages from all such precedents was that they were, in one important respect, no longer subordinate to external events. If, as I suggested earlier, individuals could rally around and jointly mob other animals to drive them off and get access to resources, that activity had within it the potential to become an organized, repetitive, and deliberate engagement with some potentially productive aspect of the environment. A living source of sustenance was no longer the reward for infinitely slow, adaptive evolution; it could be seized directly, once the obstacles that surrounded it had been overcome. Beginning with socially coordinated intimidation displays and close observation of the specific responses of other animals, the removal of "obstacles" around potential food sources may have become a skill that eventually became a central feature of the human ecological niche. Because scrapers, cutters, crushers, and diggers are among the more rudimentary of tools, the development of a "physical removal technology" was clearly a part of prehistoric humanities' relationship with their environment. However, the larger, governing part may have relied on the gross behavioral trait of cooperating to "remove" other animals from wanted resources. Frequently a dangerous enterprise, close cooperation would have had survival value for all members of the group. If this was the case, children, from an early age, would have had to be party to an everyday behavior that potentially penalized anything less than full participation. Yet a willingness to keep close and join in group activities is characteristic of many young primates in much less demanding circumstances than those proposed here. It will seem contradictory to suggest that the enfeeblement and dependency that is implied by neotenous traits should have been integral to such an apparently aggressive strategy. "Apparent" is the operative principle because mobbing (notably among birds and social mongooses) typically directs the energies of many small animals at a single larger target (20). Furthermore, there are few more insistently assertive creatures than juveniles seeking to satisfy their appetites. As exceptionally feeble and slow animals, early hominins must have had to develop exceptionally effective alternative defenses if they were to survive in more open African landscapes. If there was any way in which neoteny might have assisted social cohesion or the development of intelligence, then any enfeebling disadvantages must have been more than offset by the advantages.

It is with the development of intelligence that neoteny may have had its most influential role. One typically neotenous expression of this was delay until the brain had reached its adult size before it got "sealed" into its brain-box. Brain enlargement is the most obvious correlate of intelligence, so this late suturing of cranial bones is a typical delayed development mechanism and the most obvious manifestation of human neoteny. Likewise, the aptitude for learning seems to have been extended well into adulthood. Were it not for the fact that many species have rather short adult lives, the relative proportion of a lifetime spent as a learning, maturing juvenile would be longer in humans than in all other mammals. As it is, humans are quite long-lived; yet in spite of this, the social units of both early humans and later hunter-gatherers probably had an exceptionally high ratio of juveniles to adults. This arrangement could have helped blur distinctions between adults and juveniles while foraging, with many implications for the development of intelligence and (on the assumption of a linkage) an ever-enlarging capacity for dietary and technical innovation.

Childhood, with its curiosity, delight in exploration, and playful expressions of discovery and invention, may have provided the essential grounding, the precondition for technological invention. Experience is essential for the maintainance and diffusion of traditions in many species, but it is less certain that mature individuals were significant sources of innovation in early "*Homo*." Rather, inventiveness was more likely to have originated in less-experienced youngsters, in playful but ecologically relevant contexts (21) (figure 7.4).

Regardless of its inventors, one outcome of novel techniques and tools operating within particular ecosystems would have been that habitat-specific technologies and their users could, over time, become more and more tied into a particular region or ecotype. Paradoxically, the techniques and tools that were most habitat-specific were likely to have been the least adaptable to other ecological settings. In some circumstances, notably during periods of radical environmental change, less specific, more generalized tool users could have gained advantages while the ecospecialists became ever more firmly confined to their specific habitats. Thus, early human populations and their technologies might have paralleled the diversification of species, making the old metaphor of a single chain of descent even more obsolete.

There are still some latter-day Linneans who cannot conceive of hu-

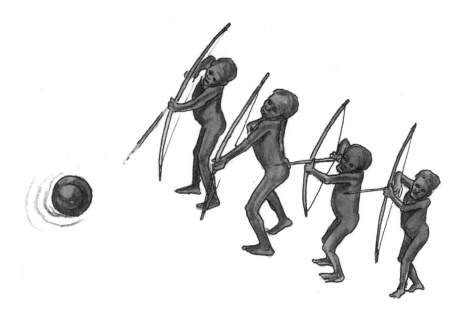

FIGURE 7.4 Small boys firing arrows at a moving target. Children practice functional skills and acquire accurate proficiency at an early age. (From an anonymous photographer. CMS bookshop, Madang P.N.G.)

mans on the same plane as horses and donkeys, with one species preferring wooded hills, another open plains—each differently adapted, yet both coexisting in close proximity. I argue that this situation, so normal for other animals, could apply in a very particular way to early humans and help explain the problem of coexistent species of *Homo* or proto-*Homo* in Africa and elsewhere. Before pursuing this line of thought, consider the nature of the evidence for the two earliest known east African species.

First, it should be pointed out that some scholars have proposed separate descent: from *afarensis* for *rudolfensis* and from *africanus* for *habilis* (22). This hypothesis would require a tree with much larger measures of both parallel and reversible evolution in teeth, brain size, and skull shape than seem plausible. The arguments that follow allow for several possible ancestries. One option is that both *rudolfensis* and *habilis* might have derived from a common ancestor originating in southern Africa, a position that could be supported by some of the features their skulls share in common. Alternatively, those similarities might be due to parallelism in two lineages deriving from different basins.

First, consider some of the limitations suffered by a putative common ancestor of the two species. South Africa is a temperate cul-de-sac; any outward expansion by a water-dependent species (and one that can be assumed to have been well adapted to its cool and peculiar climate) would have faced formidable obstacles. On three sides, there was sea; to the west, waterless deserts too. The extreme east would have been an unlikely avenue of expansion because, at that time, coastal lowlands in the eastern tropics might still have been preoccupied and dominated by ground apes. Two main routes remain that could have been available, at different times, to separate "wings" of a single parent population. Furthermore, for animals moving north into warmer, drier environments, there may have been different solutions to the problems of finding water and keeping cool.

Among the areas in which dispersing populations could have survived would have been any that resembled their cool, strongly seasonal homeland. On this count the uplands of eastern Africa were the obvious, if distant, zone for a secondary expansion. The main "upland corridor" between southern and eastern Africa is the long south-to-north mountain range that marks today's frontier between Zimbabwe and Mozambique. This is but two valleys distant from the extreme northern end of the Drakensberg Mountains. Here, we can envision an early *Australopithecus africanus* population well positioned to initiate such a northward expansion when conditions were right. In spite of lowland discontinuities, this route has, nonetheless, been traversed in both directions by numerous species of upland-adapted animals and plants. The ideal time for a successful traverse and colonization of uplands would have been during *cool* periods when entire temperature belts and their biotic communities drifted toward the tropical highlands and also slid downhill. The first Homo-like fossils, as well as the first worked stones, have been found in east Africa between 2.4 and 2.6 mya, a period that followed one of the sharpest, fastest, and most severe drops in global temperature during the Plio-Pliestocene (23; figure 6.3, p. 208). Ecological conditions that had been restricted to southern latitudes became more widespread, especially in the extensive uplands of eastern Africa. During such a climatic phase, emigrants could be said to be taking their preferred environment with them. Uplands in the very south of Africa would have remained habitable, but only for those species that were already highly adaptable or could change fast enough to keep up with the very severe cooling. Much of the community would have either died out, drifted north, or (in suitable localities) "slid" downhill into sheltered pockets. My guess is that invasion of the tropics by immediate predecessors of *H. habilis* broadly coincided with the onset of global cold.

By contrast, an alternative option for expansions in range would have been favored by very different climatic/ecological extremes and would have provided opportunities for other populations of large–brained, *Homo*-like hominins. Before the setting in of cold 2.9 mya, warm conditions would have sent more water down rivers and greatly increased the areas covered by swamps and floodplains. Here, potential corridors are permanent rivers, and habitats (whether riverbank or lakeside) are, by their very nature, narrowly linear. Hominins that needed to keep cool under much hotter conditions might have had to live under more constraints than free-ranging upland types; perhaps they relied more on shade and living in continuous and intimate contact with water. Such conditions were widespread in the East African Great Lakes region, and it was here that a homegrown hominin, in the form of *Kenyanthropus*, might have developed its own unique way of life.

If ever there was a time when aquatic or semiaquatic habits were possible in early humans, "Rudolfs," or their immediate ancestors, are the most plausible candidates. The arguments favoring a semiaquatic phase in human evolution center on some physiological peculiarities that are otherwise difficult to explain (24). For example, fat grows under the skin of most primates, but it is never as extensive as the layer covering a typical modern human. Our subcutaneous fat has sometimes been likened to the blubber of truly aquatic whales, hippopotamuses, sea cows, and seals. Our greasy skin, served by an abundance of sebaceous glands, is also peculiar and can, at a stretch, be interpreted as a waterproofing device. Our tendency to waste huge quantities of precious water through copious sweating is obviously related to keeping cool but could only develop in an animal that was seldom far from water. The fact that very young babies usually consent to immersion in water and may even swim has been cited as evidence for an aquatic or semiaquatic past. It is also claimed that humans possess a "diving reflex" in which their metabolism slows, the heart rate lowers, and long breaths can be held in spite of vigorous activity underwater. All these capabilities are absent or poorly developed in the apes. More extreme arguments claim that big brains need seafood, the low position of the human larynx is purely a design for holding breath, and our naked skin and hair tract patterns resemble those of other aquatic mammals, not other primates. If the mysterious Rudolf lineage could be shown to have passed through a prolonged phase of adaptation to river and lakesides, some of these claims will deserve more serious consideration than they have earned to date. In any event, a strictly waterside existence would have restricted the diet and demanded many specialized behaviors. Rudolfs might have had to live within a number of constraints that played no part in the life of an "upland" species of southern origin, putatively the Handy-man.

During warmer, wetter periods (as are thought to have existed up to about 3 mya), a dense network of rivers could have allowed water's-edge dwellers to expand and colonize far and wide over much of eastern and central Africa. But did they? Recent recovery of the Kenya Flat-face from 3.5 million-year-old deposits near the shores of Lake Turkana (the former Lake Rudolf) suggests that *Kenyanthropus* could now justifiably include *rudolfensis* as its descendant. It is possible that Rudolfs could derive from the migration of a very early *Australopithecus africanus*. However, differences in teeth and skull make this unlikely, so the possibility must be raised that a *Kenyanthropus platyops/rudolfensis* lineage may be of separate Eastern Rift or Great Lakes origin. The earlier emergence of this lineage from a coastal ground ape might have paralleled many of the features described in the previous chapter. After all, the higher reaches of the east African uplands share many features (particularly in relation to montane forests and lower temperatures) with the wetter parts of southern Africa. In any event, for the purposes of this exploration, the two *"Homos"*— *habilis* and *rudolfensis*—can be regarded as sharing similar evolutionary histories in separate but cooler realms of Africa. How could this choice of options relate to candidature for the status of "earliest *Homo*"?

Considering its conservative anatomy and wide distribution in the east African highlands, *Australopithecus habilis* is a good candidate for south African origins and colonization via an upland route at around 2.8 to 2.9 mya. Once established in new but favorable regions and only weakly connected with the place of origin, a nucleus of colonists could rapidly proliferate and, less rapidly, differentiate from the parental stock (which, during climatic vicissitudes, may also have continued changing). When the expanding colonists entered several such new lands, the opportunities for speciation must have multiplied, particularly where there were barriers or bottlenecks to keep subpopulations apart. In eastern Africa, fragmentation into central, eastern, and Ethiopian massifs would have encouraged diversity; an observable variety in local forms of *Australopithecus habilis* fossils could lend support to such a process.

What about *H. rudolfensis*? All we know is that this species has been found at 1.85 mya from the shores of Lake Turkana and a lot earlier, at 2.4 mya, from the shores of Lake Malawi. We can, of course, suppose that fossils separated by such big spans of time and distance will prove to be different; nonetheless, they plausibly belong to a single, very distinctive type. Clearly, lake margins were among its chosen habitats. We know that this was a new form of biped that not only used its hands but, judging from the stubs and roots of its teeth, may have had somewhat specialized teeth in its long face. Whatever it did with its teeth, that activity was probably closely coordinated with whatever it did with its hands. So far, there are no associated artifacts to hint at what sort of specialization was involved, but in its departure from a generalized dentition, 1470 seems to have adopted feeding habits that were more specialized than those of *A. habilis*. With possible emergence before 3 mya, the fossils of Lakes Turkana and Malawi could represent a lineage established enough for a large number of regional representatives to have evolved and diversified. Such fragmentation could have led to still further specialization in some regions, but *K. (H.) rudolfensis* shows enough generalized, *Homo*-like features to suggest that *less* specialized types could have developed in other regions.

The divergence of two very similar and closely related species across an ecological boundary is not without precedent, and many African taxa may have diverged during peaks of opposite climatic extremes. At such times a likely mechanism could have been the localized "engulfment" of one habitat by another (as I have suggested happened to the habitat of the ancestors of gorillas).

An example of evolution in sibling species is provided by the lechwe and the waterbuck, two valley-loving antelopes so closely related that they readily interbreed in captivity yet maintain very different life histories in the wild. Lechwe are deep-swamp specialists of extremely restricted distribution in central Africa; they come in two species, the northern Nile lechwe and the southern (Zambezian) common lechwe. The latter has different subspecies around each of the lake complexes it inhabits (a diversity evidently enhanced by the difficulties of crossing dry lands in-between the swamps, which are essentially ecological islands). The waterbuck, by contrast, recently ranged, as a single species, over all of sub-Saharan Africa wherever there was the barest minimum of surface water. Along the margins of swamps, the two species often come into close contact without hybridizing. Although the most obvious adaptations of lechwes are their elongated hooves and tolerance of continuous immersion, there are many more subtle differences between them and waterbuck. These peculiarities, which include differences in physiology and behavior, are likely to have begun during a wet climatic phase, with the effective isolation or engulfment of future lechwes (as one segment of a larger, perhaps pan-African population) within a discrete "island" of swamps. Thereafter, each adaptive refinement would have proceeded incrementally. Their sibling population, the ancestors of waterbuck (almost certainly in a different and drier region) retained more generalized habits that have demonstrably favored them under drier conditions. Once lechwe were unable to survive outside the swamps, it became inevitable that isolates would diversify still further into subspecies and species. The presence of their close relative, the waterbuck, in the intervening land was probably a major factor restricting them to swamps (25).

Is it fanciful to liken early Rudolfs to ancestral lechwe/waterbuck and envisage them as *Homo*-like pioneers spreading out through tropical Africa and perhaps beyond? Perhaps when global temperatures crashed, their sibling species the Handy-man was favored (due, perhaps, to its cold-adapted physiology and in spite of its more conservative anatomy)? It is possible that Rudolfs then became more and more restricted to the habitats that suited them, or to regions outside or beyond the range of the Handy-man. Alternatively, could a head start in evolving a larger brain and (very speculatively) easy walking have given the Rudolf lineage long-term advantages over the Handy-man? Which of the two was in an evolutionary cul-de-sac, and which one was closer to being a true human ancestor? Or were they both without issue? There are still no certain answers to such questions.

Anything that restricts movement or gene flow (whether derived from the animals' behavior or induced by local topography) can encourage divergence. Any form of separation must be decisive in shaping the fortunes of sibling populations moving out from a common source—no less for early humans than for waterbuck and lechwe, but in the former there is the complicating factor of "culture." We can be reasonably sure that hominins in wooded or scrubby uplands would have rapidly developed very different cultural and economic practices from those living along low-lying lakeshores and river levees.

The value of comparing Handy-men with Rudolfs has less to do with the details of their niches (that is a task for later, better informed generations) than with trying to answer a question that was posed earlier. In the first place, we need to think what was, until recently, the unthinkable: what permits two human species to coexist (26)? Then, given that only one species was our direct ancestor, there is the second recurrent question of what consigns the one or two other lineages to extinction without issue (27)? Part of the answer to both these questions must lie in the tendency of these culture-bearing animals to reinforce their physical adaptation to a particular region or ecosystem with appropriate techniques, and to invent the tools to go with them. The dynamics of innovation, trial and error, and a rapid feedback loop would tend to drive cultures into still further specialization until some sort of equilibrium between technology and ecology was reached. The general proposition of culture as "the human niche" is of long standing, but some years ago I explored some of the longer term, evolutionary implications of self-making in a book titled "Self-made Man" (21). In it, I redefined and multiplied that singularity into a continuous accretion of niches, many of them effectively "stolen" from other species. I also pointed to specific permutations of such stolen niches becoming characteristic of particular cultures and regional populations.

If culture has become the prime selective force shaping the evolution of modern humans, the concept should be extendable back to its beginnings in earlier human species—but it must be admitted that data get a lot scarcer. Even so, it can be guessed that sustained demographic success in any ecologically specialized niche would tend to be reinforced by selection for appropriate physiologies, anatomies, even temperaments. Thus, genetic change might travel, far behind, on the coattails of cultural change. Nonetheless, selection would eventually operate as much within parameters defined by culture as within those set by nature.

In this way, genetic changes could eventually have shadowed the cultural innovations of specialized human economies. In spite of the improved adaptedness that such changes might have brought about, such economies were ultimately vulnerable to invasions by wider ranging, less specialized cultures. These could bypass the specialists with some particularly powerful new inventions or techniques, or even through an ability to improvise new procedures to displace now obsolete routines. Such contests would have been most damaging to the local specialists during periods when their environment was suboptimal.

The deposits in which *H. habilis* have been found have yielded crude stone tools. Equally crude stone artifacts come from the same shorelines as *H. rudolfensis*, but these give no clear indication of their technology. In any case, it is impossible to reconstruct the biology of an entire lineage from one or two specimens. Contrasts between human specialists and nonspecialists are preferably deferred to better documented examples of Neanderthals and Modern humans.

Crucial dimensions of the speciation process are patterns of dispersion and the passage of enough time for genetic change to take place. Following an initial spread, there is typically separation from the parent population and the development of local peculiarities in isolated subunits; if any of these local idiosyncrasies is adaptive and successful, there may then be secondary spread. In the absence of appropriate fossils, how much time that set of events might have required is anyone's guess. That said, and accepting that the current earliest *Homo*-like fossils and the earliest traces of large-scale stone working in east Africa have been pushed back as far as 2.6 mya, the first excursions by their ancestors out of their respective homelands might have occupied quite narrow windows of time. The fact that the last known fossils allotted to *A. africanus* have been dated to as late as 2.05 mya means that these man-apes survived half a million years after the supposed departure of their *habilis* descendants. The most likely explanation is that their physiological adaptations to this southern cul-de-sac were, for a while, unassailable. The Handy-man's conservative anatomy and retention of similarities with its southern forebears also become more understandable if its initial spread was indeed due to similar physiological advantages. The fact that these similar and closely related hominins were rather commonly fossilized during the globally cool period between 2 and 3 mya also lends support to that argument. The Handy-man looks as though it was the main beneficiary of climate change, but what had happened to the ancestors of 1470 before the worldwide crash in temperatures?

Much controversy surrounds supposed triggers for the emergence of *Homo*. Changing climate is commonly cited on the implicit assumption that because the general drying out between 2.9 and 2.3 mya was associated with the first appearance of *Homo*-like fossils, it can be taken as a comprehensive explanation for our origin. The correlation looks to be true enough for *habilis*, but the species' dominance in the fossil record could prove to have been a colossal red herring if its abundance had as much to do with its physiology as with its supposed brainyness. The model presented here suggests that an entirely new way of earning a living (which I characterize as tool-assisted "niche-stealing") gave more than one type of hominin new competitiveness. It was a change that let them cross a technological threshold, which then took them into a new relationship with the environment. But why two species? One climatic extreme might have favored the development of one type of protohuman; another extreme might have favored the evolution of a second species. The spread of either species could have been facilitated by whatever climatic change favored their particular way of life, but I suspect that it was some permutation of sociotechnology that gave one or the other the long-term advantage (figure 7.5).

There is one respect in which climate might have been an important agent assisting the colonization of the tropics by *Australopithecus habilis*. Lucies and the Robust, *A. aethiopicus*, were common and widespread animals by 2.7 mya, and their established presence might be expected to have deterred incursions from *A. africanus*-like protohumans. The *disadvantages* of climatic change for the more tropical Lucies or proto-Robusts

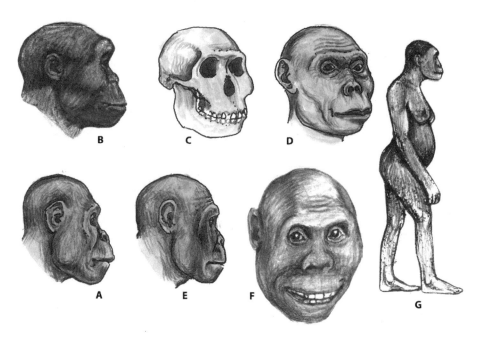

FIGURE 7.5 Reconstructions of *A. africanus* (A), *H. habilis* (B–D), and *H. rudolfensis* (E–G).

rather than the *advantages* for *habilis* could have been decisive in giving the latter a toehold. If cooler, drier conditions actively favored the latter in extensive areas that, under a moister, warmer climate, were better suited to the Lucy/Robust lineage, a cold-adapted physiology could have been a critical asset. However, the ecological success of *habilis* must also have been founded on the use of tools, an omnivorous economy, and the exploitation of resources away from the immediate proximity of river valleys. The Handy-man and Lucy/Robust lineages could each occupy their own distinct foraging zones, sharing only a common need for sleeping trees and access to water. In confirmation of this, fossil Robusts and early *habilis* are often found in the same strata and in similar riverine habitats (28). So there clearly was ecological overlap, but the former are usually more strictly riverine (29). If the precursors of Rudolfs had shared much of their overall range with the Lucy/Robusts for much longer than the Handy-man, their niche would seem to have been a rather cryptic one; in addition, they were clearly much rarer animals.

So far, the use of tools has had scant mention because I wanted first to discuss the broader ecological and social contexts and examine the separate behaviors and sensory fields in which tool using could emerge. At the broadest level of generalization, the emergence of *Homo*, whether from a southern or eastern African stock, must have been associated with a new type of direct link between hands and brain. I have proposed that it was during the ground ape phase that hands began to be released from their primary functions of locomotory grasping and supporting and levering the body. As Jolly suggested in his seed-eating hypothesis (30), habitual squatting allowed for the development of secondary functions of touch, conveying food to the mouth, and some rudimentary capacity to process food. Even at this stage, it can be supposed that new levels of "hand-processing" and hand/brain organization had been achieved.

I suggested earlier that the two main behaviors involved in the emergence of tool using were small-scale food processing (at a mainly individual level) and aggressive interactions that were primarily social. These activities operate in radically different ambits, so a corresponding divergence was likely to have occurred in the earliest contexts and functions of primitive technology. Ground apes, supposedly typical, close-up, intensive foragers, would have collected and processed much of their food within a perceptual field narrowed to the reach of their arms. Wider theaters would have surrounded aggression but probably tended to refer to rivals or threats that were within sight. A third ambit, signaled by distant calls or vistas of unknown or rarely traveled territory, would have tended to be beyond the ken of ground apes. Moving out, literally, into the pathways of other animals in more open country would have involved a challenging expansion in all motor and sensory fields (31).

Behavior and activities operate within spatial fields where particular senses provide relevant information, and particular faculties respond appropriately. Expanding home ranges into new habitats and developing novel behaviors can only take place if a basis for adaptation already exists. On such criteria, the descendants of ground apes, as portrayed, might be disqualified from ever making a living out in the open by a form of "foraging myopia." Such self-defeating short-sightedness must have been offset by behavior linked with the two roots of tool use: foraging and aggression. (Nest building can be treated as a comfort-enhancing activity not entirely divorced from subsistence.) Mobbing of other species might have provided the basis for a separate category of action, elicited by aggravating stimuli other than food but ultimately designed to secure access to it. The ultimate ecological success of *Homo* may have depended on a unique integration of separate classes of behavior, based on different senses and involving distinct forms of judgment, and each taking place in contexts that were originally separate in function. Consider some of the sensory properties of a primate's ambits as they radiate out from the individual's contact with the outside world:

1. The "arm's length" (1 m) radius. Close-up, reachable handling of food and small objects. The main linkages are between all the senses—the hands, mouth, and brain—but touch and smell are foremost during short-range contacts. Here, judgments involve properties that can be polar opposites or relative gradients; hard/soft, alive/dead, edible/distasteful, fast/slow, safe/dangerous, immovable/handleable. Each involves the recognition of simple consequences (e.g., long dead is distasteful, slow is dangerous, soft is edible); these are judgments that other animals can make but usually with the aid of preprogrammed sensory devices, specifically tuned by natural selection to guide their choices. For the ancestors of *Homo*, deficient in the preprogrammed wisdom of superspecialist species, an expanding diet depended on close observation and trial and error. Primarily the realm of immediate appetites, the main social dimensions of the "arm's length" ambit are grooming, sex, and sleep. Close contacts involve the exchange of soft, rapidly attenuating vocalizations. Where the home range and food sources are both well–known, this can be the dominant ambit.

2. Immediate surroundings (radii of 4–5 m in cover, more in open surroundings). This is the zone of immediate action in which smell and touch are of little importance but visual sensitivity to events and movement is paramount, backed up (where relevant or necessary) by auditory cues. The limbs, rather than the hands or the mouth, become the main servants of the brain's decisions. Here, judgments relate to the perceiver's own ability to move and act in space, to "get at" or to "get away." The location and behavior of conspecifics, other species (dangerous, useful, or neutral), and generalized resources (e.g., fruiting trees or nesting birds) are perceived, and the perception can be acted on. This is the main realm of immediate action and reaction, and it is the ambit in which many open-country mammals and birds spend the greater part of their time.

Likewise, this ambit appears to be the main realm for social life. With sound broadly subordinate to vision, there could be sufficient time lapse between visual and auditory signals for the effect of vocalizations (on own or other species) to be readily observed. With experience, vocalizations can be enhanced or suppressed to achieve particular ends (such as to "frighten off" one species at one moment and "be frightened off" by another). This slight dislocation in the relative primacy of one sensory channel over another may have provided proto-*Homo* with some flexibility in their use or choice of signals. A narrow temporal window between the moment of seeing an action and reacting could have provided the perceiver with an appropriate moment in which to interject vocally. Moving out of closed habitats into more open ones would have made the close monitoring of calls and countercalls easier and would have opened the way to the voice becoming a primary means of manipulating and co-ordinating group behavior (32). Mobbing displays, augmented with loud calls, could have become a routine strategy in approaching other species that obstructed access to desired resources.

3. Distant ambit (beyond the radius of immediate action to the horizon). Of all the senses, only vision and the reception or transmission of very loud calls would have been relevant to the perception and use of distant space. Thunder and the scent of rain with other audio-olfactory clues to benefits or dangers also could have galvanized early *Homo* into action. Eyes link directly to the brain, with negligeable inferences or cross-references from other senses. Here, the act of scanning would have mainly concerned perceiving the movements of distant con-specifics, enemies, prey, or patterns made by attractive vegetation types, fires, rain clouds, and the sun's cycle. Judgments would have centered on assessments of direction, relative distance, and an aware-ness of events such as the comings or goings of rain, smoke, dust, or dusk. Because the act of long-distance scanning was relatively slow, passive, and did not involve the need for immediate action, there would have been a potential for reflection and planning.

Clearly many birds and some open-country mammals have evolved very detailed spatial awareness and extremely accurate homing skills. Such skills may require learning and memory of a landscape's topography to a high order of resolution and detail, but these are superimposed on innate skills that are best developed in migratory birds. These are not events that seem to register much with the living apes, and there is noth-ing in the evolution of primates (least of all among the putative, stay-at-home ground apes) to prepare them for life in such a milieu (33, 34). If forest dwellers were to develop any skills in the perception of a wider am-bit, they would have had scant resources in the way of innate abilities. Nonetheless, open vistas might have elicited feats of spatial memory and a propensity for planning that depended more on self-constructed mod-els in the brain than on genetically coded travel plans and compasses. If such perceptions and plans were to be shared with other group members, a combination of directional gestures and arbitrary signal codes would have had to be devised.

Noting the very different sensory and functional fields that are in-volved in these three ambits serves to underline the multidimensional challenges facing the human lineage as it moved out into more exposed situations. More specifically, it draws attention to a lack of preadaptation in this "lateral expansion" that may have put them at a substantial disadvantage with many of the animals they encountered away from safely wooded riverbanks. These species may have been predators, potential prey, or competitors; but many of them, especially those with long "lineage-tenancies" in open country would have been in possession of faculties that were superior to those of the hominins in very many respects. What, if any, was the major breakthrough that allowed such a feeble and short-sighted primate to leave the shelter of riverine forests?

Before hominins could survive in the open, their ability to walk and run on two legs must have become easy and economic of energy. Their physiology must have adapted sufficiently to permit some exposure to the sun, and they might have had to tolerate some temporary dessication. In the face of strong competition, bluster alone could not have helped them find, secure, or process a wide range of different foods. Some form of diversified technology would have been essential. A more systematic and regular use of tools must have been the main innovation of proto-*Homo* and probably distinguished them from the Lucy/Robusts. Nonetheless, tools must have developed by increments, no less than biological structures. Incompleteness in the archaeological record ensures that there is very little documentation of the first appearance of artifacts, with the single exception of stone tools, which were probably of minor overall importance, particularly during the earliest periods.

If the use of tools emerged from two very different behaviors—solitary foraging and social aggression—quite different classes of tools could be expected to develop. In fact, there are technologies that do not fit too neatly into such a model, but the main groupings of primitive artifacts can be listed as follows:

1. Exploratory or extractive devices deriving from individual foraging.
   A. Wipes and Sponges. Fashioned from leaves, lichens, or chewed fiber, an extension of appetite for food or water and typically assisting in the extraction of traces of desirable food or water from crevices. In some cases, wipes may also be used to remove unpleasant substances such as feces or irritants.
   B. Probes. Includes sticks and stems, acting as extensions or miniaturizations of a finger or hand exploring a cavity or crevice. Tools of this class are occasionally made and used by apes—notably, stripped stems or twigs that are inserted into termite holes by chimps. These are primitive types of tool directly serving ap-

petite—possibly of quite marginal utility to begin with but an enduring and highly "developable" trait.

C. Levers (in the loosest sense). A variation on the probe, but more robust, simulating the pushing and pulling of hands or fingers in loosening bark or soil or unearthing roots or subterranean animals. A variant is the use of sticks to maneuver inaccessible objects within reach; very commonly used by captive apes in contrived situations and almost certainly used in the form of long-bone digging sticks by Robusts (35). Initially likely to have been a byproduct of foraging for storage organs of plants, but with many later applications.

D. Spikes. Points with a range of possible uses, from miniature variations on the probe to larger objects (e.g., found stakes, bone slivers, teeth, or antelope horns). Not seen in wild apes (but use of ready-made points can be taught). Depending on size and application, held in the fingers or with the whole hand, to impale, perforate, loosen, or degrade. A minor tool type to begin with, its later, larger morphs would become the spear, javelin, sword, and dagger; its lesser morphs, the awl and needle.

E. Blades. Grating or sawing edges on found or flaked stone, bone, shell, or wood that acted like larger, stronger fingernails or even teeth. Can be learned by apes, from humans, toward directed but contrived ends (36). A type of tool with very numerous applications for the processing of animal carcasses, but probably a minor category to begin with.

2. Hammers. Stones, blocks of wood, large animal bones, or ivory used to crush or split objects open. The most direct way of forcing access to well-protected foods enclosed within shells, cases, capsules, carapaces, or pods too hard to be broken open by teeth or hand. In most applications, this is unambiguously derived from solitary foraging techniques that are also used by apes and other nonhuman animals such as Capuchin monkeys, mongooses, and sea otters. It permits a huge expansion in diet when applied to nuts or mollusks and has remained a major tool category up to the present because it is so well adapted to the hand's grasp and the arms leverage (37). Hammering can also have a rarer aggressive connotation in which the arm's action simulates fist pummeling.

3. Clubs. Branches or bones wielded as flails, staffs, staves, or poles, mostly in defense or attack, both against other species and during fights. An artifact unambiguously associated with aggression, observable in chimpanzees directed at enemies or (in a less directed form)

as part of a generalized blustering display. In its later developments, this category overlaps in interesting ways with spikes and missiles.

4. Missiles. Beginning with sticks, debris, and stones tossed at a rival or enemy, simple "throwing" is common among chimpanzees and has been reported in other higher primates. The behavioral origins of throwing are undoubtedly associated with aggression. It is a class of implement that would have been of the greatest utility to early hominins, and the throwing of missiles has remained a major human skill. In its directly manual applications, it serves to dislodge fruit or prey; is used in hunting, homicidal, and playful activities; and is a preferred activity for testing motor aptitude. The applications of missiles, once the mechanical priciples were understood, had the potential to culminate in Roman "ballista" (and, with very different sources of power, interplanetary rockets).

5. Cordage and Wraps. Originating in vines, grasses, plant stems, spiderwebs and the sinews, skin, or tissues of animals. The rudiments of a feeling for cordage, as a binder, are sometimes perceptible in the nest-making of great apes. A less obvious awareness of cordage in apes is their occasional self-drapery with leafy vines (or, in captivity, ropes and rags) usually over the shoulders, neck, or head. All apes have a highly developed sensitivity to the strength, flexibility, and swinging potential of natural vines and lianas because of the frequency with which they have to trust their weight to them. The harvesting of many forms of grapelike fruit also makes foragers familiar with the tensile properties, drawing or holding power, and portability of laden and entangling vines.

The development of cordage is linked with the concept of wrapping, and both must have preceded the regular use of any tools that were sufficiently useful not to be thrown away or stored at their site of manufacture. The precedence of cordage and wrapping derives from the fact that a sharp edge is prone to cut its owner and cannot be carried permanently in the hand. Tool-attachment sites could have been from wrist or elbow bracelets, shoulder strings, necklets, and waist belts. The earliest preferred sites are likely to have been the neck and elbow. An important and early component of human technology, cordage developed ever more diverse uses during human evolution but is scarcely ever preserved in the archaeological record excepting tangs. At rather late stages of cultural evolution, perforated beads attest to its presence in the form of necklace string.

6. Containers. Receptacles for solid or liquid materials, of both vegetable and animal origin; likely to have been marginally later in de-

velopment than most of the preceding artifacts. Apes easily learn to use ready-made cups and jugs, and wild animals manipulate tough leaves in maneuvers that blur the difference between wrapping and containing. Containers mimic the mouth and the cupped hand, but their main early emergence would have been as by-products of appetite. Numerous foods occur naturally within "containers" of skin, rind, bone, shell, or woody material, so any tendency to carry the results of foraging to a common base would have utilized such properties. Wrapping or binding containers to make them more portable would have set up an early functional link between containers and cordage. Such a connection would also have emerged as a by-product of foraging for animals or plants that combined both qualities rather than as a contrived assemblage of separate components. Containers became crucial artifacts with the need to carry food back to a distant home base.

A majority of these artifacts can be devised from materials familiar to apes, and as numerous recent studies (and the tool-type summaries just given) have shown, apes employ, or can learn to use, a wide variety of tools (38). Many of the main incentives to turn raw materials into artifacts have been explored in some detail in my previous book, *Self-made Man*, but in common with all other books on the subject it could only discuss a central question in terms of a hopelessly poor archaeological record. Posed as an ecological problem, we need to know the point in hominin ancestry when their use of tools put them at enough of an advantage over other species to initiate a type of supraecological dominance that permitted almost continuous expansion in range.

It was not simply a matter of outcompeting other species as foragers, scavengers, or predators. It was a qualitative change in the methods by which hominins could extract energy from their environment. Ultimately, human technology would be able to extend its influence over all ecosystems, but its beginnings seem to have involved a share-out among several species of hominin, each somewhat of a techno-specialist. However, in gradually increasing the number of techniques and modifying them to suit seasonal or local changes in the environment's resources, early hominins laid the foundation for an ever-expanding number of techniques. Each of these could extend the range and depth of its user's "reach."

One analogy might be the greater efficiency of orb-webs as compared with the direct stalking of prey by spiders. The web extends the spider's reach in a manner analogous to the extension of many aspects of an or-

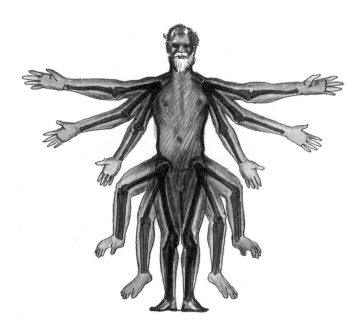

FIGURE 7.6 Self-portrait as a twelve-limbed primate. Proportions of human and ape (chimpanzee) compared with intermediate lengths interposed. (These correspond very closely with the proportions of *Praeanthropus*.)

ganism's biology beyond its immediate boundaries (figure 7.6) (39). Like spiders, hominins could trap, snare, or net their prey; but the latter could also ambush, excavate, expose, entice, corral, hook, spear, preserve, or contain a steadily enlarging range of food types. As with the many species of spider, the adoption of different techniques by different species must have been favored by many selective forces. Each type of hominin living in a peculiar environment with a peculiar range of food options would have been affected. Not least among these selective forces would have been those that bore down on the hominins' physiology, anatomy, and susceptibility to different diseases.

Because all techniques were ultimately capable of being applied to all situations in all habitats, inventions such as fire, weapons, shelters, vehicles, clothes, medication (the list goes on) had universal application with the right fine-tuning. Local and specific adaptations were eventually rendered redundant by the sheer reach of the technology itself. From the start, the elaboration of technology and techniques mimicked the elaboration of natural adaptations; it was this property that caused every expansion of range or scope to be at the expense of another species. This process may have begun with the first emergence of hominins, when their immediate impact on the environment was probably still comparable to that of other primate species such as baboons.

When did cultural factors begin to tip the balance? Can we identify the beginning of a specifically human strategy for energy extraction? The only existent physical evidence for some such momentous change is the first currently known appearance of stone tools (in Ethiopia, eastern Africa, at about 2.6 mya). However, tools seldom get buried in the hands of their makers. (Even the consistent occurrence of crude stone tools in strata that also yield the Handy-man has been challenged when the bones of Robusts also occur in the same deposits.) Given that apes use tools, there can be no doubt that *all* hominins did too. They cannot be proved to be tool dependent in the way that later humans demonstrably were, but it is very probable that the Robusts had a stereotyped technology. What is not at all likely is that Robust techniques were capable of rapid and appropriate elaboration to a variety of different ends. Robust technology would have been of a specialized and limited type. It may have been devoted to quite specific tasks, notably the excavation and processing of subterranean foods, and was probably closely associated with peculiar manual and perhaps oral/dental manipulations of food. Like spiders, different hominins might have had species- or population-specific techniques that reflected the challenges of local food types or ecologies.

Breaking human evolution down into small functional increments alters perspectives on this question. So long as erect posture, free hands, and tool use could be lumped together as a packaged approximation of humanity, any australopithecine was a fair model. If the first hominin lineages emerged from different basins, at different times, and with different attributes, as is argued here, there was likely a threshold in the relative reliance on technology that distinguished our own ancestral line from all others. Each lineage may have shared bipedalism, manipulativeness, and some technological aptitude; yet the permutations differed, each following its own trajectory within its own distinct ecological niche.

In the highly contentious field of human origins, there was rare but transitory agreement during the 1980s that *"Homo" habilis* was the earliest human and a likely precursor to the lineage that eventually gave rise to modern humans (40). The demonstration that *habilis* was an incomplete biped has contributed to the erosion of that consensus, and precursors to *rudolfensis* now have almost as strong a claim to our ancestry—one of the many uncertainties obscuring our efforts to retrace our evolutionary path. Today, the earliest uncontested precursor is *Homo er-*

gaster, the "Work-man" (a species that many regard as the African version of *Homo erectus*). That species and its collaterals are subjects for the next chapter.

REFERENCES

1. Stanley, S. M. 1992. An ecological theory for the origin of *Homo. Paleobiology* 18: 237–257.
2. Johanson, D. C., and M. A. Edey. 1981. *Lucy: The Beginnings of Humankind.* New York: Granada.
3. Clarke, R. J., F. C. Howell, and C. K. Brain. 1970. More evidence of an advanced hominid at Swartkrans. *Nature* 225: 1219–1222.
4. Johanson, D. C., and B. Edgar. 1996. *From Lucy to Language.* New York: Simon and Schuster.
5. Diamond, J. 1991. *The Rise and Fall of the Third Chimpanzee.* New York: Random House.
6. Mayr, E. 1969. *Principles of Systematic Zoology.* New York: McGraw-Hill.
7. Wood, B. A., and M. Collard. 1999. The human genus. *Science* 284: 65–71.
8. Strait, D. S., F. E. Grine, and M. A. Moniz. 1997. A reappraisal of early hominid phylogeny. *Journal of Human Evolution* 32: 17–82.
9. Walker, A. C., and R.E.F. Leakey. 1978. The hominids of East Turkana. *Scientific American.* 239(8): 54–66.
10. Leakey, L.S.B., P. V. Tobias, and J. R. Napier. 1964. A new species of the genus *Homo* from Olduvai Gorge. *Nature* 202: 308–312.
11. Johanson, D. C., F. T. Massao, G. G. Eck, T. D. White, R. C. Walter, W. H. Kimbel, B. Asfaw, P. Manega, P. Ndessokia, and G. Suwa. 1987. New partial skeleton of *Homo habilis* from Olduvai Gorge, Tanzania. *Nature* 327: 205–209.
12. Leakey, R.E.F. 1971. Further evidence of lower Pleistocene hominids from East Rudolf, North Kenya. *Nature* 231: 241–245.
13. Susman, R. L., and J. T. Stern. 1982. Functional morphology of *Homo habilis. Science* 217: 931–934.
14. Bromage, T. G., F. Schrenk, and F. W. Zonneveld. 1995. Paleoanthropology of the Malawi Rift: an early hominid mandible from the Chiwondo Beds, northern Malawi. *Journal of Human Evolution* 28: 71–108.
15. Walker, A. C. 1981. Diets and teeth. Dietary hypotheses and human evolution. *Philosophical Transactions of the Royal Society, London* B 292: 57–64.
16. Vrba, E. S. 1988. Late Pliocene climatic events and human evolution. In *Evolutionary History of the "Robust" Australopithecines,* ed. F. E. Grine, 405–426. New York: Aldine de Gruyter.
17. Huxley, J. 1932. *Problems of Relative Growth.* New York: Dial Press.
18. Shea, B. T. 1989. Heterochrony in human evolution: the case for neoteny reconsidered. *Yearbook of Physical Anthropology* 32: 69–101.
19. Hamilton, W. D. 1971. Geometry for the selfish herd. *Journal of Theoretical Biology* 31: 295–311.
20. Kingdon, J. 1977. *East African Mammals. An Atlas of Evolution in Africa,* Vol. 3A: Carnivores. London: Academic Press.

21. Kingdon, J. 1993a. *Self-made Man. Human Evolution from Eden to Extinction?* New York: John Wiley & Sons.

22. Bromage, T. G., and F. Schrenk. 1995. Biogeographic and climatic basis for a narrative of early hominid evolution. *Journal of Human Evolution* 28: 109–114.

23. Denton, G. H. 1999. Cenozoic climate change. In *African Biogeography, Climate Change, and Human Evolution*, ed. T. G. Bromage and F. Schrenk, 94–114. New York: Oxford University Press.

24. Morgan, E. 1982. *The Aquatic Ape. A Theory of Human Evolution.* London: Souvenir Press.

25. Hardin, G. 1960. The competitive exclusion principle. Science 131: 1292–1297.

26. Tattersall, I. 1986. Species recognition in human paleontology. *Journal of Human Evolution* 15: 165–176.

27. Wood, B. A. 1993. Early Homo: how many species? In *Species, Species Concepts, and Primate Evolution*, ed. W. H. Kimbel and L. B. Martin, 485–522. New York: Plenum.

28. White, T. D. 1988. The comparative biology of "robust" *Australopithecus*: Clues from context. In *Evolutionary History of the "Robust" Australopithecines*, ed. F. E. Grine, 449–483. New York: Aldine de Gruyter.

29. White, T. D., and J. M. Harris. 1977. Suid evolution and correlation of African hominid localities. *Science* 198: 13–21.

30. Jolly, C. J. 1970. The seed eaters: a new model of hominid differentiation based on baboon analogy. *Man* 5: 5–26.

31. Peters, C. R., and R. J. Blumenschine. 1996. Landscape perspective on possible land use patterns for early Pleistocene hominids in the Olduvai Basin, Tanzania: Pt. II. Expanding the landscape models. *Kaupia* 6: 175–221.

32. Cheney, D. L. and R. M. Seyfarth. 1990. *How Monkeys See the World.* Chicago: University of Chicago Press.

33. Garber, P. A. 1996. On the move: how and why animals travel in groups. In *On the Move*, ed. S. Boinski and P. A. Garber. Chicago: University of Chicago Press.

34. McGrew, M. C. 1992. *Chimpanzee Material Culture: Implications for Human Evolution.* Cambridge, UK: Cambridge University Press.

35. Brain, C. K., and P. Shipman. 1993. The Swartkrans bone tools. In *Swartkrans. A Case's Chronicle of Early Man*, ed. C. K. Brain. *Transvaal Museum Monograph* 8: 195–218.

36. Marzke, M. W. 1997. Precision grips, hand morphology and tools. *American Journal of Physical Anthropology* 102: 91–110.

37. Marzke, M. W. 1986. Tool use and the evolution of hominid hands and bipedality. In *Primate Evolution*, ed. J. G. Else and P. C. Lee, 203–209. Cambridge, UK: Cambridge University Press.

38. Whiten, A., and C. Boesch. 2001. Chimpanzee cultures. *Scientific American* January 2001: 61–67.

39. Dawkins, R. 1982. *The Extended Phenotype.* Oxford: W. H. Freeman.

40. Stringer, C. B. 1986. The credibility of *Homo habilis*. In *Major Topics in Primate and Human Evolution*, ed. B. Wood, L. Martin, and P. Andrews, 266–294. Cambridge, UK: Cambridge University Press.

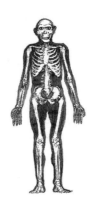

CHAPTER 8
On Going Far with Fire
Africans Go Abroad

Northwest Africa or Arabia the most likely regions for the evolution, in isolation, of *Homo ergaster* (1.8–1 mya), a member of the *H. erectus* complex (1.8–0.3 mya in Asia). Evidence for the use of fire by *H. ergaster/erectus* as early as 1.6 mya. Association of Erects with the Achulean stone tool industry. The emergence of "archaic" or Heidelberg humans in Africa or Eurasia about 1 mya and their evolution into Neanderthals in northern temperate habitats. Tool-assisted specialization a potential throughout hominin and human evolution.

Hannibal's war elephants and their forced march up through Spain and over the Alps have provided an exotic footnote in Mediterranean history and a welcome diversion for countless bored primary school pupils. Few stop to ponder where Hannibal got his elephants from before he shipped them over the Straits of Gibraltar. Today, the pastoral landscapes of densely populated Morocco and Algeria may seem an unlikely habitat for wild African elephants, yet their relatively recent existence in the Atlas Mountains illustrates an important

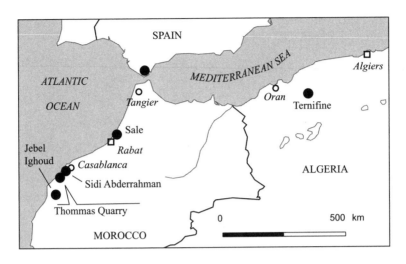

FIGURE 8.1 Map of northwestern Africa illustrating localities where discoveries of *Homo erectus* and archaic humans have been made.

pattern in African biogeography—and one that may have a direct bearing on the evolution of modern humans.

Capturing rain off the Atlantic, the mountains in this northwestern corner of Africa once offered a choice of habitats ranging from nearly subalpine through montane forests to woodlands, savannas, steppes, and a wide, fertile coastal plain. During favorable climatic cycles through much of the last few million years, there were temporary ecological connections with tropical Africa across the high ground, and mountain chains that run obliquely across the Sahara from northwest to southeast and less directly along the Mediterranean coast to the Nile valley and its sources (1). The Nile valley also provided a narrow corridor between the Mediterranean coast and tropical Africa. At other times, there were contacts with Europe and western Asia. As a result, the Atlas region supported a unique mixture of tropical African and Eurasian flora and fauna, with rhinos, zebras, giraffes, hippopotamuses, buffaloes, various antelopes, and gelada baboons; it also included a predominantly Mediterranean fauna such as deer, Eurasian carnivores, and Barbary macaques (2). During arid cycles of climate, Atlas biota would have been totally cut off from their parent populations in tropical Africa, and the desiccating influence of the Sahara would have been more pervasive even than it is today.

Fossils of hominins are very fragmentary and, so far, have been found only in a scatter of Pleistocene deposits in the Atlas region (figure 8.1) (3). Even so, late and inconclusive as they are, these fossils signify the presence and survival of yet another tropical mammal, an early human, close to the margins of the Sahara. Although their precise identity is controversial, these fossils certainly confirm that a new form of human had adapted to very dry, open conditions.

I return to these fossils later, but given that proven "early" fossils should be our first guide to the arrival of a new sort of human, we must look elsewhere—while acknowledging along the way that these humans may have been widespread and common long before they turned up as fossils.

The earliest dated recovery of this new type of hominin from eastern Africa was about 2 mya (4), where the species suddenly appears—as Alan Walker, the leading expert on this species has remarked—"without a past" and fully formed (5). Labeled "Work-man" (*Homo ergaster*) in Africa and "Erect-man" (*Homo erectus*) in Asia (many authorities regard them as a single species), this tall, slender human had made startling changes in proportions, especially when compared with the diminutive and still rather apelike *Homo habilis* of the previous chapter.

Up to the present, *habilis* has been the preferred model for an ancestor to the "Work-man" and other Erects, but the possibility of another ancestral lineage should be kept in mind. At present, the most realistic alternative is the Rudolf lineage, but this is plausible only if an early, less specialized form had found its way up into the northern reaches of Africa.

Because the adaptations of equatorial species inhibit their colonization of temperate regions, the speedy exit, spread, and success of the *ergaster/erectus* lineage out of Africa is strongly suggestive that this was a northern, not a tropical species. Such a conclusion immediately narrows the options as to where they could have evolved. Fossils (albeit late ones) make the Atlas Mountains a strong contender, but Arabia is another possibility.

So far, my sketch of our ancestors and their evolutionary travels has followed a sequence, moving along, jerkily, toward a gradually more recognizable portrait of humanity. The trail has been faint and fragmentary, leaping chasms of empty space and time, the clues being all-too-rare fossils helped along by genetic inferences and tentative reconstructions drawn from the geography and dying ecologies of contemporary Africa.

In the previous chapter, the trail ended abruptly with something of a precipice. In spite of the temporal gap being very narrow, a much more substantial gap in knowledge and fossil history separates the Handy-man and Rudolfs from the Work-man (and, by implication, our own more certain line of descent). The clear implication is that the fossil record is sampling three species that were already widespread (and sufficiently abundant to get fossilized) but offers no guide as to places of origin—rather, a few distribution records on a largely blank map. In the case of *rudolfensis* and *habilis*, later records could well be retrieving two species close to the end of their evolutionary lives, while *H. ergaster* was just beginning. This is not an unusual situation. For example, the modern Kongoni, a long-faced grazing antelope is known, from fossils, to have evolved fast and very recently. Yet it coexists with older members of the same lineage, gnus and topis, while archaic, relict species (bontebok, blesbok and hirola) survive as rarities in remote enclaves (6).

Perhaps the fossil Rudolfs and Handy-man were the equivalent of hominin bontebok and hirola, while the Work-man was the Kongoni of his time. Certainly the latter's modern anatomy suggests he was standing on our side of the canyon that separates us from the apes, australopiths, Rudolfs, and Handy-man. The coexistence, for a while, of several hominin species is only anomalous if we insist on the concept of a single line of descent. Species that adapt to peculiar regional conditions can often export their advantages to quite distant places, and there is no reason to suppose that hominin species were any different (especially those with a tried and tested but relatively specialized technology). The important questions of how and when the Work-man evolved are not entirely separable from the geographic question of where such a distinctive animal could have evolved. It is a question that returns us to the Atlas Mountains and Arabia.

New, unknown fossils have been bridging chasms between older, better known fossils ever since the search for our origins began. Perhaps, then, we can hope that present gaps may soon be plugged with new discoveries. New fossils have often come from unexpected quarters, sometimes revealing some of the secrets of "fossils without a past." So is there any fossil evidence in support of the northwest of Africa as a place of origin for *H. ergaster*? In 1953, Camille Arambourg described three lower jaws from Ternifine in the foothills of the Algerian Atlas (3). These resemble east African *H. ergaster* jaws but are probably later in age. Other fragments of early *Homo* are also known from the Morrocan Atlas and, in spite of some being too recent to occupy an ancestral position, illustrate that this was prime habitat for them (7). To evolve adaptations to dry, open country, the Work-man would have had to suffer a period of prolonged isolation from other hominins, in an appropriately discrete habitat. There were few, perhaps no, places in tropical or southern Africa that could have provided such a combination; however, northwestern Africa (or possibly Arabia) offered both. The alternatives fall between very early members of the Handy-man or Rudolf lineages. Spreading into one or both of these regions, this hominin evolved in isolation from other, more tropical species. An ancestor, resembling the Handy-man, can be envisaged crossing the Sahara during a benign period when mountain massifs provided a route between the southeast and northwest ends of the Sahara. By contrast, an alternative, but less plausible, ancestor, perhaps an early Rudolf, might have followed the Nile valley and spread along the coasts of the Mediterranean and Red Sea.

Before pursuing the geography of its possible origins, consider what sort of human the Work-man was. The species was first named *Homo ergaster* by Colin Groves and Vratislav Mazak in 1975; their evidence was a lower jaw from East Turkana, Kenya, that had well-worn, rounded molars that were so small they resembled those of some modern humans (8). The following year, Alan Walker and Richard Leakey described a new, nearly complete female skull that clearly belonged to the same species (9). However, it was only in 1984 that a near-complete skeleton was found, one of the most significant fossil finds of the century (10).

Most of our knowledge of *H. ergaster* comes from this momentous discovery, the detailed study and description of which was organized and coordinated by the evolutionary anatomist Alan Walker (11). The remains of this tall, slender young male revealed, for the first time, the existence of a true long-distance walker, with limb proportions, body–size, and build virtually indistinguishable from our own. Its cranial capacity, relative to its size, was modest: 850 to 900 cc (compared to about 1500 cc for comparably sized Moderns, 775 cc for *H. rudolfensis*, and 510 to 680 cc for *H. habilis*, which was a smaller animal). When the considerable differences in body sizes are taken into account, the brain/body ratios of *ergaster* and the earlier so-called "*Homo*" species scarcely differed. In other words, the Work-man's body had been transformed into something approximating a modern athlete, but in spite of some cranial remodeling there was meager measurable change in the relative size of the brain (12).

The cranium was lozenge shaped, with a short, vertical face hitched to the front; the shallow dome of the cranial vault extended back to meet the flattened skull base in a distinctively pointed bun at the rear. There was the trace of a keel along the top of the cranium, while the sides were flattened to accommodate temporal muscles more extensive than those of modern humans but much less developed than in *Australopithecus*. Whereas the base of the skull was still flat in *habilis* and *rudolfensis*, there were signs of flexion in *H. ergaster*—and this anticipates one of the more cryptic yet significant traits of Moderns. Among the many structural readjustments associated with this buckling of the skull's "chassis" is depression of the larynx such that it lies lower in the neck. It has been argued that such repositioning of the voice box was essential to improve

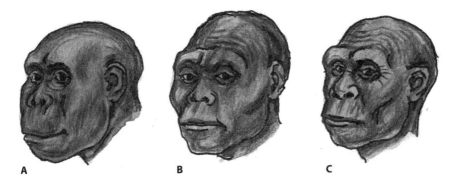

FIGURE 8.2 Reconstructed heads of three near contemporaries. A. *"H." rudolfensis*. B. *"H" habilis*. C. *H. ergaster (erectus)*.

controlled production of sound—one of the prerequisites for speech. An unexpected and well-hidden detail has been discovered in the inner ear, where the semicircular canals (which are important for balance in fully bipedal activities) closely resemble those of modern humans and are unlike the more apelike canals of *Australopithicus africanus* (13). This detail illustrates that the aquisition of bipedalism is an incremental process. It also demonstrates that standing up in ground apes would have been, literally, only the first step toward developing a faculty that has involved a succession of adaptive refinements spun out over several million years.

There was no trace of what we would call a forehead; instead, there was a shallow depression behind rounded brows that were thinly boned (like the rest of the skull, but projected well forward above the root of the nose. Unlike its flat-nosed predecessors, the Work-man had crimped-out margins around the nasal cavity revealing the existence of a projecting nose. The adaptive significance of this detail is thought to concern the humidification of dry air (14), which would certainly represent a striking innovation. If the apelike noses of early hominins were adapted to breathing the moist air of shaded forest and their apelike limb proportions betrayed a continuing dependence on trees, here was further confirmation that trees and shade were no longer preconditions for the survival of an advanced human (15, 16). In this respect, it might be argued that the existence of a protruding nose is better anticipated by the narrowed nasal margins of *rudolfensis* than by the very flat broad nose of *habilis*. On the other hand, the Work-man's short face is more like the latter's than the clipper-like muzzle of Rudolfs (figure 8.2).

The chest was flattened fore and aft, as in modern humans, losing all trace of the apish conical form. Between rib cage and pelvis, there was a

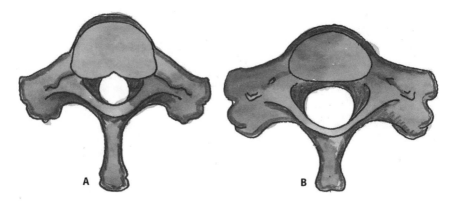

FIGURE 8.3 Vertebrae of *Homo ergaster* (A) and *Homo sapiens* (B) to show the greater diameter of the spinal cord in the latter. (From Walker, A. C., and R.E.F. Leakey, eds. *The Narioko-tome* Homo erectus *Skeleton*. Cambridge, MA: Harvard University Press.)

visible "waist," indicating that movements in the upper part of the body were fully independent from those in the lower. These novel abdominal proportions also suggest a smaller gut, which in turn implies a diet with less bulk and a high energy and protein content. Here, say several authorities, were the first efficient, regular, and systematic human hunters (17).

The central cores or "bodies" of the lumbar vertebrae were broad and circular, with flat, weight-bearing surfaces that enhanced balance and strength in the vertebral column. This feature is shared by modern humans, but there were two respects in which the Work-man's vertebrae differed: the spines were consistently longer, and the canal that enclosed the spinal cord was consistently narrower (figure 8.3) (18). This latter detail implies many fewer nerve cells serving the chest muscles and diaphragm. The smaller diameter of the spinal canal has been interpreted as indicating that the Work-man was "illinguate" (a word coined by Alan Walker to describe a human species without a spoken language) (5). Studies of speech production have shown that getting the rhythms and tones of speech right calls for fine control of air within the chest. This control is exerted by a "fan" of multiple-headed muscles, known as the Serratus complex (19). These muscles originally served quadrupedal animals to attach the top of the shoulder blades to the ribs of the thorax (acting as a sort of weight-bearing sling). Going bipedal effectively released these muscles from a predominantly suspensory function, permitting them to be adapted to other roles. In this case, the bellows action of ribs and diaphragm is intimately associated with improved communication and, ultimately, has a direct bearing on the cultivation and development of intelligence.

In this connection, an important aspect of hominin hunting that has had scant attention is the benefit, for a hunter, in having the ability to deceive prey through the mimicry of signals made by other animals. Calling up antelopes or birds with vocal lures is a familiar practice for numerous hunting societies, and the ability to imitate calls of all sorts is well developed in modern humans. It is frequently elaborated into onomatopoeic songs and games that involve still further manipulation of the breath, voice, tongue, lips, and mouth. The roots of these deceptively simple skills must lie deep in our hunting past. That it took 4 million years or more for a "borrowed" side effect of bipedalism to become enlisted in such significant and different adaptations as speech and song illustrates the mosaic nature of adaptive change and the role of "increments" in building up new structures and faculties, bit by bit from components that were designed for quite different purposes.

The pelvis of Work-man was actually narrower than in modern humans, a detail that not only confirms the advantages (for balance and movement) of a highly localized and stable center of gravity above the two legs, but also suggests that the small-headed babies of Work-women could pass through a narrower canal at birth than our own large-headed ones (20). Where the upper femurs angled inward to form ball-and-socket joints with the pelvis, there was a long "neck," providing extra leverage from the upper thigh muscles and a longer swing to the stride—another indication of new and exceptional skills at walking and running (21). It is intriguing to speculate as to whether they might have won all the track events had they been able to compete in our Olympic Games. I think that is unlikely, at least for the long-distance events, because they were substantially closer to their forest-dwelling ancestry than modern humans, and were likely to have made less complete physiological adaptations to extremes of heat and cold. In spite of their ability to move about in dry, open habitats, they probably resembled many other dry-country species in preferring to be active in the cooler mornings and evenings. Nonetheless, it can be predicted that the Work-man had made substantial physiological improvements on his predecessors—advances that would provide the basis for still greater tolerances in his descendants: of cold in Neanderthals, and of heat in some populations of Moderns (figure 8.4).

More than one explanation can be made for the differences in body proportions and foot structure between the athletic, modern-shaped Work-man and the longer armed, more australopithecine-shaped Handy-man. One can suppose that from a post–A. africanus ancestor, the Work-man lineage so improved its adaptation to a wholly terrestrial way of life that, for a quick retreat, fast runs became a viable alternative tactic to

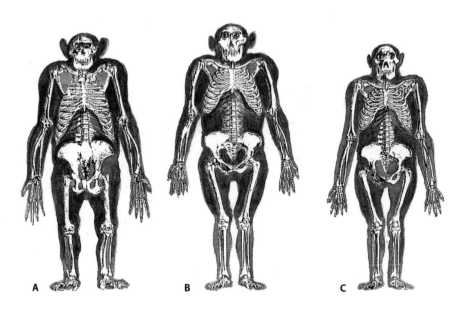

FIGURE 8.4 Body proportions of seven species of hominid. A. Chimpanzee. *Pan troglodytes.* B. *"Australopithecus" afarensis.* C. *A. africanus.* D. *"Homo" habilis.* E. *H. neanderthalensis.* F. *H. ergaster.* G. *H. sapiens.*

climbing the nearest tree. Larger size, too, might have made a difference in dealing with some predators, particularly if the forceful hurling of missiles was combined with clumped and noisy group responses. These changes do imply that the Work-man's ancestral habitat was not only less woody, but that trees were often too sparse or too small to be reliable refuges from predators. In my view, a more important implication is that resources were more scattered and therefore required substantially more walking (an assumption that I, and many others, have mistakenly applied to much earlier phases of human evolution). Were Habilines the Work-man's immediate precursors, their anatomy and behavior would have had to be very extensively overhauled to achieve truly efficient and sustainable walking, running, and self-protection out in the open. Indeed, their limbs show fewer adaptations to life away from trees than those of some of their Robust contemporaries. In the absence of any skeletal material to show what sort of body they had, it is impossible to speculate about Rudolf locomotion.

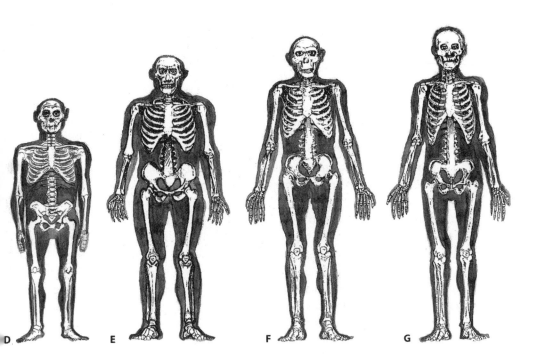

D   E   F   G

I have already suggested that if hominin evolution began with the separation of newly bipedal animals into distinct river basin populations,
one of the incidental effects might have been a near monopoly of resources accessible to the first lineage to break out of its basin. I tentatively
identified the Lucies and their descendants as that lineage. It is therefore
conceivable that preoccupation of more open environments by Lucies
and Robusts might have constrained early humans from moving out of
wooded country. Obviously, if an early hominin population happened to
become isolated in a poorly wooded region without any competing australopiths, such constraints would have disappeared. On current evidence,
australopiths never made it beyond the southern margins of the Sahara.
Colonization of northwest Africa or Arabia might have been eased if the
colonist already possessed a tolerance for lower temperatures, something
as possible for a form with southern as for one of East African highland
origins. Meanwhile, the Lucy lineage, by originating in warmer latitudes
and at lower altitudes, remained less tolerant of cold. If this was the case,

the soils most likely to yield fossils that link early *habilis* or *rudolfensis* with *ergaster* will be those of the north African littoral, with Arabia as a possible runner-up.

As was pointed out earlier, there are two major schools of thought on early human origins; one, the "Lucy School," puts *Praeanthropus* ("*Australopithecus afarensis*") on the main line of human evolution and seeks the link back through *garhi*, effectively bypassing *africanus* (22). The "*africanus* School" derives *habilis* directly from *africanus* and supposes that *habilis* and *ergaster* also belong to the same lineage (23). Now, with the discovery of *Kenyanthropus platyops*, a third possibility has arisen. *Rudolfensis* may be an independent lineage and could be considered an offshoot of our own line. It could be argued that *habilis* remained too conservative for too long to be a viable ancestor for *H. ergaster* but this line of descent gains strong support from the discovery of a strikingly *habilis*-like skull among the European Erects excavated at Dmanisi in Georgia. Thus *habilis* remains an adequate model of the earliest and most conservative type of *Homo* and was probably not all that different from the very earliest members of the *H. rudolfensis* line. There can be some confidence that further exploration will eventually uncover more fossil links to resolve the arguments.

As more hominin remains have been discovered, so phylogenetic trees have proliferated, all based on different permutations of existent fossils. The earliest of these consisted of no more than three or four names strung along a straight line. Trees have always tended to be rooted in whatever were the currently known earliest hominin fossil. Thus, "Lucy School" trees began with the "early walker," "*A. afarensis*" being firmly identified as the ancestor. The status of "earliest hominin" has been usurped again and again by subsequent discoveries of still earlier fossils. In spite of these cycles of redundancy, the possibility that the Lucies and Robusts may represent their own distinctive tree rather than the main trunk of human evolution has had few takers so far. The less popular "*Australopithecus africanus* School" offers a tree that not only implies human descent from inferior walkers but a confusing picture of *two* species at the very base of the *Homo* tree trunk. The earlier, impoverished trees assumed homogeneity in human evolution, and scholars of that era tended to lump many of the characteristics we recognize today as uniquely human—upright stance, brain enlargement, temperature regulation, naked skin, cooking, peculiar diets (some even claim monogamy!)—into large packages, implying that they all evolved together. All manner of conjectural trees, based on ingenious permutations, continue to decorate contemporary discussions of human evolution.

It will be clear enough by now that my piecemeal approach to hominin evolution includes the possibility of a "slow start" for the *Homo* lineage (relative to the *Praeanthropus/Paranthropus* group, which I regard as belonging to a separate, more precocious sibling lineage). I am suggesting that a mosaic of anatomical features—hands, feet, limb proportions, teeth, and brain—all evolved in separate, sequential progressions, even if each individual change was essentially unpredictable. Further, I am suggesting that the most radical transformation could be ascribed to an outlying population suffering a relatively late but lengthy isolation in a distant, dry, open habitat. This habitat was substantially more demanding than anything to be found in the tropics, but it may have had the great advantage of no competing hominins.

If the Work-man emerged in an advanced condition and made up for a "lost past" (by suddenly appearing, fossilized, in a wide scatter of localities), what was the secret of success?

It was not relative brain size because *Kenyanthropus rudolfensis* and *Australopithecus habilis* had comparable brain-body ratios. This is not to say that there couldn't have been qualitative changes in intelligence, but simply measuring brain volumes does not suggest major change (unlike the huge contrasts between medium-brained *ergaster* and, say, giant-brained *neanderthalensis*).

It was not walking per se: several australopiths had long been accomplished walkers, although their shorter legs and somewhat gangly, long-armed bodies would have been no match for the Work-man for pacing over long distances, still less for running.

It was not the use of stone tools alone, because crude stone tools of the "Oldowan" type are recorded from various parts of eastern and southern Africa more than half a million years before the Work-man's appearance there. Most tool occurrences have been associated with the Handy-Man or, less securely, with Rudolfs and even late Robusts.

It is possible, however, that another sort of tool provided the Work-man with an unassailable advantage: fire.

Although their earliest fossil appearances in east Africa date to about 2 mya, one *H. ergaster* from the Swartkrans Cave in South Africa (dated to about 1.5 mya) hints at fire making by sharing its bed of sediment with a pile of charred bones (24). There is further evidence of fire, with mixes of tinder and fire woods, from around Lake Turkana; these specimens have been reliably dated to 1.6 mya (25) The association of *ergaster* with fire is of exceptional interest because scholars have long ascribed the reduced size of human teeth to cooking and food processing, and it is the small cheek teeth that most immediately distinguish *H. ergaster*. Indeed, one of the most significant changes between *rudolfensis/habilis* and *ergaster* is the proportional decline in the relative size of the molar teeth. The more recent discovery of small-toothed *Kenyanthropus platyops* and its link with *K. rudolfensis* weakens that argument. These new discoveries imply that diminished molar teeth might have been an earlier trait in one or more hominin lines.

Relationships between fire, softer food, and smaller teeth are not clear-cut, because the teeth of some specimens show so much abrasion and damage that their consumption of tough foods seems certain. It may be amusing to speculate that small-toothed hominins in fire-wracked landscapes might have discovered how much easier it was to chew the victims of bush fires than their free-ranging, less tasty, raw equivalents. Could it be that their dental inferiority speeded their use of fire? Notwithstanding such asides, the rooting of the teeth and the extrapolated size of chewing muscles are consistent with much-diminished biting forces compared with most earlier hominins. There are many potential uses for fire, most of them explored in some detail in my earlier work (26), where I suggested that effective control of fire was an innovation of the Erects and began before 1 mya. Firing food, or cooking, was a "tool" that neutralized bacteria and toxins, released nutrients, and allowed a vast expansion in the food base by making indigestible material edible. Thus the main way in which fire impinged on human gross anatomy was in its use for food processing. Persistent fire-softening and tool use could have permanently diminished the challenge of very hard foods, as teeth no longer had to be heavy-duty shredders or crushers. Had there been a functional connection between smaller teeth and fire, there would have been the very strong implication that much time must have passed before selection for smaller teeth could be manifest. In this case, the earliest control of fire would have had to substantially pre-date the first appearance of *H. ergaster*, implying that the actual discovery was made by the species' still unknown precursor in its still unknown place of origin. In fact, there could have been an even closer relationship between small teeth, an early use of food-processing tools, and perhaps a deliberate choice of softer foods in a small-toothed lineage. As it is, there are independent environmental reasons for linking fire making with the Work-man. If the Atlas region was this species' region of origin, it can be remarked that (in common with some other Mediterranean environments) fires triggered by lightening are an annual hazard. In the very dry Machia bushlands, many plants are adapted to fire, indicating a long history of predictable seasonal fires. Hominins living in such environments would be more intimately familiar with fire than those from less combustible vegetation zones. If the extraordinary success of humans on their first excursions (both within and outside of Africa) is to be explained, not only was use of fire a likely ally, but the possibility that it began to be used before 2 mya has to be borne in mind.

Until very recently, specimens of *H. ergaster* were restricted to the African continent and were treated as the African form of Asian *H. erectus*. This picture changed in 1991 and 2000 when skulls scarcely distinguishable from east African *H. ergaster* were excavated at Dmanisi in Georgia. The skulls have been reliably dated to 1.7–1.75 mya and they include a small adolescent that resembles a *habilis* in several features, including, notably, a cranial capacity of only 600 cc. The skulls are associated with more than a thousand stone artifacts of the Oldowan, or Mode I culture. (The tools most closely associated with *later H. erectus* belong to the Achulean culture, with their earliest appearance dated to 1.5 mya [27].) The practice of sinking *H. ergaster* in *H. erectus* can be justified by many real resemblances and by the fact the Asian Erects were the first fossils of primitive, small-brained humans to be discovered; they were used as a model for later discoveries to be compared with and named after (figure 8.5).

A single, thick-boned cranium with heavily inflated brow ridges was first discovered in Java, Indonesia, by Eugene Dubois in 1891. In describing this fossil as *"Pithecanthropus erectus,"* Dubois was not only fulfilling his prediction about the existence of "ape-men," he was also using a name that had been previously coined by Ernst Haekel (29) many years before the discovery—a name that expressed confidence in the reality of human evolution. Many other fossils of this species have since been recovered from Java, some more complete and one with massive cheek bones as well as the characteristic bar of inflated bone above the eye sockets. Frequent fractures in these thick Asian skulls, sustained while their owners were still alive, suggest an exceptionally violent life (up to 7 or 8 breaks in some skulls). The most likely origin of such wounding is other members of the same species; it is possible that ritualized clubbing contests (known to be necessary initiation ordeals in some recent modern human societies) could have been an influence. Selection for thick skulls could operate only if such practices were widespread and sustained over many generations. Bizarre as it may seem, such cultural skull battering should not be entirely dismissed as an explanation for the anomalous thickening and enlarged brow ridges. If it could ever be proven, such behavior would be one more dramatic instance of human evolutionary "self-making."

The Erects may have entered southeast Asia as early as 1.8 mya (30) and occupied the region for at least 1.5 million years. The skulls of other

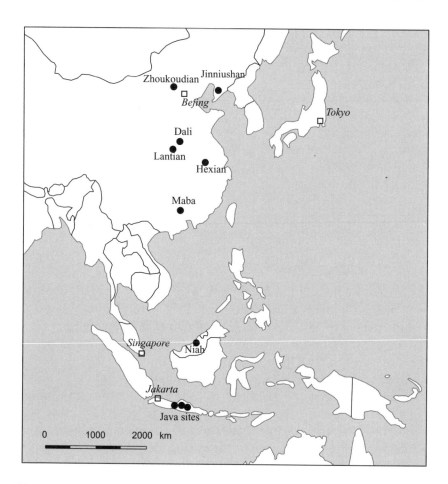

FIGURE 8.5 Map showing the locations of major sites of human fossil discoveries in eastern and southeastern Asia.

Asian humans were first discovered in the 1920s in the caves of "Dragon Bone Hill" or Zhoukoudian, near Beijing. First described as "Pekin Man" or *"Sinanthropus pekinensis,"* these fossils were soon reallocated to *erectus,* in spite of having less heavily built skulls. Today, there is a trend to return the name *sinensis* to these Chinese specimens, although their affinities with *ergaster* and *erectus* are not disputed seriously (figure 8.6). The differences between Chinese and Javan fossils are the clearest evidence for differentiation into distinct populations (31). Still further taxa may yet emerge as new *erectus* fossils are found in other parts of Eurasia. The "lumpers" prefer to retain *erectus* as a single specific designation for all

FIGURE 8.6 Three (species? subspecies?) of *erectus*. A. *H. ergaster*. B. *H. pekinensis*. C. *H. erectus* (Java). (In part after De Beer, G. 1964. *Atlas of Evolution*. London: Nelson.)

the members of what is clearly a single radiation, while the "splitters" seek more recognition of differences that involve enormous distances in space and a time span going back at least 1.8 million years (32). From the perspective of understanding our own evolution, it becomes important to identify which branch of this far-flung diaspora might have given rise to *Homo sapiens*. Again, there are fierce controversies, some backing Asian origins, others African, while others suppose that we belong to a global melting pot that began with the travels and travails of Erect humans (33).

The rationale for separating thin-skulled *H. ergaster*, as a species, from thick-skulled *H. erectus* is not entirely clear-cut. For example, one African *ergaster* skull (from Olduvai) has massive *erectus*-like brows. There are several ways of interpreting evidence that is culled from a mere handful of fossils, with each interpretation raising different questions. Was the Olduvai specimen merely a sample of the variability of the African parental population? Did this polymorphism produce the particular morphological type that passed through some narrow bottleneck on its way eastward? Did Erects get back into Africa? (most unlikely) In support of the first proposition, hominins have been consistently prone to producing gracile and Robust types—a tendency that may throw light on many puzzling fossils. Even today, our own species is extremely varied; it is on such diversity that natural selection works, sometimes very fast. Numerous recently discovered fossils suggest that the Work-man is not only the likely source for Erects, but that a later Africa-based population of the same lineage gave rise to Moderns.

If, after evolving there in isolation, the Work-man emerged from northwest Africa or Arabia only around 2 mya, it is quite possible (if a dry-land connection was present at the time) for an African and Eurasian dispersal to be roughly contemporaneous.*

There are several reasons for believing that modern humans could not have evolved directly from later Erects. One reason concerns the less flexed base of Erect skulls (34). This is not a single, insignificant trait but rather involves a whole complex of bones with networks of blood vessel and nerve canals and angles of muscle attachment that are difficult to modify (whereas other more superficial areas of the skull are much more plastic and subject to variation in proportions). The very derived state of thick-boned Erect skulls cannot easily be converted back into those of thin-boned Moderns without effectively reversing the complexities of the skull base (as well as several other anatomical characteristics). So the more likely option is that our lineage arose from an *ergaster* population that had remained in Africa. (An alternative, but less likely, possibility is that another as yet unknown *ergaster*-derived people, from western Eurasia, found their way back into Africa and gave rise there to *H. sapiens*.)

The abruptness and rapidity with which Eurasia was colonized by *Homo* begs the question of why no earlier hominin had succeeded. The Miocene apes that long, long before found their way into Africa from Eurasia arrived as four-limbed animals. For hominins to find a way back 8 million years later must have involved more than simply walking out on two legs; otherwise Lucies or Robusts would have beaten *Homo* to it by at least a million years. Instead, the spatial and ecological constraints of remaining tied to woodlands in the vicinity of permanent rivers seems to have kept the "tropic-origined" Lucy lineage in Africa, whereas the *Homo* lineage—a more temperate-adapted "thief of niches"—could as readily steal them in Eurasia as in Africa. If an animal originating in the temperate south could establish itself in the temperate northwest of Africa (or Arabia) and there learn to tend fire, its ability to colonize other temperate regions (as well as tropical Asia) is somewhat more understandable.

Unlike Lucies and Robusts, *Homo* seems to have crossed a new threshold where biological adaptation was overtaken by cultural adaptation as the mechanism for ecological success. The metaphor of a threshold may be misleading inasmuch as accretions of innovation must have been developed stage by stage. Nonetheless, the use of fire must represent quite the largest leap toward a "cultural animal." Alterations in anatomy con-

*The discovery, at Longgupo, China, of fossil teeth and jaw fragments (dated to 1.9–1.7 mya) has been taken by some to imply that Erect humans might have evolved independently in Eurasia from still more primitive ancestors. In fact, these fragmentary finds are consistent with an early *H. ergaster* being the most plausible model for the immediate ancestor of *H. erectus*.

tinued to be meaningful manifestations of adaptive change, but direct adaptations to local environments were subordinate to—and, to an increasing degree, a consequence of—culture. The preeminence of culture became most marked in *H. sapiens*, but regional differentiation in later populations of *ergaster* and *erectus* suggests that both species eventually made important adaptive accommodations to local conditions.

Whatever those accommodations were, the fact that *Homo erectus* was the first hominin to be found fossilized at both ends of the Eurasian landmass around 1.8 mya implies an extremely quick and efficient dispersal and an ecological success that must have depended on at least one major technical breakthrough. It is difficult not to conclude that one reason for that success, especially across a wide span of relatively cool ecoregions, was the possession of a new and flexible tool-mediated relationship with the environment, in which control of fire was a major key to the Erects' success.

The link between *H. ergaster* and fire may be much more extensive than has been appreciated (35). What are commonly called "secondary grasslands" are in fact made by two principal agencies: one is grazing, browsing, and trampling by large or numerous herbivores; the other is fire. It is also possible that globally dwindling CO_2 levels have progressively favored grasses, which can take up this gas at very low levels. (Primary or edaphic grasslands grow in much more restricted sites, such as seasonally or permanently waterlogged soils or in sites with very shallow, fast-draining, or drought-prone soils, above impenetrable substrates or in deserts.) African habitats are thought to have shown some opening up about 5.5 mya, presumably in response to drier climates. However, recent research suggests that the development of extensive savannas is much more recent, on the order of 1 mya (36, 37). If fire became a general tool of *Homo ergaster* and their descendants, perhaps these new people were instrumental in making grasslands even more prevalent (rather than responding passively to expanding rangelands, as they are so commonly portrayed). Thus, the spread of fire-climax grasslands may not have preceded this burgeoning population of highly successful hominins but could have followed hard on their pyromaniac heels! Humans are undoubtedly niche stealers, but in this instance their activities may have vastly extended secondary grasslands, creating new niches for grassland and fire-climax species—above all for themselves.

For all species of animals and plants, it is unique techniques of making a living that define a species and its ecological niche. In the case of *Homo*, techniques represent an interplay between behavior and actual technology in the shape of tools. Unfortunately, the majority of tool types have left no record. Apart from rare (and hotly contested) hints of fire at fossil sites, the physical evidence for human technological history comes down to stone tools that have been allocated to a succession of named stone industries. Each industry broadly replaced what went before with ever more refined manufacturing techniques, but each lasted for a substantial time as a recognizable class of artifacts (although sudden eruptions of advanced techniques hint that the skills needed for an industry long predate or anticipate their general expression). These industries have been categorized in four major groupings. The first, the Oldowan or Mode I culture, is currently dated from about 2.6 to 1.3 mya but has been recorded from Atapuerca in Spain as late as 780,000 years ago. The second culture, the Achulean (1.4–0.3 mya) was succeeded by the Middle Palaeolithic (from about 0.25 mya to the last few thousand years), which in turn gave way to the refined stone knapping of the Upper Palaeolithic (mainly since about 35,000 years ago). The last two cultures need not concern us here (see figure 9.3, pp. 306–307).

The best known artifact associated with the Erects is the Acheulean hand axe. This pear-shaped, bifaced, blade-edged tool has been the source of much speculation about its primary functions. There is no sign of waisting or modification to allow for binding to a haft or handle, but the axes are often palm sized or somewhat larger and seem well shaped to be gripped by a single hand. The weight and size are typically well suited to absorbing the force of work throughout the hand, and the long blade cuts more efficiently and blunts more slowly than small flake tools that, by contrast, exert more strain on the wrist and fingers.

The most surprising thing about this tool is the fact that it appeared quite suddenly about 1.5 mya in Africa, where it seems to have been invented by *H. ergaster*. Unlike the beginning of the Middle Stone Age industry, which seems to coincide with the emergence of wholly modern humans and implies a fundamental change in mentality, the Acheulean hand axe seems more like a conventional "invention" by an already deft flaker of stones. Whatever its functions, the ability to make this distinctively shaped tool spread rapidly throughout Africa and eventually to parts of Asia and Europe. Furthermore, the tool seems to have gone on being used by archaic humans long after *H. ergaster* was extinct. Among the speculations about function, the oldest is that these were literally axes, used to chop down trees and cut up carcasses; numerous tests have confirmed their utility in butchering large animals, More equivocal is their effectiveness on tree trunks. Acheulean hand axes have also been envisaged as a portable, general-purpose tool that could be resharpened and made available far from natural sources of knappable material. Another theory, jocularly known as the "killer frisbee hypothesis," envisages them being thrown like a spinning discus to disable prey; tests have proved this to be possible. The symmetry and "unused" technical perfection of many specimens have suggested to some scholars an aesthetic dimension to the tool.In addition, as I remarked before, the need to display a fine grasp of techniques or design must have conferred prestige or elicited approval from fellows wherever tools and technology were concerned (26). Humans, like chimps, are aware that "expertise" confers power, and the display of a good command of skills would have been a prime tenet of prehistoric education from the earliest times. It is, therefore, possible that the social knapping of hand axes might have become an occasional "recreational" activity, conducted in a spirit of mild rivalry. This use would have been possible only in localities where the raw materials for hand axes was particularly abundant; at least two such sites are known, both densely sewn with hundreds of hand axes. Like many other, poorer hand axe sites, Isimila, near Iringa, Tanzania, and Olorgesailie in the Kenya Rift Valley are on the pebbly floor of broad valley bottoms. Usually thought to be "factories" for a wider distribution, these two sites might also have acquired some sort of ritual significance because some of the axes are gigantic, several times as heavy as the palm-sized "functional" models.

It is important always to remember the numerous techniques that must have coexisted, or pre-dated the knapping of stones: the making and use of cordage, skins, and wrapping materials; and the manufacture of various wooden and bone implements. It is extremely rare for these to survive for long, yet all of them are predicated through the existence of stone tools (which require ties or suspenders, wrappers, hafts, or shafts, while others are demonstrably designed to process various less durable materials into important tools or artifacts). We can be confident that all early *Homo* used a wide range of implements (38). We can be equally sure that there were periodic technical innovations that then spread widely. In spite of such probabilities, the Achulean stone tools of the Erects and Work-men show minimal signs of innovation for long periods of time (39).

Yet there is one piece of intriguing evidence that implies very substantial innovation. There is now good evidence that by 0.8 mya, Erects on the Indonesian archipelago had developed watercraft: their stone tools, blunted flake blades, well-worked cores, and cobble choppers turn up on the island of Flores, together with the remains of fossil elephants, tortoises, and other animals (40). Either Erects had developed watercraft, or some peculiarity of their mainland economy rendered them prone to some freak of natural rafting. Discovering how to make boats or rafts after a million years of living on the shores of islands is some confirmation that regional populations of early *Homo* living in highly distinctive habitats were capable of inventing technologies appropriate to their particular environment. At present, there is no sign of island hopping being widespread among Erects, so it is possible that any wider application of sea travel may have had to await the arrival of wholly modern humans, who were better equipped to disseminate knowledge of their own innovations (and, in the case of watercraft, perhaps the inventions of their Erect human precursors as well). The apparent uniqueness of the Flores Landing could imply that a crucial difference between Erects and Moderns—and one that had far-reaching consequences—was the ability of Moderns to communicate beyond the confines of a localized group and to quickly grasp diverse applications for technical inventions.

The impact of Erects on both island and mainland faunas seems to have been substantial (41, 42). It seems likely that they were effective hunters, which, in turn, implies some skills in coordinating group activity. The extreme durability of Erect populations (over some 1.8 million years) and the conservative nature of their surviving artifacts offer an extraordinary contrast to the much shorter career of almost continuously innovative *Homo sapiens*. Because Erects had many resemblances with Moderns, this contrast emphasises that relatively small shifts in how faculties are ordered can have far-reaching effects. Erects were not only exceptionally successful humans, but their "limitations" (from our own perspective) seem to have been accompanied by more ecological stability than our own species has been able to achieve. Such stability may have had more to do with the resilience of ecological communities and the tradition-bound nature of some Erect economies than any "virtue" in their exploitation of resources. Nonetheless, the fact is that they survived, as a distinct lineage, at least eight times as long as Moderns have been around (so far!).

How long humans have been around, and how varied they can be are questions that belong to an enduring controversy: do the Erect humans of Eurasia feature in our direct ancestry? Essentially, the assertion that they do dates back to a time when *Homo erectus* skulls provided the only physical evidence for an early type of human. Also, for a variety of reasons that seemed quite plausible at the time, Asia was once considered the most likely home of primitive humans. The idea that Asian Erects are directly ancestral to Moderns is therefore grounded less in the contemporary spectrum of fossils than it is a reflection of academic conservatism or inertia (and, in some cases, odd forms of regional chauvinism). The sur-

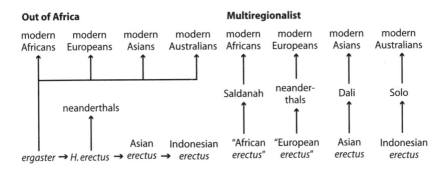

FIGURE 8.7 Two models for the origins of modern humans. While the molecular evidence mainly supports "Out of Africa," some genetic mixing may have left traces of earlier genotypes.

viving academic tradition that thinks modern people descended directly from Erects also asserts that Chinese Erects are ancestral to modern Chinese people, that Javan Erects gave rise to Australian aboriginal people, and that contemporary Africans are the separate descendants of African Erects. These ideas grew out of 19th-century efforts to match a mere handful of poorly known fossils with the stereotypes of living populations or "races" that were current at the time. It is a school of thought that still has some adherents—and its most influential recent proponent was Carleton Coon, who sought our ancestry and its dispersion among the Erects and rather vaguely labeled the process of modernization "parallel evolution." Its contemporary label is "multiregionalism," supporters of which believe that once Erects had become established throughout Africa and Eurasia, their regional populations remained relatively immobile and, in the course of nearly 2 million years, gave rise to modern "races." The resemblance of all Moderns humans is explained as a gradual absorption and spread of useful genetic traits from neighbors, combined with the development of parallel adaptations, such as brain enlargement, as part of a vaguely modernizing trend (figure 8.7).

Supporters of this school point out that it is possible for a favorable trait or clutch of traits to emerge within an established gene–pool, and for such traits to diffuse throughout that population's distribution. In effect, they argue, it is traits, not species, that have been replaced over the last 2 million years. It is true that selection against earlier traits and for some newer ones might, theoretically, be able to transform an entire population, but the process is very dependent on an unimpeded flow of genes. New genes cannot easily infiltrate a genome that is widely scattered and already specialized, or well differentiated, region by region. For traits to spread to the point at which they become universal, they must be decisively favored by selection all over the range.

The alternative "Out of Africa" school of human origins—now overwhelmingly supported by research from many sources, and supported here—identifies Africa as a recent, rather than an ancient, region of origin. It interprets the evidence from genes and fossils alike as pointing to Moderns being a younger replacement, rather than a modified older species. The biogeographic interest and strength of this school is its recognition that the numerous respects in which we differ from our ancestors, whether arrived at through the diffusion of traits or the replacement of species, must have had a place or places of origin. That all other human and hominin species have become extinct (some relatively recently, and some apparently quite abruptly) is fact, as is the late appearance, beginning in Africa, of recognizable Moderns as fossils. That said, it must also be acknowledged that many permutations of "replacement" could embrace traits or species, and the two are not easily untangled from either fossils or genes.

The multiregionalists claim that Eurasia and Africa spawned different favorable traits, the influence of each spreading out very slowly in ever-expanding circles to reach and benefit people in all continents. "Out of Africanists" counter that an isolated population (in possession of traits that truly conferred decisive advantages) emerged from Africa to then replace, at a much faster pace, related ("parent," "sibling," or "conservative") populations.

That the Erects were a distinct type of human is now scarcely contested. The global longevity of Erects (as a species or species group) presents a contrast with the very brief career of wholly modern people. This contrast also applies to another enigmatic guild of fossil humans sometimes referred to as "Heidelbergs" or *Homo heidelbergensis*; these are treated by others as subspecies of *Homo sapiens* and confusingly known as "Archaic *Homo sapiens*," "Archaic Moderns," and more colloquially as "middlebrows." It has been suggested that they derived from a second late excursion of *ergaster*-like humans out of Africa and that their main innovation was an increase in brain size that brought them within the range of modern humans. Alternatively, they could have arisen within Eurasia. First named in 1908 from a massive lower jaw excavated near Heidelberg in Germany, this species is now represented by numerous remains from a wide scatter of European sites. Specimens are currently known from about a half to a quarter of a million years in Eurasia (43) but have plausible precursors in Africa from about 1 mya (44) and 0.6

A   B

FIGURE 8.8 African and western Eurasian descendants of the Heidelbergs. A. The Kabwe, or Broken Hill human. B. Reconstruction of a late Neanderthal face.

mya (at Bodo in Ethiopia) (45). A fragment of upper jaw and orbit dated to about 800,000 years ago has been retrieved in Spain and judged by its discoverers to be an intermediate between the Work-man and the Heidelbergs; it has been given the name *Homo antecessor*—yet another claim to directly ancestral status!

In western Eurasia, fossils of "archaics" are sufficiently numerous to illustrate step-by-step evolution into a more specialized cold-adapted form, the celebrated Neanderthal humans. On present evidence, these were absent from eastern Asia, although archaic equivalents may be represented by a 200,000-year-old skull from Jinniushan and others from Dali and Chaohu (46). It is possible that the long-established Erects inhibited or substantially delayed the penetration of eastern Asia by the later Heidelberg guild.

The Neanderthal brain was as big, or larger, than that of modern humans but had a different shape. Long, flat, and loaf-like, it was encased in a massive thick skull, tucked behind a receding forehead and deeply overhung eyebrow ridges. Skeletons were also massive, with short, bowed limbs, a peculiarly elongated pubic bone, long backs, and rotund, barrel-shaped chests—proportions not found in any modern people (47). Such odd looks derived ultimately from taller, longer-limbed, but much smaller brained humans of the "Work-man" group. Neanderthal ancestors may have began to differentiate from European Heidelbergs about half a million years ago, eventually becoming the indigenous people of western Eurasia. By 130,000 years ago, unmistakably Neanderthal fossils, their cave sites, and their Mousterian stone tool industries occur all over Europe (figure 8.8).

Neanderthals seem to have lived in small groups, and the bones of elderly and disabled individuals imply that families took care of their weaker members. It is possible that they buried their dead. As might be predicted in a cold environment, ash from their fires is found in many sites, yet there is, as yet, no sign of the deliberate construction of stone hearths or pits.

Tools were handled in peculiar ways; hints of this lie in the very thick, knobby finger bones and the bowed, short forearm. However, it was the way in which arms and hands, as prime agents of human technological skills, linked up with a third agent, a dental clamp, that Neanderthals were unique. This specialization gave them their diagnostic "muzzlec" consisting of massive jaw muscle attachments and big front teeth (48). It is known from the state of the teeth that the jaws were systematically used as a clamp (49). Crowns are rounded and heavily worn, and their surfaces are scored from back to front (even in very young individuals). Mainly animal tissues were clenched, shredded, and cut by a combination of manual pulling, biting, and slicing with stone blades (which sometimes slipped to chip the teeth).

Why should "clamping" develop in a single regional population of humans? Does the ecology of their cool, northwestern province provide clues? Are there enough traces left of their feeding behavior, or subsistence strategy, to explain their very unmodern facial anatomy? Northern summers offer a similar range of habitats and resources to those found further south, so the focus can be narrowed to winter, when the need to find predictable sources of meat (supplemented, perhaps, by roots, nuts, barks, buds, and fungi) would have been crucial to survival. In common with other meat-eating animals, Neanderthals would have had to live near the winter concentration grounds of their food animals. The debris of their middens and hearths suggest many regional differences, but wild cattle, horses, goats, deer, pigs, mammoths, and rodents appear to be the most frequently eaten animals. But were these all hunted? Competition for such prey would have been intense; big cats, wolves, and hyenas would have been as adept at catching and killing herbivores as their modern equivalents in Africa today.

Ice Age winters may have killed as many, if not more, victims as drought does on the plains of Africa, but there would have been far fewer rivals for the available meat. It was not just the thick wool on dead rhinos that discouraged jackals, or the lack of thermals that would have deterred vultures; rather, it was the fact that corpses would have set solid within hours of death. Bodies could have stayed that way for up to 6 months, especially if crude forms of ice houses were built around the larger bodies to keep them frozen and protected from other scavengers.

Deeply frozen cadavers would have accumulated in predictable localities, such as dangerous fords in the path of reindeer migrations, rocky ravines where horses or bison regularly stampeded, valley funnels where ibex and sheep came off the mountains, or boggy lands crowded with herds of mammoths and aurochs. After the survivors and their mobile hunters had moved onward, these stores of meat could be used with great economy and efficiency by only one scavenging species—Neanderthals— because by using fire, they could thaw out meals, bit by bit, whenever needed.

Anyone who has wrestled with a deeply frozen carcass will know that brute force is necessary to move it, let alone break into it. Imagine every meal demanding that sort of unmannerly struggle, and the Neanderthal's herculean body build becomes easier to explain. All but the most cold-resistant individuals, with the strongest teeth and jaws, would have been weeded out by natural selection. It is no surprise, then, that Neanderthals appear to have aged fast and suffered abnormal damage to bones and teeth. In summer, their life may not have been very different from that of their contemporaries in Africa or tropical Asia. What defined them as a special type of human seems to have been a unique combination of brawn and technology, directed at withstanding the stresses and exploiting the opportunities of winter.

The stratagem must have developed during the very severe Ice Age, which lasted from 180,000 to 130,000 years ago. Fifty thousand very bitter winters were no trivial ordeals for an originally tropical primate to adapt to and withstand. Even so, this alliance between fire, stone tools, face-clamps, and the fists of Goliath was no more than northern elaboration of a much older and well-established mode of existence. It implies that Neanderthals, their predecessors, and their cousins (our own ancestral line) had a common ecological niche—that of the tool-assisted scavenger (even if this was a strongly seasonal tactic).

Both technology and physique must have played their part in this extraordinary and well-tested adaptation. Unfortunately, physiological adaptations do not fossilize, but body proportions imply mass-surface ratios. It is to be expected that superior tolerance for cold, lack of light, and general stress may have helped Neanderthals resist competition from less-specialized humans. They seem to have held their own against their southern neighbors, modern humans, for at least 60,000 years but were in retreat by about 40,000 years ago. The lurid racial battles portrayed by dramatists and filmmakers need not be invoked to explain their demise; simple demography might have been enough (50). Neanderthal decline was assured once the Moderns began to press north (having devised their own strategies for coping with winter—and probably after stealing a few

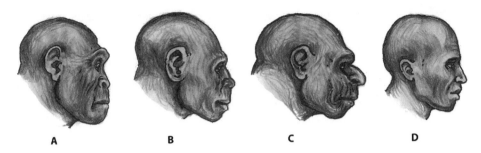

A   **B**   **C**   **D**

FIGURE 8.9 Profiles of "Workman," *Homo ergaster (erectus)* (A); Archaic, *H. heidelbergensis* (B); Neanderthal, *H. neanderthalensis* (C); and modern human, *H. sapiens* (D).

from their competitors). Smaller social units, a lower overall density, and slower or less-successful breeding would all have prejudiced Neanderthal survival (47). In their ultimate extinction, Neanderthals may have played out a recurring theme in human evolution—biological overspecialization for a very specific ecology. Eventually, they were defeated by technological innovations in the hands of biologically less-specialized competitors. In this instance, the Neanderthals had come to incorporate, in the most literal sense, their own bodies in the ecological strategy that had ensured their survival for more than 100,000 years within a temperate geographic region. When more independent and flexible techniques arrived, Neanderthal "self-integration" into a super-specialized mode of earning a living seems to have proved fatal.

Although the western Eurasian Heidelbergs progressively diverged toward Neanderthals, the anatomy of the earlier forms seems to have derived from African precursors of the *H. ergaster* type, possibly from an emigration out of Africa that was at least a million years later than that of the Erects (figure 8.9).

In Africa itself, there are fossil skulls that are sometimes called "African Heidelbergs" or "Archaic Sapiens." A particularly well preserved Heidelberg from Greece, "The Petralona Skull" (from about 350,000 ago) resembles some skulls from Africa. Such evidence reinforces the idea that there was some gene flow between Africa and western Eurasia long after the Erect excursion but well before the emergence of fully modern humans.

Lumping the African skulls with the Eurasian Heidelbergs leaves the geographic origins of Moderns an open question; proposals for both Eurasian and African origins have been argued. However, in spite of many uncertainties, most lines of evidence point to an African origin. The genetic evidence, in particular, seems to support the emergence, somewhere in Africa, about quarter of a million years ago, of the specific lineage that we call "anatomically modern humans," or *Homo sapiens sapiens*, our own common ancestral line.

It is customary, in books that address the evolutionary origins of Moderns, to lay out the array of diverse opinions and theories and advertise the conflicts that surround this most emotive of subjects. For some protagonists, particularly those that think we are all modified Erects or Neanderthals, this is not the end of a chapter, but merely a halfway pause in the final story. When it comes to the controversies that surround our acquisition of evolved speech, rational intelligence, "consciousness," and demonstrable creativity, there are entire books devoted to the polemics and summaries of the current state of play (51). Leaving aside claims for divine intervention, some currently fashionable arguments have tended to narrow the discussion to the choice of promising genes as made by females choosing males to mate with, and vice versa (52–54). According to these debates, the answers must lie somewhere in the exciting arena of how early humans chose, seduced, bribed, cheated, deceived, protected, or conserved their actual or would-be mates. In the account that follows, there is some acknowledgement for the role of mate choice, but I see its influence as indirect. It is, of course, axiomatic that males and females have differing interests in reproduction and go about the mating game in different ways, but I see the role of natural selection less in charismatic foreplay or bedazzled climaxes than in the survivorship of young humans growing up in environments where it was no longer the quality of specialized adaptations that was selected for, but a technical and inventive mentality within a technique-dependent society (55, 56). Rather than mate choice, I see the quality of mothers rearing their children and the quality of their offspring as the prime forces that selected for sustained and sustainable intelligence and the evolution of wholly modern humans.

The emergence of our own lineage as an entirely new form of human, and the dispersal of this, my larger family, will complete my self-portrait in the next chapter.

REFERENCES

1. Butzer, K. W. 1973. Past climates of the Tibesti Mountains, central Sahara. *Geographical Review* 63: 395–397.
2. Kowalski, K., and B. Rzebik-Kowalska. 1991. *Mammals of Algeria*. Warsaw: Ossolineum.
3. Arambourg, C. 1957. "Recentes decouvertes de paleontologie humaine realisees en Afrique du Nord francaise". In *Third Pan-African Congress on Prehistory, Livingstone, 1955*, ed. J. D. Clarke. London.

4. Leakey, R.E.F., and A. Walker. 1989. Early *Homo erectus* from West Lake Turkana, Kenya. In *Hominidae. Proceedings of the 2ⁿᵈ Int. Congress of Human Palaeontology*, ed. G. Giacobini, 209–215. Milan: Jaca book.

5. Walker, A. C., and P. Shipman. 1996. *The Wisdom of the Bones: In Search of Human Origins*. New York: Alfred A. Knopf.

6. Kingdon, J. 1982. *East African Mammals. An Atlas of Evolution in Africa*, Vol. 3D. Bovids. London: Academic Press.

7. Rightmire, G. P. 1994. *100 years of Pithecanthropus: the Homo erectus Problem*, ed. J. L. Fransen. *Courier Forschungsinstitut Senckenberg* 319–326.

8. Groves, C. P., and V. Mazak. 1975. An approach to the taxonomy of the Hominidae: gracile Villafranchian hominids of Africa. *Casopis pro Mineralogii a Geologii* 20: 225–246.

9. Leakey, R.E.F., and A. Walker. 1976. *Australopithecus, Homo erectus* and the single species hypothesis. *Nature* 261: 572–574.

10. Shipman, P. 1992. Human ancestors' early steps out of Africa. New Scientist 133: 24.

11. Walker, A. C., and R.E.F. Leakey, eds. 1993. *The Nariokotome* Homo erectus *Skeleton*. Cambridge: Harvard University Press.

12. Rightmire, G. P. 1986. Stasis in *Homo erectus* defended. *Paleobiology* 12: 324–325.

13. Spoor, C. F., B. A. Wood, and F. Zonneveld. 1994. Evidence for a link between human semicircular canal size and bipedal behavior. *Journal of Human Evolution* 30: 183–187.

14. Weiner, J. S. 1954. Nose-shape and climate. *American Journal of Physical Anthropology* 12: 1–4.

15. Franciscus, R. G., and E. Trinkaus. 1988. Nasal morphology and the emergence of *Homo erectus*. *American Journal of Physical Anthropology* 75: 517–527.

16. Wolpoff, M. H. 1968. Climatic influence on the skeletal nasal aperture. *American Journal of Physical Anthropology* 29: 405–423.

17. Walker, A. C., and R.E.F. Leakey. 1978. The hominids of East Turkana. *Scientific American* 239(8): 54–66.

18. MacLarnon, A. 1993. The vertebral canal. In *The Nariokotome* Homo erectus *Skeleton*, ed. A. Walker and R. Leakey, 359–390. Cambridge: Harvard University Press.

19. Kingdon, J. 1971. *East African Mammals. An Atlas of Evolution in Africa*, Vol. 1. London: Academic Press.

20. Falk, D. 1991. Breech birth of the genus *Homo*: Why bipedalism preceded the increase in brain size. In *Origine(s) de la Bipédie chez les Hominidés*, ed. Y. Coppens and B. Senut, 259–266. Paris: CNRS.

21. Fedak, M. A., B. Pinshow, and K. Schmidt-Nielsen. 1974. Energetic cost of bipedal running. *American Journal of Physiology* 227: 1038–1044.

22. Johanson, D. C., and M. A. Edey. 1981. *Lucy: The Beginnings of Humankind*. New York: Granada.

23. Tobias, P. V., ed. 1985. *Hominid Evolution Past, Present and Future*. New York: Alan Liss.

24. Brain, C. K., and A. Sillen. 1988. Evidence from the Swartkrans cave for the earliest use of fire. *Nature* 336: 464–466.

25. Shipman, P. 1991. A journey towards human origins. *New Scientist* 19: 45–47.

26. Kingdon, J. 1993a. *Self-made Man. Human Evolution from Eden to Extinction?* New York: John Wiley & Sons.

27. Gabunia, L., and A. Vekua. 1995. A Plio-Pleistocene hominid from Dmanisi, East Georgia. *Nature* 509–512.

28. Shipman, P. 2000. Doubting Dmanisi. *American Scientist* 88(6): 491–494.

29. Haeckel, E. 1868. *The History of Creation, or the Development of the Earth and Its Inhabitants by the Action of Natural Causes. A Popular Exposition of the Doctrine of Evolution in General, and That of Darwin, Goethe, and Lamarck in Particular* (translated by E. Ray Lankester). New York: D. Appleton.

30. Swisher III, C., G. H. Curtis, and T. Jacob. 1994. Age of the earliest known hominids in Java, Indonesia. *Science* 263: 1118–1121.

31. Bilsborough, A., and B. A. Wood. 1986. The nature, origin and fate of *Homo erectus*. In *Major Topics in Primate and Human Evolution*, ed. B. Wood, L. Martin, and P. Andrews, 295–316. Cambridge, UK: Cambridge University Press.

32. Groves, C. P. 1989. *A Theory of Human and Primate Evolution*. Oxford: Clarendon Press.

33. Thorne, A. G., and R. Raymond. 1989. Man on the Rim. The peopling of the Pacific. Sidney: Angus & Robertson.

34. Aiello, L., and M. C. Dean. 1990. *An Introduction to Human Evolutionary Anatomy*. London: Academic Press.

35. Oakely, K. P. 1961. On man's use of fire, with comments on tool-making and hunting. In *Social Life of Early Man*, ed. S. L. Washburn, 176–193. Chicago: Aldine.

36. Cerling, T. E. 1992. Development of grasslands and savannas in East Africa during the Neogene. *Palaeogeography, Palaeoclimatology, Palaeoecology, Global and Planetary Change* 97: 241–247.

37. Sikes, N. E. 1999. Plio-Pleistocene floral context and habitat preferences of sympatric hominid species in east Africa. In *African Biogeography, Climate Change, and Human Evolution*, ed. T. G. Bromage and F. Schrenk, 301–315. New York: Oxford University Press.

38. Keeley, L. 1980. *Experimental Determination of Stone Tool Uses*. Chicago: University of Chicago Press.

39. Rightmire, G. P. 1993. Did climatic change influence human evolution? *Evolutionary Anthropology* 2: 43–45.

40. Morwood, M. J., P. B. O'Sullivan, F. Aziz, and A. Raza. 1998. Fission-track ages of stone tools and fossils on the east Indonesia island of Flores. *Nature* 392: 173–176.

41. Hooijer, D. A. 1955. Fossil Proboscidea from the Malay Archipelago and the Punjab. *Zool. Verband Mus. Leiden* 28: 1–146.

42. Coon, C. S. 1962. *The Origin of Races*. New York: Knopf.

43. Tattersall, I. 1993. *The Human Odyssey: Four Million Years of Human Evolution*. New York: Prentice Hall.

44. Abbate, E., A. Albianelli, A. Azzaroli, M. Benvenuti, B. Tesfamariam, P. Bruni, N. Cipriani, R. J. Clarke, G. Ficcarelli, R. Macchiarelli, G. Napoleone, M. Papini, L. Rook, M. Sagri, T. M. Tecle, D. Torre, and I. Villa. 1998. A one-million-year-old *Homo* cranium from the Danakil (Afar) Depression of Eritrea. *Nature* 393(6684): 458–460.

45. Johanson, D. C., and B. Edgar. 1996. *From Lucy to Language*. New York: Simon & Schuster.

46. Chen, T., Q. Yang, and E. Wu. 1994. Antiquity of *Homo sapiens* in China. *Nature* 368: 55–56.

47. Trinkaus, E., and P. Shipman. 1993. *The Neanderthals*. New York: Alfred A. Knopf.

48. Brace, C. L., H. Nelson, and M. Korn. 1971. *Atlas of Fossil Man*. New York.

49. Trinkaus, E., ed. 1989. *The Emergence of Modern Humans: Biocultural Adaptations in the Later Pleistocene*. Cambridge, UK: Cambridge University Press.

50. Stringer, C. B., and C. Gamble. 1993. *In Search of the Neanderthals*. Thames & Hudson: London.

51. Kohn, M. 1999. *As We Know It. Coming to Terms with an Evolved Mind*. London: Granta.

52. Whiten, A., and R. W. Byrne, eds. 1988. *Machiavellian Intelligence*. Oxford: Oxford University Press.

53. Buss, D. M. 1994. The strategies of human mating. *American Scientist* 82: 238–249.

54. Hrdy, S. B. 1993. Sex and the mating game. In *Reinventing the Future*, ed. T. A. Bass, 7–25. Reading, MA: Addison-Wesley.

55. Gibson, K., and T. Ingold, eds. 1992. *Tools, Language and Intelligence*. Cambridge, UK: Cambridge University Press.

56. Isaac, G. 1976. Stages of cultural elaboration in the Pleistocene. In *Origins and Evolution of Language and Speech*, ed. R. S. Harnard, H. D. Steklis, and J. Lancaster. New York: New York Academy of Sciences.

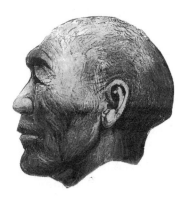

CHAPTER 9

On Being a Self-made Human

The Modern Diaspora

An African Archaic human or Heidelberg, the Buia skull (from the Eritrean coast, about 1 mya) represents the first fossil evidence for a braincase of modern aspect. Genetic evidence suggests the emergence of a minuscule population, popularly represented as the offspring of an "African Eve" between 250,000 and 300,000 years ago. Earliest actual fossils of Moderns in eastern Africa at about 130,000 to 120,000 B.C. Estimated dates for the emigration of Moderns out of Africa range from 200, 150, 140, 100 to 65 thousand years. Worldwide dispersal the ultimate triumph of bipedalism? Significance of Y chromosome genes for timing events and tracing dispersals. Evidence for a dispersal out of Africa to east Asia/South Pacific and back again. "Self-making" and "niche-stealing" fundamental traits of modern humans.

Self-portraits are notoriously susceptible to self-delusion, self-congratulation, or swollen-headed self-aggrandizement. In naming us all *Homo sapiens*, "wise humans," Linnaeus was maintaining this tradition. Darwin, instead, broke this mold when he concluded his portrait of human evolution with the reminder that, for all our noble qualities—benevolence, sympathy, godlike intellect, "all these exalted powers"—we still bear the indelible stamp of our lowly origin (1).

The greater part of this book has been devoted to examining some of the many evolutionary increments that preceded, accompanied, and succeeded the special achievement of getting up onto two legs from the particularly lowly occupation of grubbing around on the forest floor.

With the discovery of more and more extinct two-legged primates, we have been forced to cede one of our exalted powers to an ever-enlarging family of upstanding hominin cousins and ancestors. With erectness shared (and actually the only measure of separation between hominins and apes), what other crude measure of our stature remains? The Handyman made tools more than 2 mya (2); later, Erect humans, tall and strong, reached islands on watercraft (3); massive Neanderthals even had larger brains (4). That these hominins could communicate with one another is certain, but the degree to which they could talk and think as I do is still unknown and must in some senses remain unknowable—so what is left to ennoble me? What distinguishes my self-portrait from that of my premodern ancestors?

The best that can be said is that my physique is more delicate, my forehead is higher, and my teeth are smaller. Compared with their apishness or their hulking muscularity, I look less like an adult and more like one of their children—a descendant that has "failed" to grow up.

This is no aberration: it is the culmination of a trait that runs through almost every stage of our evolution. Of vertebrates, mammals have the most prolonged immaturity; of mammals, primates take this the furthest; and among primates, apes spend the longest time preparing for adulthood. Hominins continued the trajectory until modern humans, protected from natural vicissitudes by innumerable self-made artifacts, could almost be said to be permanently immature! Our technology plays a role analogous to maternal protection in that it detaches us from many of the disciplines of the environment. From this perspective, the ultimate legacy of bipedalism and emancipated hands is that we have prolonged childhood and created an all-embracing technological parent.

What are we to make of a primate that shows only moderate differences in the size and weight of males and females but, especially among females, retains such juvenile features as high foreheads and a strong predilection for play? Some consequences for the retention of juvenile characteristics into adulthood, neoteny, have been discussed in relation to earlier species of humans. The physiological and developmental mechanisms are reasonably well known, but the ways in which selection might bring about such a peculiar and, on the face of it, retrogressive outcome are less certain (5, 6). That selection can target particular age grades is manifest in many animals—most obviously among insects and amphibians, in which there are great morphological differences between larval, pupal, and adult forms. In these animals, development, feeding, or activity is able to proceed throughout an individual's life because the ruses that outwit predators are specific and appropriate to each of the animal's particular stages. The dangers faced at any one stage are of a different order from those faced at the next stage; thus, more tadpoles die, and from different predators, than do frogs. The most familiar manifestations of selection operating through differential mortality is in those animal societies whose males compete ferociously for females and for which there is natural selection for massive bodies or elaborate jousting equipment in a brutally self-culled male class. Where males are subordinate to females, the latter tend to be larger; and in the rare species where females cede caring for the young to males, dominant females are generally both larger and more colorful.

Thus, selection for a particular sex or age class tends to be expressed in morphological structure. If differential mortality operated on nonadult humans to enhance our neotenous tendencies, what form might it have taken?

Whatever it was that killed or spaced children out—be it accidents, disease, stress, starvation, infanticide, or lactation-controlled fertility—the likely result was fewer, higher quality children. The selection may not always have been exerted directly on children or juveniles; it could just as well have been the product of parental behavior, more specifically the differential success of mothers in rearing the next generation (7). As with any other animal, a mother's altruism can be seen as the expression of her drive to protect the survival of her own genes. Even so, the behavior and qualities of the children in eliciting such prolonged and intense mother-love must have played a decisive role; in other words, there must have been selection for children that remained "lovable" to their mothers. The discussion need not be confined to mothers, however; it is safe to assume that males protected, or at least participated in the defense of, females and youngsters against external dangers. As with mother-love, this behavior can be interpreted as a long-evolved interest in the survival of the males' own genes—or, at the very least, the genes of close kin (gorillas scarcely differ from humans in this respect). In groups or bands that were mostly made up of closely related individuals, altruistic behavior could as well be described as "kin selection" as "reciprocal altruism." The extreme helplessness, slow growth, and long lives of the young demanded sustained emotional responses to their needs in *all* members of the group, and humans could not have afforded the fast turnover of membership and infanticidal propensities of such social species as lions. We can envisage long-lasting, relatively stable social groups in which intragroup behavior might have rendered some classes more vulnerable than others but which were mainly threatened from outside by predators, enemies, diseases, and accidents.

As yet, there is no direct evidence for the detailed population structure or demography within groups of prehistoric foraging humans that might allow us to work out their average child mortality. Still further from the realms of possibility is an ability to learn how children lived and died during the long, slow ascent of Archaic humans. A rare hint, and one that permits comparison with chimpanzees, is a population pyramid drawn up for an extinct Amerindian society (figure 9.1) (8), which illustrates that most deaths were among the very young and the very old, and that survivorship for both sexes of adults was probably high once adulthood had been reached. The pyramid also demonstrates that because the proportion of a lifetime that was spent as a nonadult was longer than for any other mammal, the ratio of nonadult members in any single social group would have remained relatively high compared with that of other mammals. Up to a third could have been less than fully adult (*in spite* of high mortality or low birth rates), which would have given this class some sort of proportional role in determining the day-to-day functioning of the group. It would be wrong to see nonadults as mere hangers-on, tagging along on the heels of the decision-taking adults; they must have been active participants in the overall foraging success of the group (9). The brighter and better the children were at communicating among themselves and with adults, at collecting food and keeping out of harms way, the better the chances for overall survival.

A high proportion of nonadults defines what might have been special about human evolution. It also hints at where the power of natural selection is targeted and is most influential. In thinking that selection favors smart kids and loving mothers, I depart from most other students of human evolution, who have concentrated on sexual selection and mate choice in their search for primary selection mechanisms. We know the elementary arithmetic of sex: female eggs are rarer, more precious, and expensive than male sperm; most females breed, and some males do not. In those species whose males compete most ferociously, the males look and behave very differently from females, and only a tiny fraction of the males breed. Human sexes are clearly different enough, but there is no

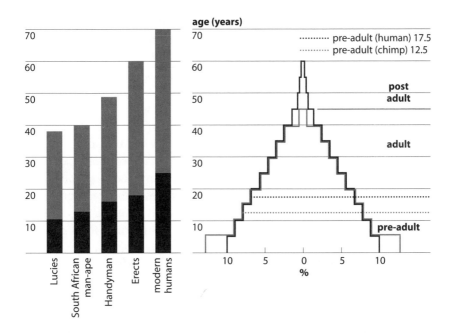

FIGURE 9.1 Left: Crude estimates of the growth period in hominins expressed as a ratio of estimated life span. (After McNamara, K. 1997. *Shapes of Time*. Johns Hopkins.) Right: Estimated age pyramids for prehistoric hunter-gatherer populations (Libben, 750–1000 A.D.). (From Howell, N. 1982. Village composition implied by a paleodemographic life table: the Libben site. *American Journal of Physical Anthropology* 59: 263–269.) Comparison with estimated age classes (in a savanna/forest mosaic habitat) at Mahale, Tanzania. (From Nishida, T., ed. 1990. *The Chimpanzees of the Mahale Mountains: Sexual and Life History Strategies*. Tokyo: University of Tokyo Press.) The relative ratio of infant, juvenile, and subadult classes represents about 49% in chimpanzees and 56% in humans. Adult ratio of population: chimp 51%, human 41%. Post-adult ratio of population: chimp 0%, human 3%. *Note*: None of these ratios is exact; the figures serve to illustrate a more general feature of hominins.

evidence that only a minority of early modern human males bred once they had achieved adulthood. Sexual selection was not a process of ruthlessly pruning out a high proportion of the males. Quite the contrary: in multimale groups, there should have been greater rewards for harmonious relations than in singlemale social systems, and the preeminence of children in the group may well have favored a dampening down of male-male aggression. A social system devoted to nurturing and involving a high proportion of nonadult members is likely to be more tolerant across the board. Where humans may have differed most from other primates could be in the filtering of social behavior through the varied demands of a tool-assisted economy. Every individual's activity must have demanded responses at so many cognitive levels that it is difficult to imagine the group foraging effectively without a great deal of social tolerance and an appropriate communication system—a system that could address social relations simultaneously with technical and strategic decision making. I see the period between conception and reproduction of the conceived (15 to 20 years) as passing in a very peculiar environment. This setting not only embraced an ecosystem and its food resources, as it would for any animal, but all contact with that environment was mediated through tools and the other people in the group. It is within that peculiar environment that we must seek the secrets of our ecological dominance and the special properties of our minds: the two are connected. In the real life of a prehistoric human, there would have been an unbroken continuum between an animal appetite for food in all its diversity; the context in which food was found (i.e., ecology); the ways in which food could be obtained (i.e., tools and techniques); and the social context in which the food was gathered, shared, and consumed (i.e., the group or "society"). All the peculiarities of human thought or mentality, language, and social awareness must have arisen within that continuum (10), and it is an intellectual conceit to dismiss subsistence and tools and lift out social relations as the "prime mover" in the development of human minds. Other animals have complex societies, but the facts that tools and techniques are the interface between modern humans and nature (11) and that their effective use to harvest resources requires complex communication skills are differences of overriding significance.

These differences are significant because tools are not separable from the technical procedures and mental planning that are used in their employment, from the ecological or food-getting contexts in which they have been devised (12), or from the social cooperation that enhances their scope and range. The use, practice, learning, and rewards of technical expertise and skills in obtaining food exist as much in their ecological setting as they do in a social one, but the public display of competence in an ever widening repertoire of techniques must have been a ubiquitous and continuous preoccupation among prehistoric peoples. For any single individual, significant social status would have attended articulate displays of competence in the gathering of food or the protection of resources.

Remember, too, that all foraging would have occurred in response to a rapidly changing calendar of food availability. These seasonal changes would have demanded versatility and good judgment as to when to switch from one food or one set of skills to another. Furthermore, a prime difference between the ecological niche of other animals and the "niche-stealing" specialization of humans was the continuous expansion of food sources. Novelty was a continuous, not an intermittent, property of hu-

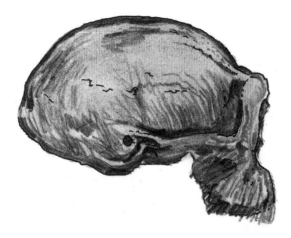

FIGURE 9.2 The Buia skull. (After Abbate, E., A. Albianelli, A. Azzaroli, M. Benvenuti, B. Tesfamariam, P. Bruni, N. Cipriani, R. J. Clarke, G. Ficcarelli, R. Macchiarelli, G. Napoleone, M. Papini, L. Rook, M. Sagri, T. M. Tecle, D. Torre, and I. Villa. 1998. A one-million-year-old *Homo* cranium from the Danakil (Afar) Depression of Eritrea. *Nature* 393[6684]: 458–460.)

man foraging not only because food sources were changing by season (and being augmented as more niches were "stolen") but also because there was always a class of youngsters that were learning how to survive on these resources for the first time. This learning was not innate for the food nor for the technique, and it would have been a continuous challenge—for both teacher and the taught—to articulate and occasionally improve skills (13, 14). Such skills were "autorewarding" in that both participants, sometimes the entire group, ended up with the prize of extra items of food.

A more detailed consideration of what made Moderns different is given later. Return now to the physical peculiarities of childlike Moderns and the fossil evidence for the first appearance of this sort of human.

There are social, developmental, and general evolutionary dimensions to neoteny, but among its more obvious morphological attributes is an enlarged brain. So far, the earliest evidence for the beginnings of a *Homo sapiens*–like swollen head comes from a 1 million-year-old fossil skull from Buia (figure 9.2), in what is now the Danakil desert, on Eritrea's Red Sea coast (15).

Although this skull clearly resembles *H. ergaster* in many respects, including the overall size of its brain, the proportions of the cranium suggest a developmental pattern that anticipates that of modern humans. It may not seem much to go by, but earlier *Homo* skulls, especially Asian ones, have prominent keels along the tops of their skulls, whereas the

Buia one has a flatter, broader top. To appreciate the significance of this change, consider the two main forces that shape the cranial vault of a skull. Bone fragments, like pottery shards or bits of ostrich egg, give little sense of the processes that shape brains, bird's eggs, or water pots. Imagine, instead, the soft, growing material of a young brain as if it were a maturing fruit or the *left* hand of the potter, pushing *outward* from within the enclosing rind or clay. Were that the only force, all heads would look more like fruit or water pots than they do; but there are external forces at work too, modeling the bones from the *outside*, like the potter's *right* hand. In this respect, the Buia skull may mark an interesting and subtle switch in the relative importance of inner and outer influences. Asian Erects had a fan of thick temporal chewing muscles that were attached high up on the sides of the skull; these, like the face and brow, would have had their own role in dragging out the final outlines of an adult's head. During early growth, expanding temporal muscles would have counteracted, however subtly, from the outside, the inner pressures exerted by a rapidly growing brain. Extrapolating from other animals with large temporals, it is possible that these muscles could have influenced the "keeling" of the crown that is so characteristic of thick Erect human crania. Whatever the exact explanation for a cranial keel, its absence in the thinner Buia skull could suggest that an extracranial influence might have begun to give way to intracranial forces ultimately traceable to an expanding brain.

To date, after Buia, the next earliest "Archaic sapiens" comes from Bodo, in Ethiopia, dated to about 600,000 years ago. This gigantic skull, the largest of a human in fossil history, has some resemblances with the much later Petralona skull from Greece (mentioned in the last chapter). But the Bodo skull differs from the European and Buia skulls in the area of jaw attachment and in the conformation of the forehead—structures that anticipate, in still more respects, the arrangement in *Homo sapiens* (16). This skull was found with numerous Achulean tools and suffered "cleaning" (or perhaps butchery) from stone tools at the hands of its fellows. Another Archaic, once considered to be an African Neanderthal, is the undated "Broken Hill skull" from Kabwe, Zambia, which is a startlingly modern, but heavily browed, foreheadless skull with a cranial capacity of nearly 1300 cc (17; figure 8.8, p. 283). Yet another skull combining primitive and modern characteristics is the "Hofmeyr skull" from the Eastern Cape in South Africa (18).

This scatter of Archaic humans from the Horn to the Cape, spanning more than half a million years, suggests a single, long-lived lineage. In spite of its small brain, the Buia skull implies continuity within the continent, even, perhaps, the existence of a single, widely distributed gene pool. Fossils showing "intermediate" features raise long-acknowledged problems in defining where *Homo ergaster* (or *H. erectus*) ends, where *Homo sapiens* begins, and where intermediates such as *H. heidelbergensis* or *H. antecessor* fit in. Are we vaguely more than a million years old, or is there some rubicon for our emergence as wholly modern humans? The fossil record unambiguously hints at some continuity for modern-looking humans over the last million years; yet there is genetic evidence for a sudden break in this apparent continuity, and its causes are the subject of much speculation.

A revolutionary insight into the peculiarity of modern human evolution has emerged from a new, very oblique approach taken by geneticists at the University of California in San Diego (19). They compared the genetic diversity of African apes and modern humans by selecting a small region of mitochondrial (mt) DNA where mutations are thought to accumulate at a steady rate. The most significant conclusion to emerge from this comparison was that there was more mt DNA variation in the 55 chimpanzees belonging to a single test group than in the entire human population. Gorillas show equivalent measures of ancient diversity to those found in the chimps. This evidence proves that our lineage has been through a very extreme form of genetic bottleneck. The single sample of DNA that has been retrieved from a Neanderthal cranium has helped narrow the timing of this bottleneck, because the sequence differs from that of Moderns in ways that put it outside our own range of variation. At some stage, after our divergence from Neanderthals—and most plausibly at the very start of our emergence as a distinct species—our own bloodline ancestors in Africa were reduced to a mere handful, robbing their descendants of the genetic diversity that had once belonged to a formerly wide-ranging and presumably adaptively variable population (20).

The depth of genetic separation between the surviving ape species and various human populations has been meticulously quantified by Satoshi Horai and his colleagues (21). Using mt DNA sequences, Horai and his team measured mutation rates in four ape species and three modern human populations. Modern humans showed a maximum range of 18 mutations, whereas the two chimpanzee species (Common and "Pygmy") were about 90 mutations apart. When chimps were compared with humans, they were nearly 200 mutations apart; with gorillas they were more than 250 apart; and with orangutans, over 500 mutations separated them from their African brethren. Once more, the clustering of chimps with humans rather than with other apes was confirmed while also demonstrating the skimpiness of diversity in humans. There are other lines of

genetic evidence to support the idea of a relatively sudden and surprisingly late emergence of modern humans from a minuscule population,
popularly represented as a single, symbolic "African Eve" (22). Genetic
impoverishment is masked by individual as well as regional variations in
physical appearance. That this is nothing new is shown by substantial
variation in nearly fifty 15,000-year-old skulls from a single cave in Algeria. Individual differences are apparent even earlier, for example in skulls
from Jebel Gafzeh in Israel (90,000–115,000 years ago).

The discovery that modern humans went through a genetic bottleneck
in Africa *after* diverging from Neanderthals has fundamental relevance
for our understanding of how and when humans spread over the globe
(21). It also poses certain questions: Why did the contraction in numbers
take place? Where dd it take place? Speculation on possible causes has
centered on exterminating agencies such as pandemics, volcanic explo-
sions, starvation, climate change, intrahominid war, or some rather unusual form of isolation. Whatever the agency, it must have been rather selective, because there is no evidence so far for other organisms in Africa
suffering a comprehensive setback. Such evidence makes the direct influence of climate change or some global catastrophe unlikely explanations.
Likewise, in the light of modern humans' later success, it is difficult to see
how their subsequent good fortunes could emerge after being victimized
by other hominins. Indeed, an exclusive sense of cultural identity may
have been a crucial characteristic of Moderns both at species and local,
purely cultural levels. Such exclusivity may well have been an important
factor in preserving the genetic identity of Moderns whenever they came
into contact with other humans (23). This cultural exclusivity could have
been of particular significance at times when contact with sibling Archaics could have represented a genetic threat to their fledgling identity.
During such confrontations, the protagonists would have been self-de-
fined in terms of their economies and the languages and technologies
that served them (not that people would have self-consciously defined
themselves, just that their artifacts, including language, had become inseparable from how they lived and survived as human animals). These
cultural barriers, not genetic incompatibility, would have been the primary obstacle to genetic exchange (24). The extinction of all other types
of humans hints at endemic intolerance in our species; closer to home,
there are enough instances of genocide *within Homo sapiens* to make active hostility toward other cultures one of our major characteristics. In
the face of massive evidence that this has been a recurrent theme in our
history, some multiregionalist ideologues continue to reject its significance for evolutionary history. Indeed, contesting that intolerance is any

part of our makeup (on supposedly ethical grounds) has served to prolong the multiregional/African Eve controversy that was mentioned in
the last chapter. It is also a legacy that is difficult to reconcile with the
suggestion that our ancestors might have begun their existence as the *victims* of an all-out fight, only to become the ultimate victors over their unknown persecutors (25).

Two preconditions might have been necessary for a tiny band of technologically oriented Moderns to emerge from one corner of Africa to take
over the continent (and eventually the world). One is a relatively prolonged period of isolation. This is a problematic option for advanced,
mobile humans in a land mass that is as singularly continental as Africa;
but some degree of regional isolation could have been caused by an unusually severe environmental change or upheaval, reinforced by an epi-
demic or starvation. The second precondition could have centered on a
decisive role for technology. If the ancestral Moderns were equipped with
peculiarly appropriate tools and techniques to cope with the vicissitudes
of their environment, they might have been able to maintain physical
separation from other humans long enough for cultural barriers to arise.
If it came to the point where such barriers became impregnable, human
evolution could be said to have become "self-made," because both obstacles *against* and facilitators *for* gene flow did not so much derive from the
environment as they were self-generated.

If we are to envisage possible contexts for the emergence of Moderns,
some degree of environmental challenge seems likely, but there is, as yet,
no hint as to *what* that challenge might have been. Likewise, with the
causes of that challenge unknown, thinking about *where* our precariously
vulnerable little band of ancestors might have survived poses innumerable questions—most being impossible to address. Whatever the explanation, whatever the agency, and wherever the scene, the consequences
may be more relevant for us, the descendants of this catastrophe, than its
causes; nonetheless, the latter are of absorbing interest and worthy of
some speculation.

Consider, first, the paradox of an extreme genetic contraction followed
by a truly astonishing global expansion. Widely distributed and long-lived
genomes (such as those of chimps and gorillas) have supposedly conferred
an increased ability to adapt to the whole spectrum of ever-varying conditions to which their long histories and formerly wide ranges have exposed
them. By contrast, a constricted genome brings with it the well-known
dangers of inbreeding. Diminished variability in a tiny parent gene pool
also implies reduced viability when conditions change, whether that
change is due to the migration of genes or the passage of time. What

could possibly explain how the offspring of a theoretically disadvantaged population of humans overcame this adaptive handicap to become one of the most successful and widely distributed of all mammals?

One interpretation concerns the very nature of adaptation. Specific gene variants are known to confer resistance to environmental extremes or to help combat equally specific diseases. More variants increase the probability that some individuals will survive almost any catastrophe. But variants need time to accumulate because they depend on relatively rare mutations. Yet the earliest Moderns, unlike chimps and gorillas, had been robbed of their variation by the recency of the bottleneck they had been through. How could their genes make up for the time they had not had? How could the supposed handicaps of an impoverished genome be offset? At least part of the answer could lie in these early Moderns' application of novel tools and techniques to many problems that were formerly the province of genetic adaptation. For example, many animals have evolved the ability to detoxify poisonous plant foods. By contrast, Moderns have overcome a variety of these poisons by peeling, drying, boiling, or soaking. Other potential foods are protected by physical barriers that humans can breach with a variety of tools. Most animals can only evolve resistance to parasites over very many generations, whereas modern humans (and some apes, notably chimps (26, 27) have learned to apply or swallow plant or mineral deterrents that act as parasite-specific as genetic defenses. These shortcuts to open up well-protected resources or thwart potential enemies can only give their devisors a substantial advantage over less ingenious relatives. Even greater advantages accrue to a species that can easily and quickly spread knowledge of such techniques through language and demonstration. In this respect, small anatomical changes in the laryngeal region might have been critical (28, 29).

Inherited abilities at solving technical and intellectual puzzles undoubtedly existed among the Archaic humans that we know were widely distributed all over Africa for nearly a million years. Yet one interpretation of the evidence suggests a sudden concentration of genetic "quality" in a single, highly localized minuscule population. The quality of a very particular set of genes may have made up for the handicap of losing overall diversity within a large gene pool. The amount of genetic change required to differentiate Moderns from Archaics need not have been large (as little as 10 to 20 mutations). The main alterations could have been changes in the timing and expression of established genes, which could have been sufficient to set Moderns apart. The genes that came together and expressed a set of new faculties would have been of the same basic quality as those that serve our intellectual and technological prowess today. A possible implication of a lottery-like convergence of such genes is that Moderns might have come into existence with the equivalent of a genetic Big Bang! Before this event, the genes for a Plato or a Darwin were literally inconceivable. Given the right intellectual climate (impossible, of course), both sages had as much chance of arising immediately after that "Bang" as now.

With the hindsight of subsequent global success, are there any clues as to where, within Africa, modern human ancestors might best have acquired a uniquely effective set of genes—and, perhaps, developed the mindset, language, and skills that underpinned that success?

In previous chapters, I have argued the case for various forms of hominin developing in isolation on ecological islands. It may seem inappropriate to extend that argument to a time long after Erect humans had found their way right across Asia to the farthest reaches of China and Indonesia. For Erect humans onward, was it not already too late for serious constraints on movement to come from the environment?

An assumption that geography no longer shaped modern human evolution and dispersal ignores many very real and continuing limitations on advanced types of humans (30). The most obvious of ecological constraints was the need for fresh water, but a lack of water would have been only one of several limitations that were vastly amplified by climatic vicissitudes. It would be false to visualize early humans spreading far and wide over all landscapes at all times, and we can never picture the depths to which once-flourishing populations might have sunk when virtually all conditions conspired against them. Thus, the relatively rapid spread of early humans must have been as much assisted by good climatic cycles as inhibited by unfavorable ones, but paleontological records of either extreme are still rare or absent. Among unfavorable conditions would have been inhospitable vegetation, hostile terrain, epidemics, or even local catastrophes such as famines, volcanic eruptions, or cometary strikes. Combinations of numerous competing or predatory species and endemic diseases might have excluded humans from some extensive and otherwise attractive regions.

The least obvious obstacle to free movement is the most paradoxical in that the secret of their own success—their technology—could effectively lock humans out of those habitats or regions where it was inappropriate. Prehistoric technology was not a universal tool kit: it developed and was relevant to the specifics of particular places at particular times. There was always the danger that the specificity of a regional population's tools, as well as conservatism in the techniques employed in their use, could be self-confining.

It must be acknowledged that humans in possession of an extensive culture and technology could only have been excluded from really wide tracts of habitable country by temporary or rare events. Even so, there could have been occasions when humans were so thin on the ground, or so unevenly distributed, that some marginal pockets were effectively cut off from the mainstream. At such times, it is not too fanciful to imagine an isolated gene pool under extreme selective pressure to come up with technical and intellectual solutions to difficult environmental challenges.

At present, any attempt to guess at where the Modern genotype might have originated would pile surmise on surmise, particularly given the current nonexistent fossil record. Nonetheless, observations are worth making on how the choice of regions might be narrowed. For a start, Moderns seem to be as remarkable for how long they remained in Africa as for their later global travels. The estimated dates at which Moderns left Africa range from about 100,000 to 222,000 years ago (24), while their emergence as a distinct form of human (possibly marked by a new form of Middle Stone Age technology) is estimated at between 250,000 to 300,000 years ago. At present, there is no direct evidence to link the arrival of the Middle Stone Age or Middle Paleolithic with the first anatomically modern humans, but the coincidences in timing and region make this a reasonable, if tentative, hypothesis (figure 9.3).

The gap between "emergence" and exit out of Africa carries a strong implication that the northern reaches of the continent were not prime habitat and also suggests that Moderns originated south of the Sahara. When Moderns did leave Africa, they would seem not to have colonized cooler habitats; that they were excluded from cool, dry habitats because these were already preoccupied by Neanderthals or Erects could be part but not all of the explanation. If this aversion derived from their immediate place of origin, then north Africa and Ethiopia are even less likely sources for the earliest Moderns, as are the South African Uplands. If choice is further narrowed by eliminating both rain forests and deserts from consideration, the largest blocks of remnant territory become the savannas of eastern, western, and southern Africa. Amorphous boundaries may make these regions seem rather vague places of origin for mobile, adaptable people; yet it is still possible that at the time, during appropriate cycles of climate, east Africa was better insulated by the lakes, mountains, and swamps that surrounded it than it is today. For savanna species, isolation could have been most marked during warm, wet periods when the spread of forests, both montane and lowland, would have helped plug up its otherwise rather permeable "borders." It is also possible that the diseases that flourish during warm, wet periods might have reduced local populations.

As it happens, eastern and southern Africa are the regions in which the earliest fossils have been found, and also the source of virtually all the earliest middle Stone Age tools (i.e., 300,000–180,000 years ago) that are generally considered to be the work of Moderns. (Although, as yet, there are no direct associations between stones and human bones before about 130,000 years ago.) East Africa is a particularly benign part of the continent and, from prehistoric times up to the present, has been inhabited by exceptionally diverse communities of animals and plants. These communities present as many challenges as opportunities to a primate, particularly to one that can "steal niches."

Apart from its potential for isolation by a ring fence of lakes, mountains, and forests, eastern Africa has at least two interesting and different properties that might have had a bearing on the development of Moderns. First, at the broadest level of genes and physiology, hominins had existed there longer than anywhere else, so early adaptations to specifically local climates, foods, and other factors (notably to diseases) were likely to be prevalent. The presence of such adaptations, in turn, implies a possible mitigation of one of the major disadvantages of a contracted genome: lowered resistance to disease. Second, as the main corridor between north and south, east African hominins were likely to have accumulated genes with diverse origins. If these had mixed, the area would have been a genetic melting pot (much as it is today) for diverse adaptive traits as well as cultural traditions.

Over time, the genetic diversity of Archaic humans in this region might have been particularly great. In these circumstances, it is not entirely inconceivable that the particular permutation that survived the bottleneck might have made an unusually favorable combination of genes.

So, what factors might have given an east African minipopulation of Archaics decisive advantages over other populations? Part of the answer could lie in something as simple as a series of changes in the timing of various developmental processes. Rather than depending on some random mutations for innovation, the existence of a wide choice of traits (already circulating in the population) could have permitted a form of genetic shuffling to initiate evolutionarily significant change.

The development of neotenous features that so clearly differentiates Moderns from Archaics might have depended on just such change; I contend that the genetic, physiological, and social bases for such change lay in juggling a whole suite of characteristics to prolong immaturity. One potential consequence for delaying the onset of adult characteristics is

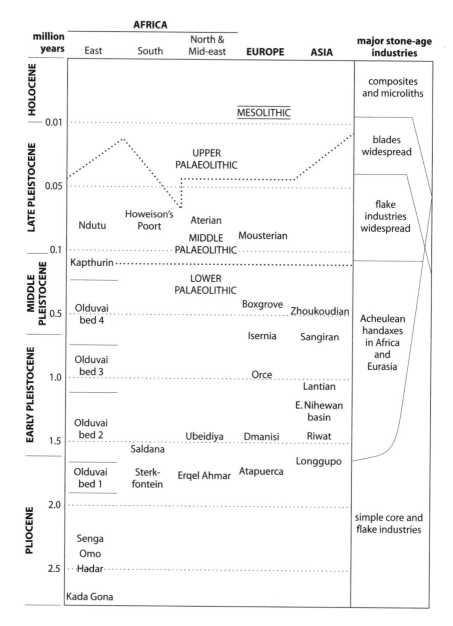

FIGURE 9.3 Palaeolithic sequences (left) and artifacts from Beds I & II of Olduvai (right, below), Upper Paleolithic. A. Tanged point/perforator. B. Burin. C. End scraper. D. Curved backed blade. E. Denticulate. F. Knife. G. Microbladelets and blade scraper. H. Five microliths. I. Bifacial lanceolate. J. Tranchet, core-axe. K. Tanged point, core scraper, core axe (Lupemban). L. Blade core (Aurignacian). M. Howeison's Poort unifacial points. N. Backed

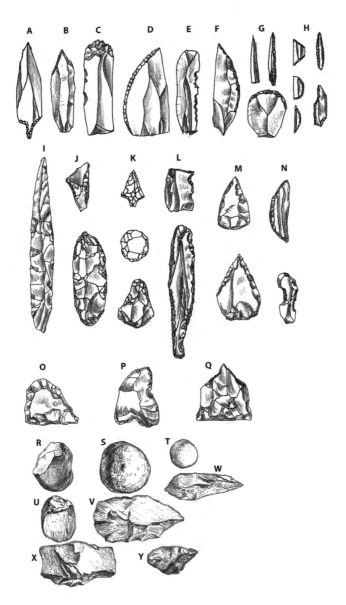

blade, notched scraper (Howeison's Poort type). O. Side scraper. P. Notch flake. Q. Borer. R. Chopper. S. Hammer/anvil. T. Sphere. U. Hammerstone. V. Hand axe. W. Pick. X. Cleaver. Y. Proto-biface. (Diagram modified after Gowlett, J. 1992. Stages of the Palaeolithic. In *The Cambridge Encyclopedia of Human Evolution*, ed. S. Jones, R. D. Martin, and D. R. Pilbeam. New York: Cambridge University Press.)

the extension of immaturity and a higher proportion of the population (or the group) being subadult. However, it is possible that this potential might have been offset, in prehistoric populations, by a disproportionate mortality within this class. There are usually substantial evolutionary repercussions when the main burden of competition and mortality falls on an age or sex class. For example, intrasexual competition selects for those characteristics that confer the most reproductive advantage on the winners; these characteristics are typically structures for sexual display, large size, or weapons for offence or defense. In the hypothetical case of human juvenile rivalry (predominantly directed at gaining adult attention and protection), one aspect of selection could have been a tranche of behaviors and aptitudes that had the immediate effect of integrating youngsters more effectively and completely into the social group. A species that depended on tools and techniques for survival would have put a high value on practical demonstrations of technical aptitude. A species that was rapidly inventing new techniques would have valued versatility and the ability of other group members, of all ages, to join in on and improvise as well as share skills.

It is known that the size of social units enlarged in Moderns so that the overall density and complexity of their society must have changed. Extended kinship structures with reciprocal but flexible obligations could have ushered in a much more intensive use of resources. The quality of relations with neighboring groups probably varied widely, but more frequent contacts with strangers should have taxed both communication and strategic skills. Children would have been an integral part of their social structures, and it is possible that ranking among children and adults first became possible and influential in supposedly enlarged social units, Neoteny-driven evolution implies numerous social consequences, among them a fine balance between cooperation and competition among children. During childhood, the safety of a vulnerable class is best met by groupings in which cooperative behavior encourages clumping. However, it is also normal for children, no less than other young animals, to compete, even while cooperating. For juvenile tool users, demonstrations of skill in the operation of artifacts become major opportunities to express prowess. Herein lie the seeds of almost continuous innovation—but also, just possibly, the emergence of social stratification (31). That unskilled children were punished in foraging societies can be illustrated by a scene witnessed during Darwin's voyage of the *Beagle*. A Fuegian child dropped a basket of seagull eggs, whereupon the boy's father dashed him against the stones again and again until, battered and bleeding, he was abandoned to die.

In these circumstances, there could have been selective advantages for "smart kids." From the juvenile's perspective, there would also have been the self-motivated desire to join in on adult activities. Wanting to share in whatever the social group was busy with must have emphasized preexistent (probably innate) impulses to share—to share activity, to share skills, and to share the wish to share! For example, there are clear indications of infants empathizing with their mother or other children well before 2 years of age (32). A self-driven compulsion toward social participation implies a need for effective communication, which must have provided powerful incentives to elaborate or improve language. There are strong indications of how inventive children can be, for example when the offspring of disadvantaged adults (whether disadvantaged or not themselves) tend to outstrip their parents in communication skills—inventing, along the way, grammatical structures that seem to be an innate component in all languages and pidjins (33).

Improving communication would have had obvious benefits for an economy that no longer relied on innate foraging behavior, because every successful harvesting technique had first to be invented and then learned and passed on as just one of many components in an extensive strategy for subsistence (34). Intelligible codes of communication may have helped small groups to join up or even negotiate with larger units and encourage an exchange of goods and ideas. As we know, even today, the exchange of anything across a territorial boundary is not easy; especially for adults, it can even be described as "unnatural." Yet once a common grammar had been evolved, arbitrary signals for simple universals could quickly be learned, especially by "smart kids" who were strongly motivated to display and share skills. If high juvenile mortality was the norm, survival into adulthood in prehistoric societies may have been unusually dependent on a sustained demonstration and performance of aptitudes, especially in efficient communication, during childhood. Prehistoric children neither sat exams nor won prizes. but they may well have had to cram in an anticipation of all the technical, ecological and social complexities of adult life, suffering penalties and rewards in the form of death versus life.

In other animals, there are limits on a youngster's openness to novelty, both in the capacity to learn and in the temporal window during which innovation is possible. In human societies, the extension of this temporal window may have had an even larger social function. Children were not just sponges for lessons from the adults and observant learners from the environment; in effect, they may have been specialists in innovation (35), a role that could have been crucial in relations between neighboring groups. Whether neighborly relations were friendly or hostile, children or childlike adults, like nerve endings, may have been effective interfaces between groups and thus a major agency in exchange.

In trying to deduce what characteristics favored the long-term survival of one minipopulation of African Archaics, it seems possible that neotenous tendencies and some small modifications in the larynx might have enhanced the ability to exchange information—something vitally necessary for a continuously enlarging repertoire of techniques. Populations without mechanisms for exchange, ones that were less willing or less physically capable of sharing innovations, would have been overtaken by those that were.)

In any competition between different populations of closely related humans, the ability to impart technological knowledge and adapt to novel situations probably had more implications for brain size than any direct facility in the use of tools. The resulting chain of learning may have mimicked inheritance as the pathway for a package of information, but its survival would have been peculiarly hazardous (36). Along this route, survival value lay first of all in adaptability, mental quickness, and an aptitude for devising, teaching, and learning new skills, including the invention of new tools. Low densities and scattered small-unit distributions would have increased the hazards. High densities and "networking" would have enhanced the chances of traditions surviving (and so getting elaborated further) as well as diffusing more widely and faster. In fact, if a fast-developing technology served by language was the main advantage possessed by Moderns, their greatest selective advantage would have lain in any social or demographic features (such as language-linked extended families) that enhanced the spread of that technology.

To convert the almost infinite inflow of information from the environment into practical programs of action (that demonstrably serve the group's interests) requires not only flexibility of response but a high order of memory. The brain must not only store an inventory of biota and phenomena but also an even larger repertoire of relevant programs to take advantage of events as they occur. Observations on the weather or the behavior of an obscure plant or animal may be irrelevant in one setting but may be the key to survival in another context, which may be immediate or far in the future.

Interposed between individual humans and the external world that surrounded them was their family, their technology, and their culture—inextricable intertwined. In terms of plain subsistence, every experience, every encounter had to be "translated" into how it could be processed to enhance survival. This would not have been a conscious, analytical procedure, but it would have depended on a very considerable store and choice of possible courses of action. It is impossible to know how a pre-historic forager "made up his mind," but an individual's choices for action might have been guided by a memory bank in which each potential "nested" in a larger framework. Such a framework would have been built around the actual technological repertoires of a particular economy as well as a knowledge of the ecological and physical properties of potential foods and tools (37). Thus, the mental framework for thinking about cord and the artifacts or toys made from it (such as snares, nets, and cats cra-dles) might have been compartmented separately from thoughts about stone and its use for hammers, knives, and axes. The challenge of making practical connections at the same time as drawing distinctions between, say, wrapping twine around a stone tool to carry it versus stripping string-bark off a stem with a blade might have challenged prehistoric people to make verbal connections between separate categories of phenomena. Likewise, the practicalities of harvesting or processing birds rather than fish, or fruit versus roots, each with its associated ecological or even culi-nary contexts, must have occupied mental compartments. These were part of a family group's shared experiences, tastes, and memories; enjoy-ing them no less than getting them required cooperation and the sharing of tastes, plans of action, and allocation of roles. Communicating such needs and tastes called for verbal tools no less than traps and snares. Such inversions of status or priority for the categories themselves might have alerted them to the ambiguities of oversimple associative name-labeling and advertised the need for labeling and distinguishing subject from ob-ject and nouns from verbs. A human child was unlike any other young animal in that its brain had to accommodate all the technological poten-tials inherent in any experience. All experiences had to be "translated," requiring a vastly enlarged memory bank of which brain enlargement was the physical expression.

Stone Age education can be envisaged as both program and process. It was a program that allowed the execution and learning of more complex techniques, each deriving from simpler ones and mediated by more com-plex verbal and conceptual structures that were, in turn, modified from the vocalizations and expressed wants of very small children. It was a process built on the need to service economies and subsistence strategies that could no longer rely on automatic responses to innate appetites. The sheer diversity of subsistence and the methods employed in harvesting that diversity called for signaling systems that could convey appropri-ately nuanced information about the many challenges that tool-medi-ated subsistence presented. At some stage, probably long before the evolution of Moderns and perhaps linked with the earliest forms of systematic hunting, charades or mimetic gestures might have been a large part of communication. During the early stages, there may have been little need for language such as we know it today, but gestural semaphore might have offered an emphatic set of visual signals that were at their most effective when backed up by vocal accompaniments. Eventually, these could be semidetached from their mimetic origins and contexts in subsistence to become systematic linguistic components. When the brain circuitry that is used during ape vocalizations is compared with that employed by humans during speech, the main alterations reflect the extra and unusual demands of spoken language. It has been suggested that at least 2 million years of continuous selection must have been involved in the brain-language interaction (38).

The ability to cooperate in larger groups and to diffuse knowledge is generally thought to have contributed to the long-term triumph of Moderns over Neanderthals and Erects (39). Part of this success may be attributable to language and to some of the side effects of neoteny discussed earlier, but part of the success story may have been little more than an extension of *H. ergaster* strategies. For example, I have described these earlier humans manipulating elements and properties and using materials in a mimicry of natural adaptations. This behavior rapidly multiplied the number of ecological niches that could be invaded so that an increasing number of animal species had a new competitor encroach on at least a part of their former niche. This central human characteristic, which I have called "niche stealing," became ever more efficient with the emergence of Moderns (11).

From the beginning of the Middle Stone Age onward, more versatile tools also suggest greater flexibility and adaptability in exploiting a range of resources greater than that used by any previous species. One corollary of Moderns taking an ever larger variety of both traditional and new food types (as well as expanding territories) was that other types of humans, no less than other animal species, had a new intruder and had to share their means of subsistence, if not become prey. When conditions were less than optimal for any species of competing animal (including earlier types of hominins that suffered the disadvantage of impoverished or stereotyped tool kits), competitiveness was reduced because there were fewer fallback alternatives. During droughts or famines, the balance would tip still further against the survival of competitors, because "niche-stealing" (or outright predation) effectively deprived those species of their living.

Turning from the generalities of modern emergence to the specifics,

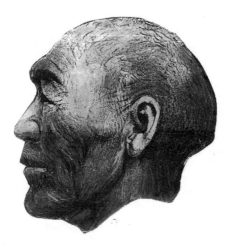

FIGURE 9.4 Reconstruction of one of the earliest relatively complete Modern skulls, from Laetoli. (From Kingdon, J. 1993. *Self-made Man. Human Evolution from Eden to Extinction?* New York: John Wiley & Sons.)

where have the first reliably identified and dated specimens of wholly modern humans been found? The discovery and dating of Modern fossils in prehistoric Africa is relatively recent, and many of the supposedly earliest fossils are very fragmentary. The first relatively complete skulls come from Omo (Kibbish) in southern Ethiopia (40) and Laetoli (Ngaloba) in Tanzania (41) from about 130,000 to 120,000 years ago (figure 9.4). Were the middle Stone Age to be proved a reliable marker for the existence of Moderns, and assuming correct dates for these specimens, they may actually sample a period that could be anything between 28,000 to well over 100,000 years into the Moderns' career. Although it is known that Moderns coexisted with Neanderthals and Erects in Eurasia, we still know nothing about whether Moderns coexisted or competed with earlier populations or species of humans or hominins within Africa. The diffusion of early people or of their cultures *within* Africa is also unknown. The ground is only marginally firmer when it comes to Moderns leaving Africa, but the options for a pattern of dispersal narrow dramatically, and there are useful observations to be made about the ecology of human environments and the factors that might have shaped the patterns of their dispersal. In the context of a book about bipedalism and its legacies, walking out of Africa to the ends of the Earth could be seen as the ultimate triumph; however, its course and timing are far from clear. In reconstructing this dispersal, genetic differences between living people become a useful guide.

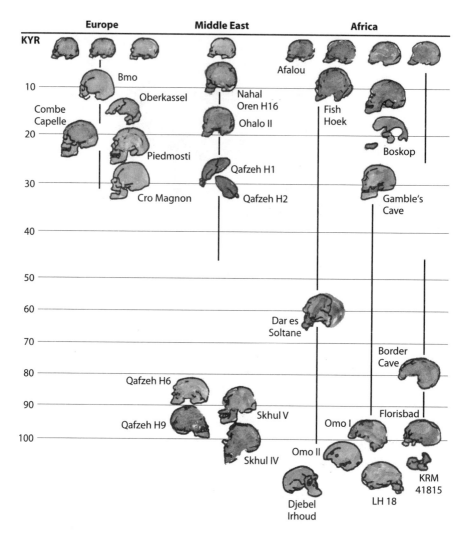

FIGURE 9.5 The nature of the evidence; a selection of key skulls of Moderns arranged by lo-cality and age. (After Foley, R. A. 1995. *Humans before Humanity*. Oxford, UK: Blackwell.)

The evolutionary significance of genetic differences found in living pop-ulations of humans is deduced from disparities in their DNA. More recent migrations and mixings may greatly obscure the picture, but past splits be-tween formerly single populations nearly always have a geographic con-text. The most accessible and frequently tested of these geographic differ-ences concerns the contrast between Africans and non-Africans.

Genetically, modern Africans have proved to be the most diverse peo-ple on Earth. Such diversity is due partly to migrations and mixing, but it

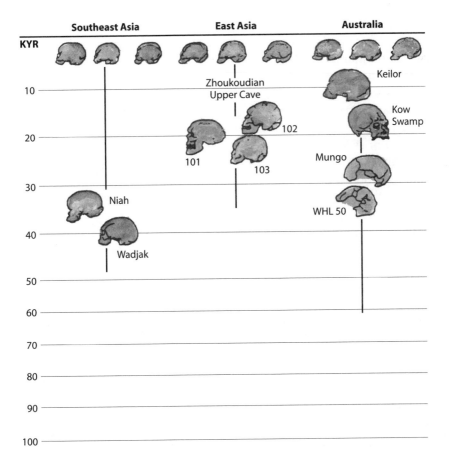

is first and foremost the product of time; the continent has been inhab-ited longer than any other. It has also been argued that this genetic diver-sity could be explained by population numbers having been consistently higher in Africa than anywhere else, but evidence for this is hard to come by. Two recent estimates agree on a primary African–non-African split at 190,000 to 222,000 years ago (42, 43). Other estimates (of 156,000, 140,000, 100,000, and 65,000 years) may have produced younger dates partly because they ignore subsequent gene flow back into Africa. How-ever, a recent estimate that does take account of genes returning to Africa dates a major backflow into Africa at about 57,000 years ago and the ear-liest exit at 135,000 years ago (44). In the end, it may have to be geology, rather than genes, that will finally arbitrate over this contentious, signifi-cant, but ultimately datable event (figure 9.5).

Even so, and allowing for the current shortcomings of molecular clocks, genetics has a special authority because it can compare the genes of living Africans with living non-Africans; it can demonstrate their ultimate common origin but also reveal something of the history of their immediately disparate ancestors.

Efforts to date the emergence of living modern from extinct humans on the basis of genes suffer from great margins of error compounded by considerable uncertainty that this was a single event. One study based on a portion of the Y chromosome came up with 270,000 years ago (45), a date that has some correspondence with the first appearance of Middle Stone Age tools but has been superceded by more recent, younger estimates (46). As a result, there is still no clear date for that mysterious event: "the emergence of modern humans."

The Y chromosome has proved a boon to the study of human dispersal because it has few functions beyond determining male sex and, being confined to males, can be used to trace male descent (47). More significantly, for the geneticist, it has, relative to other chromosomes, rather few genetic differences and remains unchanged through the generations except for very rare DNA mutations (48). These mutations are invaluable as genetic markers because they show structural changes that are directional, allowing gene trees to be constructed. Not only can trees be constructed, but the trees exist in space and time, so the presence of a marker gene in a contemporary man could, in theory, be traced back to ancestors in a specific place at a specific time. In practice, detecting the history of human population dispersals depends on gathering large numbers of gene samples from different parts of the world and working out relative frequencies of the gene clusters that are known as haplotypes.

A dramatic picture of human dispersal out of Africa has emerged from an analysis of polymorphisms on the Y chromosome by Michael Hammer's team at the University of Arizona at Tucson (49) (figure 9.6). The ancestral haplotype (labeled 1A) is found only in Africa, where it is rare but found at its highest frequency among the Khoisan of southwestern Africa. At a time tentatively dated at around 200,000 years ago, a single mutation on this gene must have arisen in one man living in Africa. His descendants found their way out of Africa, and this haplotype (labeled 1B) is the commonest and most widespread of the ten major groups identified by the Hammer team; it accounts for more than half of their global sample, including a majority of south Asians and virtually all Australian aborigines. A further mutation of 1B in Asia (labeled 3G) is found in a wide scatter of eastern Asiatic men. Currently dated at 60,000 years, the mutation might have been older. Surprisingly, a further mutation of 3G, generating the haplotype 3A, has been found in only a small but wide

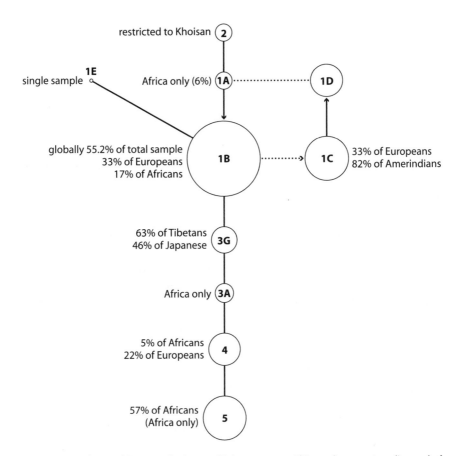

FIGURE 9.6 Polymorphisms on the human Y chromosome. This work suggests a dispersal of Moderns from an African source population through Eurasia to Australia and eventually to the Americas. It also suggests that the 3A, 4, and 5 polymorphisms, all found in contemporary Africans, derive from a Far Eastern source. (From Hammer, M. F., T. Karafet, A. Rasanayagam, E. T. Wood, T. K. Altheide, T. Jenkins, R. C. Griffiths, A. R. Templeton, and S. L. Zegura. 1998. Out of Africa and back again: nested cladistic analysis of human Y chromosome variation. *Molecular Biology and Evolution* 15[4]: 427–441.)

scattering of African men, suggesting that some of their ancestors came from Asia. Two further haplotypes, both derivative of 3A, account for some two thirds of the contemporary Africans sampled, demonstrating that the majority of living Africans owe part of their inheritance to Asian ancestors.

How can this genetic excursion into Eurasia followed by a return to Africa be explained? Traditional histories of human migration were forced to work backward from a better informed present toward the ever greater unknowns of prehistory. Now, several sources of information, including genetics, are able to retrace events in sequential order, most notably reconstructing the spread of Moderns out from their African homeland. It is a perspective that begins with small numbers and simple directions and moves, step by step, toward a far-from-simple pattern of multiple eddies back and forth. The return of genes from Asia to Africa may have been but one of these eddies, but it illustrates many important features of the nature of the human diaspora. I attempted to retrace this extraordinary saga, from ecological and biogeographic clues, in a book that predated most of the discoveries that depended on these technical innovations in genetics. Discovery of the 1B haplotype has confirmed that a mere handful of individuals crossed the narrow neck of land between today's Egypt and the Eurasian land mass. The dominance of this gene in southern Asia and the south Pacific hints that the initial dispersal of Moderns, once they got out of Africa, was exceptionally rapid and extensive, especially across the tropics. However, the variant's blanket coverage of a vast swath of country can give no indication of how the initial dispersal must have proceeded by stages.

Modern humans that had occupied the African savannas for anything between 28,000 to more than 100,000 years could be expected to have a strong preference for the habitats that most resembled those they had inhabited in Africa. On this count, India would have been the most immediately attractive and significant block of habitable territory and the area where a significant population could most rapidly proliferate (11). Also, the clearly defined borders of the subcontinent would have soon served to contain a single, very large population unit. Research on the M group of mtDNA complicates any simplistic picture of a single wave of emigrants out of Africa (50). The virtual absence of this marker in the Levant as well as its presumed spread from Ethiopia through southern Arabia to India, suggests a later wave (65,000–60,000 years ago) presumably assisted by watercraft across the Bab el Mandab Straits (51). At such a late date, it is debatable whether movements were eastward or westward.

Confirmation of a very early Afro-Indian connection comes from the Middle Stone Age tools (uncertainly dated at 163,000–120,000 years) that come from many sites across India and can scarcely be distinguished from their African equivalents (52). Unfortunately, fossils of early Moderns still remain to be found. A single "Indian Heidelberg" has been found in Narmada, as have Acheulean tools assigned to Erects, but the lives and fates of Moderns as well as of the populations that preceded them in India are still completely unknown and present a challenge to Indian archaeology.

Expansion further east might have seen a pause at the Ganges delta, but rain shadows within the Malaysian and Indonesian archipelago support patches of wooded grassland that might have proved attractive to incoming Moderns. These were, perhaps, a minor habitat for a minor population because savannas would have existed as islands in a sea of initially uninhabitable tropical rainforest. More significantly, river mouths and seashores had the potential of offering a major habitat for a major population because the resources of the intertidal zone are immense for any species that can subsist on them. Fossil Erects attest to their having inhabited the Indonesian archipelago for more than a million years before the arrival of Moderns, and their presence on the island of Flores suggests that Erects had developed enough of a marine culture to get there by watercraft. What relations might have existed between Moderns and Erects must be pure conjecture at this point, but it seems possible that the well-tested, if simple, technologies of Erects might have provided Moderns with many shortcuts to adaptation as they moved into new habitats in a new region. Techniques can be appropriated through imitation or conquest; appropriation of genetic adaptations can be achieved only through interbreeding (which can compromise species-specific advantages for the descendants of *both* parent species). Hybridization between Moderns and other, older types of humans, may well have occurred quite frequently but seems not to have left any functionally significant genetic legacy in most Modern populations (although traces of a non-Modern inheritance may eventually be detectable). Meanwhile, multiregionalists continue to claim resemblances between the skulls of Australian aborigines and those of Java Erects, also between Europeans and Neanderthals (53, 54)!

The possibility that appropriations from Erect culture might have helped speed up modern penetration of tropical southeast Asia has obvious implications for the invasion of Australia. Fossils from reliably dated strata have already proved that Moderns were in Australia by 63,000 years ago (55), but there are other clues to suggest a substantially earlier date. Dominance of the 1B gene in Australia shows that all of the founding stock of modern aborigines (possibly a very small number of men) shared the original "out of Africa" haplotype. Although this similarity could be taken as consistent with early invasion, it could equally well be pure chance.

Today, marine-based economies are such a familiar and worldwide phenomenon that their prehistoric roots are seldom given a thought. If it is obvious that an early form of marine economy delivered the ancestors of aborigines to Australian shores, it is equally obvious that later elaborations of marine culture took Melanesians to the many islands of Melanesia and, much later, Polynesians to the central Pacific. Less obvious is the possibility that marine economies might once have been so dominant, demographically, in southeast and east Asia and the south Pacific that they provided springboards for still later colonizations, even of continental interiors. Hammer's data on the frequency distribution of 3G haplotypes hints that this could be so. Because this otherwise rare morph has given rise to three other haplotypes that are now widespread, its origins must be relatively ancient. The fact that 3G is very common in Japan (56) suggests that the Japanese may have conserved part of the genotype of the most ancient inhabitants of the islands, a people named after their culture, Jomon. The Ainu people of Hokkaido and Sakhalin Islands are commonly identified as contemporary descendants of the Jomon, but their most ancient ancestors may well have been much more widely distributed in eastern Asia, especially close to the coast. Hammer regards 3G as a possible marker for the Jomon heritage of the Japanese. Given our present knowledge, the only other locality where this gene is common is among Tibetans, among whom it is possessed by 63 percent of the tested population! This statistic is difficult to make sense of until it is remembered that mountains often conserve what are otherwise conservative populations that also happen to be in possession of specialized adaptations. High altitudes, by challenging physiology in numerous ways, give immediate advantage to the first individuals to acquire appropriate adaptations (such as, say, genetic mutations that improve oxygen metabolism) (57). In this way, the descendants of a single early colonist can proliferate, providing a demographic base for still further selection and, over time, the preservation, over an entire mountain block, of a single lineage. Just as selection may favor one lineage in the mountains, it can eliminate sibling stocks in the lowlands, giving rise to sharp genetic differences along geographic or ecological discontinuities.

A thin scatter of 3G across a wide area of southeastern Asia, with two large pockets on its geographic margins, is consistent with a formerly more widespread occurrence. The ancient population that first colonized Tibet could have arrived there, possibly over many generations, via any of the great rivers that drain the Himalayas. However, a special interest must attach to those that empty into the Bay of Bengal: 3G's three derivative haplotypes are found mainly in Africa, and their ancestral carriers almost certainly inhabited the shores of the Indian Ocean (figure 9.7).

Exactly how and when this gene complex found its way back into Africa cannot, as yet, be deciphered with any certainty from either genes or fossils. However, it is almost impossible to envisage the westward flow going overland through already inhabited territory, as its earlier, unimpeded, east-flowing counterpart had done. By far, the most likely route was by steady diffusion along the coasts and islands of the Indian Ocean.

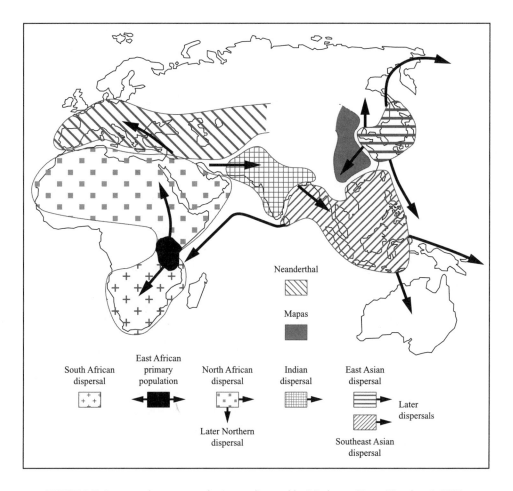

FIGURE 9.7 Suggested sequence of primary dispersal by Moderns. (From Kingdon, J. 1993. *Self-made Man. Human Evolution from Eden to Extinction?* New York: John Wiley & Sons.)

Its vectors were shore dwellers that I have dubbed the "Banda Strandlopers." The Banda Sea, west of Papua New Guinea, lies near the center of an island realm that stretches from the Bay of Bengal to the Solomon Islands or Fiji and comprises many thousands of equatorial islands, most of them well watered and habitable. The term strandloper (borrowed from a recent population living on the Namib coast) describes a prehistoric economy in which the people exploited coastal and estuarine habitats. The name does not imply a monolithic race but rather practitioners of a subsistence economy that was based on seashore and littoral resources, especially shellfish. Their technology would have centered on the capture and processing of mostly small aquatic animals and plants within the reaches of the tides, but terrestrial foods, especially shore-loving plants, were un-likely to have been abandoned. At a very early stage, the strandlopers must have invented (or copied from Erects) the first significant human vehicles: simple watercraft, probably outrigged canoes or rafts, initially propelled by poles or oars. (I explored many of the details of watercraft development in my earlier volume [11].) By making it possible to reach islands, watercraft would have increased the potential for genetic isola-tion; it would also have increased the possibilities for complex intercon-nections between groups that were no longer separated by a journey on foot but instead by boat travel over the sea.

Banda Strandlopers could be of special interest to human genetics be-cause they illustrate links between a technologically specialized way of life and biological adaptation. Exploiting the intertidal zone, activity would have been determined by the timing of the tides (especially where mollusks were the main food). For a modified forest ape, the major haz-ard of such an extraordinary way of making a living was solar radiation and heat. Women and children, as major foragers, would have been ex-posed to punishing levels of radiation when the tides coincided with midday, and it can be predicted that selection for heat and radiation re-sistance would have been intense. I contend that the results of that selec-tion are evident in the physiology, physiognomy, and dark complexions of numerous contemporary peoples and predict that it may one day be possible to reconstruct the beginnings of this adaptation to super-stress. In most other parts of the then-inhabited world, prehistoric Moderns would have been free to rest or seek shade during the heat of the day be-cause few nonmarine food sources require foraging at midday. In suffer-ing selection as a direct result of their economy, the Banda may not have been unique, but the strandloper strategy for subsistence is unique in the predictability of solar radiation as a major selective force. The physiology of melanin in the skin suggests that the original color for Moderns was light brown, allowing protection through tanning in summer (or in sun-nier habitats) and paling in winter (or in northerly latitudes). Extremes of permanent superpigmentation (in people of Banda inheritance) and de-pigmentation (in people with some Baltic ancestry) are both derivative conditions. A peculiar property of melanin physiology in females is known as "melasma," in which the skin of women darkens after the first pregnancy; this follows an initial lightening during puberty. Such fine-tuning suggests that many processes may be involved in skin chemistry (58) and that only further research will reveal all the evolutionary impli-cations inherent in our main radiation-filtering mechanism. Nonetheless, it is predictable that human skin, hair, and internal thermoregulation would have had to make structural adjustments to the super-stresses of the strandlopers' existence. Superpigmentation is a convenient diagnostic marker for a major adaptive event in prehistory, but it is likely to be augmented by the discovery of many others in the near future. What is already clear is that the Banda strandloper hypothesis and the distributions of people in the Indo-Pacific region suggest a mechanism that helps explain the geography and dispersal route of Hammer's "back-to-Africa" saga of the genes.

Given the vast extent of the Banda realm and the relative ease with which its linear habitats could be colonized and traveled along, it seems possible that there might have been a brief period during the earlier years of expansion when there might have been some sort of genetic commonality. After postulated beginnings on the shores of southeast Asia, the Banda would soon have expanded in different directions from many centers of population. The patterns for some of these early population groupings may soon be comprehensible through the application of new genetic techniques. In any event, opening up the resources of shoreline and marine ecosystems would have represented one of the triumphs of modern humans' ingenuity and inventiveness. Such a change in diet and economy would have had enormous demographic consequences: at the peak of its initial expansion, Banda-type economies in tropical seas of the Indo-Pacific might have supported one of the largest human populations on Earth. In southeast Asia, the legacy may be hard to discern today; there have been so many overlays, as one people after another expanded and contracted in a complex pattern of boom and bust, movement and stasis, resulting in ever more complex genetic laminations. Nonetheless, there may be many hidden inheritances from that one-time demographic advantage that may, one day, reveal themselves to genetic sleuthing. On the western bounds of their Indian Ocean range, much more complex legacies have begun to be revealed in studies of Y chromosome polymorphisms by the Hammer team. Of the three haplotypes that budded off from the east Asian 3G, the most derived ("5") is exclusive to Africa, where it accounts for 57 percent of Hammer's African sample (49). The two intermediate derivatives are also African, but their occurrence in southern Europe suggests a more tentative incursion that has had less impact on an earlier matrix of ex-African haplotypes.

One connection between bipedalism and selection for resistance to heat or radiation stress concerns running during the heat of the day in very hot, exposed environments. Banda strandlopers may have had scant need for athletic prowess, but in equatorial Africa conjoining resistance to heatstroke with athleticism may well have represented a final chapter in a progressive improvement of gait under extreme conditions. Other peoples may have had equally long strides and stamina, but the ability to sustain long-distance running under a punishing sun must have represented a measurable advantage for runners that were deeply pigmented.

Northern Europeans illustrate a rather different way in which technology can induce adaptive changes that justify labeling Moderns as "Self-Made Humans." As the ice retreated after the last glaciation, reindeer, horses, bison, deer, and salmon invaded vast new territories around the Baltic Sea; not far behind them were their human predators. However, this was no easy frontier. Like the Neanderthals before them, northern European people had to accommodate to unprecedented levels of cold, a lack of light, and a restricted diet in winter. In making this accommodation, "natural" selection for physiological adaptation was essentially subordinate to and consequent on technology, because it was fur garments, fires, and fishing equipment that made the northern migration possible in the first place. With summers shorter and winters longer, warm clothing, effective shelter building, and efficient hunting techniques were not enough to protect their users from levels of physiological stress that were way beyond the margins of tolerance for an originally tropical animal.

The first Baltic hunters must have suffered intense selection for survival in a hostile environment; but unlike the Banda strandlopers, their enemy was not too much sun and heat, but too little. In northern winters, people tend to suffer vitamin D deficiency because it is sunshine falling on the skin that produces the chemical reaction to synthesize this essential vitamin (59). Among the lethal consequences of too little vitamin D are rickets and a failure in young women to bear healthy children. Physiologists have concluded that paling of the skin favors vitamin metabolism so it must be significant that the densest distribution of the least pigmented people on Earth coincides closely with the former boundaries of the northern European ice sheet.

The mutation or mutations whereby skin, hair, and eyes became depigmented must have arisen in a population that was strategically positioned in time and place. The time was probably just before the first retreat of the glaciers, and the place was somewhere in the vicinity of today's Denmark. Passage through a narrow window of time or opportunity or the opening of a new frontier creates ideal conditions for a genetic bottleneck—in this case, for a highly distinctive population to move into an entirely uninhabited region. Similar bottlenecks have been posited for new arrivals in south and central America, and minimal genetic variation in some contemporary Amerindian groups would seem to confirm this pattern.

We may never be absolutely certain when people developed superpigmented or depigmented skin or hair because colors, like the skin, hair, and eyes that they tint, fail to fossilize. Nonetheless, the circumstances are highly suggestive for both Baltics and Banda Strandlopers. Likewise, it is uncertain when and why people became "naked" but selectively "tufted" on top of the head, over the brows, in the groin, and under the arms. In spite of attempts to link hairlessness with the evolutionary act of standing up, much variation in the hairiness of Moderns could be taken as evidence for rather recent origins. There are at least four popular theories trying to explain hairlessness. The first concerns keeping cool in grasslands; the second sees vestiges of an aquatic phase; the third finds aesthetic reasons; while the fourth suggests that less hair may help combat skin disease and parasites. I favor the last explanation and think that reducing the purchase for diseases would have begun to be important when people living in tropical environments began to stay in encampments for relatively long periods and at relatively high densities. Although skin disease might have been a risk for Erects and Heidelbergs, I suspect that it was Moderns that created the squalid conditions under which diseases would have begun to flourish, because their ever more intensive use of resources and greater sociality encouraged more densely populated and less temporary encampments. Imagine the buildup of human waste, parasites, and disease-carrying animals—all major hazards for the health of people in such encampments.

That bare skin evolved in the tropics to deter parasites was first suggested in 1874 (60). Darwin (1) doubted the magnitude of this hazard, mainly because he was thinking primarily in terms of arthropod infestations. The real threats were more likely to have been bacterial, fungal, and viral infections sheltered by a soiled hairy cover and provided with a purchase via glands attached to the abundant roots. Darwin pointed out the anomaly of hair on the head being a logical protection against the sun while hair "at the junction of all four limbs with the trunk" was positioned to be least exposed to the sun. A similar apparent anomaly is that head and axial hair are just as likely to be sources of infection yet seem to be peculiarly resistant. Possibly, localized secretions inhibit infection.

At this point, a return to the peculiar distribution of skin glands in modern humans is in order (see also chapter 2). The main gland types are sebaceous, apocrine, and eccrine (there are also tear-producing lacrimals, secretory surfaces producing wax inside the ears, and, of course, female mammary glands). Of all these, sebaceous glands seem to have the most functional relationship with hair. Exuding into the follicle, they oil the hair shaft. That they are particularly dense around the anus, mouth, and eyes suggests that they probably help inhibit infection. In humans, sebaceous glands develop during puberty, when they become large, active, and numerous. By contrast, they are smaller and fewer in African apes. The hyperdevelopment of a hair-serving gland in a vestigially haired primate seems paradoxical. Perhaps the gland's antibiotic functions have been underrated by medical science because the glands' protective action, however expressed, is so integral to a healthy skin that it has remained unremarked and undefined. The apocrines, like sebaceous glands, develop in humans only at puberty and, even then, remain restricted to nipples, genitals, armpits, and ears (in spite of being generously distributed over the skins of other mammals and primates). This peculiar distribution can be explained if human babies have retained some sensitivity to scent in their attachment to their mother's breast. Likewise, human courtship (which still relies on cues that are partly olfactory) could have favored the selective retention of glands that release their scented, milklike fluids in response to sex. Health is not normally compromised by the presence of apocrines in the groin and armpits, but their overall decline in humans may well signify the supression of pores that could have been potential avenues for infection.

The third set of skin glands, eccrines, are not associated with hair but excrete minute traces of antibiotic, urea, and salt diluted in water. It is this watery fluid that is normally labeled as "sweat" in humans ("sweat" in other animals is typically produced by the apocrine glands). The primary function of "sweating" in humans is unambiguous: it is an effective method of cooling. Just as important, eccrine sweat is the vector for an antibiotic protein called dermicidin that is known to be effective against at least four common species of bacterium (61). The switch from apocrine to eccrine sweating can be regarded as a species-specific adaptation and improvement of our immune system. Eccrine glands also respond to a wide variety of psychological stimuli (mostly excitement and fear), which is consistent with their evolutionary origin in the pores, palms, and pads of soft-footed and climbing mammals. Dampening of the digits presumably improved grip and traction during any sort of emergency and, if other mammals share eccrine antibiotics, served to protect tissues from wounding or bruising. In a uniquely intimate relationship with an adult lioness, Joy Adamson (62) felt her pet's paws become damp whenever a loud noise or unusual event made the animal "nervous." Many of us have experienced similar dampening of our palms, but our eccrines are not confined to our hands and feet; they are spread liberally all over the body. African apes also have a generous share of eccrines, mixed in with much the same number of apocrines. Our own sweat system is unique in its wholesale switchover from apocrines to eccrines. Indeed, eccrine sweating has been described, together with speech and bipedalism, as one of the truly unique characteristics of humans (63). This peculiarity has been invoked as evidence for an oceanic aquatic past on the argument that it developed to shed excess salt (64). Instead of an imaginary marine challenge to the very earliest hominins, the multiplication and spread of eccrines looks more like another physiological response to a typically modern, human-induced problem. I think it was simply elaboration of a preexistent trait that served to protect and moisten bare skin through much finer, less easily infected pores. As group numbers rose and campsites became less temporary, modern humans were increasingly exposed to infection. Messy, contaminated pelts were selected against, favoring progressively less hair. This in turn called for a cleansing and cooling system that was more appropriate to a predominantly bare skin. Copious sweating on a relatively hairless body not only cools it, it dilutes the concentration of potential contaminants with a dilute bath of antiseptic. Because extravagant sweating demands frequent drinking, the use of water (in both its pristine and sweated-out forms!) has become our main method of cooling off. In evolutionary terms, copious sweating may have helped shift body hygiene from a waterproof coat and "dry fur grooming" to the "wet skin cleansing" that is now characteristic of most people. Although the discussion dwells on sweat and hair, here is an aside: eyebrows. The visual and expressive value of eyebrows is obvious, but this could be subordinate (or at least equal) to another function: brow hairs help trap and divert sweat that might otherwise run into the eyes and interfere with vision.

While the superdevelopment of eccrine glands in modern humans is, I think, a late development linked with temperature control and hygiene, these glands may have played a more subtle role in hominin evolution. So long as ape forelimbs supported the front end of a heavy body, the knuckles, if not the hands, were tough and calloused. Dry, thickened skin is not unhealthy, but skin that is dry and inflexible provides poor traction and is much less sensitive to touch. For example, people often lick their fingers if they want to make their fingertips stickier, as when leafing pages or sorting cards. Faintly damp, softened fingertips are much more sensitive than dry, calloused ones, and I have already remarked that the manufacture of many prehistoric artifacts must have depended on enhanced fingertip sensitivity as well as manual dexterity. Eccrine glands would have joined textured fingertip skin and Meissner's corpuscles as preexistent but essential components in the elaboration of manual skill.

Hairless humans may have partially compensated for the loss of an external layer by developing an internal one. An extensive subcutaneous layer of fat in modern humans has been likened to blubber and ascribed to a marine aquatic phase. However, insulation against cold can be as effective against cold air as against cold water. Darwin marveled at the naked, body-painted Patagonian Amerindians. One, when asked how he could stand ice-cold winds without clothes, had a brief answer: "me all face."

Darwin favored sexual selection as the prime force behind our peculiar hair patterns. This is not inconsistent with the ideas that have just been outlined. If a bare skin is more likely to stay healthy, an incidental benefit for an observant and visually oriented species is that its display, especially in individuals of breeding age, can advertise youth and good health. This supports the aesthetic argument, and visual signaling might have been an important side effect of hairlessness, but is unlikely to be the primary explanation. If the mitigation of disease and parasites was the main reason for hairlessness ,it would be another instance of "self-making" in that selection for physical change was in response to a set of environmental conditions that were essentially self-inflicted, if not self-made.

A less clear-cut relationship between culture, disease, and adaptive change might have attended the lighting of fires in caves, shelters, or dwellings. In those areas where people lived their nights out in smoky enclosures, it could be predicted that some resistance to smoke-induced diseases of the respiratory tract might arise. Although there may be hints that some contemporary peoples are better able to tolerate smoke than others, there is, as yet, no evidence for an evolutionary dimension. However, there can be little doubt that smoke-sensitive individuals would have been at risk in many prehistoric encampments and might even have succumbed. Their deaths, like those of skin-ulcered campers, would have expressed selection for survival in a man-made environment.

Studying how local populations seem to have adapted to self-made stresses quickly suggests reasons for the uniqueness as well as the diversity of "self-made" modern humans. Portraits that acknowledge the true breadth of that diversity require entirely new perspectives on the geography and history of our dispersals out of Africa (and back again). Telling the story of those dispersals will lie very much in the hands of geneticists, but theirs will be lifeless portraits if they are not set in the context of innumerable local encounters with challenging habitats. The geneticists will identify small bands of prehistoric adventurers passing through narrow genetic bottlenecks during their dispersal all over the globe; but it will be in the details of their adaptation to the far north, mountains, humid forests, and scorching seasides that we will develop a new respect for our own diversity. Above all, the exploration of physiological and physical diversity may help us reach a more truthful understanding of those much misunderstood entities we call "races."

The diversity we see in Moderns has been the combined outcome of very recent and very ancient events. As I tried to explore in my earlier book, a better understanding of our evolved nature will be an ever more pressing imperative if we are to make the best of it rather than submit to its more destructive traits. It may also help us search for a more truthful reflection of that stranger who stares back at us from the mirror. I have come a long way from my ground ape self, but on the pathway to self-knowledge there is still a long hard walk ahead.

REFERENCES

1. Darwin, C. 1871. *The Descent of Man and Selection in Relation to Sex*. London: John Murray.
2. Leakey, L.S.B., P. V. Tobias, and J. R. Napier. 1964. A new species of the genus *Homo* from Olduvai Gorge. *Nature* 202: 308–312.
3. Morwood, M. J., P. B. O'Sullivan, F. Aziz, and A. Raza. 1998. Fission-track ages of stone tools and fossils on the east Indonesia island of Flores. *Nature* 392: 173–176.
4. Trinkaus, E., ed. 1989. *The Emergence of Modern Humans: Biocultural Adaptations in the Later Pleistocene*. Cambridge, UK: Cambridge University Press.
5. Morgan, E. 1995. *The Descent of the Child: Human Evolution from a New Perspective*. Oxford: Oxford University Press.
6. Whiten, A. 1994. Grades of mind reading. In *Children's Early Understanding of Mind: Origins and Development*, ed. C. Lewis and P. Mitchell, 47–70. Hillsdale, NJ: Erlbaum.
7. Zihlman, A. 1981. Women as shapers of the human adaptation. In *Woman the Gatherer*, ed. F. Dahlberg, 75–120. New Haven, CT: Yale University Press.
8. Howell, N. 1982. Village composition implied by a paleodemographic life table: the Libben site. *American Journal of Physical Anthropology* 59: 263–269.
9. Bogin, B. 1990. The evolution of human childhood. *Bioscience* 40: 16–25.
10. Renfrew, C. 1987. *Archaeology and Language*. Cambridge, UK: Cambridge University Press.
11. Kingdon, J. 1993a. *Self-made Man. Human Evolution from Eden to Extinction?* New York: John Wiley & Sons.
12. Mellars, P., and C. Stringer, eds. 1989. *The Human Revolution: Behavioural and Biological Perspectives on the Origins of Modern Humans*. Edinburgh, UK: Edinburgh University Press.
13. Whiten, A., and R. W. Byrne. 1991. The emergence of metarepresentation in human ontogeny and primate phylogeny. In *Natural Theories of Mind: Evolution, Development and Simulation of Everyday Mindreading*, ed. A. Whiten, 267–281. Oxford, UK: Basil Blackwell.

14. Welford, A. T. 1968. *Fundamentals of Skill*. London: Methuen.
15. Abbate, E., A. Albianelli, A. Azzaroli, M. Benvenuti, B. Tesfamariam, P. Bruni, N. Cipriani, R. J. Clarke, G. Ficcarelli, R. Macchiarelli, G. Napoleone, M. Papini, L. Rook, M. Sagri, T. M. Tecle, D. Torre, and I. Villa. 1998. A one-million-year-old *Homo* cranium from the Danakil (Afar) Depression of Eritrea. *Nature* 393(6684): 458–460.
16. Conroy, G. C., C. J. Jolly, D. Cramer, and J. E. Kalb. 1978. Newly discovered fossil hominid skull from the Afar Depression, Ethiopia. *Nature* 307: 423–428.
17. Johanson, D. C., and M. A. Edey. 1981. *Lucy: The Beginnings of Humankind*. New York: Granada.
18. Day, M. H. 1989. *Guide to Fossil Man*. 4th edn. London: Cassels.
19. Stone, A. C., R. Bonner, and M. Hammer. 1998. Y chromosome diversity in *Pan troglodytes*. AAPA abstracts of contributors to Dual Congress.
20. Horai, S., T. Gojobori, and E. Marsunaga. 1987. Evolutionary implications of mitochondrial DNA polymorphism in human populations. In *Human Genetics, Proceedings of the 7th International Congress*, ed. F. Vogel and K. Sperling, 177–181. Heidelberg: Springer-Verlag.
21. Horai, S., K. Hayasaka, R. Kondo, K. Tsugane, and N. Takahata. 1995. Recent African origin of modern humans revealed by complete sequences of hominoid mitochondrial DNAs. *Proceedings of the National Academy of Science USA* 92: 532–536.
22. Cann, R. L., M. Stoneking, and A. C. Wilson. 1987. Mitochondrial DNA in human evolution. *Nature* 325: 31–36.
23. Stringer, C. B., and C. Gamble. 1993. *In Search of the Neanderthals*. Thames & Hudson: London.
24. Stringer, C., and R. McKie. 1996. *African Exodus*. London: Jonathan Cape/ Pimlico.
25. Brookes, M. 1999. Apocalypse then. *New Scientist* 14 August 32–35.
26. Huffman, M. A., T. Nishida, and S. Uehara. 1990. Intestinal parasites and medicinal plant use in wild chimpanzees: Possible behavioral adaptation for the control of parasites. Mahale Mountains Chimpanzees Research Project Ecological Report No. 72.
27. Wrangham, R. W. 1995. Relationship of chimpanzee leaf-swallowing to a tapeworm infection. *American Journal of Primatology* 37(4): 297–303.
28. Lieberman, P. 1984. *Biology and Evolution of Language*. Cambridge, MA: Harvard University Press.
29. Lenneberg, E. H. 1967. *Biological Foundations of Language*. New York: Wiley.
30. Ruff, C. B. 1994. Morphological adaptation to climate in modern and fossil hominids. *Yearbook of Physical Anthropology* 37: 65–107.
31. Soffer, O. 1994. Ancestral lifeways in Eurasia. In *Origins of Anatomically Modern Humans*, ed. M. and D. Nitecki. New York: Plenum Press.
32. Astington, J. 1993. *The Child's Discovery of the Mind*. Cambridge, MA: Harvard University Press.
33. Chomsky, N. 1968. *Language and Mind*. New York: Harcourt Brace Jovanovich.
34. Bickerton, D. 1990. *Language and Species*. Chicago: University of Chicago Press.
35. Pinker, S. 1994. *The Language Instinct: How the Mind Creates Language*. New York: Morrow.

36. Pinker, S., and P. Bloom. 1990. Natural language and natural selection. *Behavioral and Brain Sciences* 13: 707–784.

37. Barkow, J., L. Cosmides, and J. Tooby. 1992. *The Adapted Mind: Evolutionary Psychology and the Generation of Culture.* Oxford, UK: Oxford University Press.

38. Deacon, H. J. 1989. Late Pleistocene palaeoecology and archaeology in the southern Cape, South Africa. In *The Human Revolution: Behavioural and Biological Perspectives on the Origins of Modern Humans,* ed. P. Mellars and C. Stringer, 547–564. Edinburgh, UK: Edinburgh University Press.

39. Trinkaus, E., and P. Shipman. 1993. *The Neanderthals.* New York: Alfred A. Knopf.

40. Day, M. H., and C. Stringer. 1982. A reconsideration of the Omo Kibish remains and the *erectus-sapiens* transition. In *L'Homo Erectus et la Place de l'Homme de Tautavel Parmi les Hominides Fossiles,* Vol. 1, 814–846. Première Congress Internationale de Paléontologie Humaine. Prétirage; CNRS.

41. Magori, C. C., and M. H. Day. 1993. Laetoli hominid 18: An early *Homo sapiens* skull. *Journal of Human Evolution* 12: 747–753.

42. Bodmer, W. F., and R. McKie. 1994. *The Book of Man. The Quest to Discover Our Genetic Heritage.* London: Little Brown.

43. Ruvolo, M., S. Zehr, M. Von Dornum, D. Pan, B. Chang, and J. Lin. 1993. Mitochondrial COII sequences and modern human origins. *Molecular Biology and Evolution* 10(6): 1115–1135.

44. Zhivotovsky, L. A. 2001. Estimating divergence time with the use of microsatellite genetic distances: Impacts of population growth and gene flow. *Molecular Biology and Evolution* 18: 700–709.

45. Dorit, R. L., H. Akashi, and W. Gilbert. 1995. Absence of polymorphism at the ZFY locus on the human Y chromosome. *Science* 268: 1183–1185.

46. Shen, P. D., F. Wang, P. A. Underhill, C. Franco, W. H. Yang, A. Roxas, R. Sung, A. A. Lin, R. W. Hyman, D. Vollrath, R. W. Davis, L. L. Cavalli-Sforza, and P. J. Oefner. 2000. Population genetic implications from sequence variation in four Y chromosome genes. *PNAS* 97(13): 7354–7359.

47. Hammer, M. F. 1995. A recent common ancestry of human Y chromosomes. *Nature* 378: 376–378.

48. Hammer, M. F., and S. L. Zegura. 1996. The role of the Y chromosome in human evolutionary studies. *Evolutionary Anthropology* 5: 116–134.

49. Hammer, M. F., T. Karafet, A. Rasanayagam, E. T. Wood, T. K. Altheide, T. Jenkins, R. C. Griffiths, A. R. Templeton, and S. L. Zegura. 1998. Out of Africa and back again: Nested cladistic analysis of human Y chromosome variation. *Molecular Biology and Evolution* 15(4): 427–441.

50. Lahr, M. M., and R. A. Foley. 1994. Multiple dispersals and modern human origins. *Evolutionary Anthropology* 3: 48–60.

51. Quintana-Murci, L., O. Semino, H-J. Bandelt, G. Passarino, K. McElreavey, and A. S. Santachiara-Benerecetti. 1999. Genetic evidence of an early exit of *Homo sapiens sapiens* from Africa through eastern Africa. *Nature Genetics* 23: 437–441.

52. Misra, V. N. 1987. Presidential address. *Proceedings of the 74th Indian Scientific Congress 1–24, Bengal, Calcutta.*

53. Frayer, D. W., M. H. Wolpoff, A. G. Thorne, F. H. Smith, and G. G. Pope. 1993.

Theories of modern human origins: The paleontological test. *American Anthropology* 95: 14–50.

54. Thorne, A. G., and M. H. Wolpoff. 1992. The multiregional evolution of humans. *Scientific American* April: 76–83.

55. Flannery, T. F. 1994. *The Future Eaters. An Ecological History of the Australasian Lands and People.* Chatswood NSW: Reed.

56. Hammer, M. F., and S. Horai. 1995. Y chromosome DNA variation and the peopling of Japan. *American Journal of Human Genetics* 56: 951–962.

57. Torroni, A., J. Miller, L. G. Moore, S. Zamudio, J. Zhuang, T. Droma, and D. C. Wallace. 1994. Mitochondrial DNA analysis in Tibet: Implications for the origin of the Tibetan population and its adaptation to high altitude. *American Journal of Physical Anthropology* 93: 189–199.

58. Byard, P. J. 1981. Quantitative genetics of human skin colour. *Yearbook of Physical Anthropology* 24: 123–137.

59. Loomis, F. W. 1967. Skin-pigment regulation of vitamin D biosynthesis in man. *Science* 157: 501–506.

60. Belt, W. 1874. *A Naturalist in Nicaragua.* London: Murray.

61. Randerson, B. 2001. Killer armpits. *New Scientist* 10 Nov.

62. Adamson, J. 1961. *Living Free.* London: Collins.

63. Sokolov, V. E. 1982. *Mammal Skin.* University of California Press.

64. Morgan, E. 1990. *The Scars of Evolution. What Our Bodies Tell Us about Human Origins.* London: Souvenir Press.

CHAPTER 10

In Conclusion

Confessions of a Repentant Vandal

Summary of evidence for a diverse radiation of hominids, of which modern humans are the lone bipedal survivor. Will thinking bipeds continue to survive? The need to understand the evolution of humans as that of just one more African mammal and the only survivor of at least 18 species of hominins. The need to understand the natural world and the workings of global environments and ecosystems. A global, process-oriented database needed to study and implement sustainability in all environments worldwide. Need for niche-thieves to learn how to live on their home planet without consuming it.

In 1969, when cinemas and televisions around the world screened two astronauts bouncing and lurching over a barren lunar stage, it was not their grotesque movements that impressed: rather, it was the old frontier image of two pioneering bipeds leaving tracks on new territory. Today, photographs of the trail of their boots in the dust of the moon recalls those of the 3.5 million-year-old footprints at Laetoli. Neil Armstrong's message "one small leap for a man, one giant step for mankind" crackled through space to an incredulous audience and, on their return, Richard Nixon hailed the astronauts' 7-day expedition to the moon as "The Greatest Week since The Creation." The rhetoric of feet carrying humanity to this theatrical climax seemed to invite an atavistic flashback to a storybook beginning: "we set off on two feet and look where it took us!" In triumphantly associating his scientists' prowess with events from the book of Genesis, a story in which two people trudge out of Eden, Nixon revealed the chasm between technological "neo-knowledge" and prescientific "archaeo-knowledge."

He was not alone in turning to Biblical symbolism. Even the most Darwinian of scientists like to use metaphors from Genesis, the favorites being Eden, Adam, Eve, and Exodus (1–3).

Among many interpretations of the Biblical Genesis, the most credible symbol is of Adam as Everyboy and Eve as Everygirl, while the Garden of Eden is Childhood and Expulsion the inevitable arrival of Adulthood. In that context the Fruit of the Tree of Knowledge becomes the arrival of Puberty, but why not take one more liberty with this infinitely malleable myth? Suppose then that Eden is a sort of generic Human Childhood, where our ancestors' infinite capacity for making up stories to explain the meaning, workings, and origins of life must eventually be displaced by the realities of a Universe governed by external laws and structures, as discovered from the Tree of Knowledge: science. In this retelling, a new life outside of Eden will no longer be subject to the Divine Ordinances of our mythic past but will have to be lived with the discipline of scientifically verifiable facts! In place of supernatural comforts, Everyboy Adam and Everygirl Eve must recognize their "lowly origin" in the molecules and soils of this small Earth. The contemporary world is caught up in a similar mythic bind. The vast majority of people, fearful of where knowledge might take them, are captive to religious and educational traditions that have little or no connection with the scientific and technological culture on which, for good or ill, their future survival now depends.

In that future, our generation might be of little more significance than relay runners for the genes of our descendants, were it not for the intellectual and environmental legacy that we leave behind. Each generation inherits liabilities as well as benefits from their predecessors' occupancy of a world that becomes progressively more shop-soiled and second- or third-hand. In the immediate future, liabilities threaten to overwhelm assets. There are many reasons for such a sorry legacy to our descendants. Of those that are relevant to the thesis of this book, the most obvious is the absence of a coherent biological vision of what it is to be a human being. As Richard Dawkins puts it, "Intelligent life on a planet comes of age when it first works out the reason for its own existence" (4). Another liability is the lack of any consensus for a disciplined role in this extraordinary mammal's relationship with nature. In calling for more scientifically oriented policies, Helena Cronin remarks that "our evolved minds are designed to help us to react appropriately to the different environments that we find ourselves in. It is thanks to our genetic endowment, not in spite of it, that we can generate our rich behavioral repertoire. Change the environment and you change the behavior. So an understanding of the evolved psychology of our species—of our motivations and desires— is vital for political action" (5). Likewise, our shallow perspectives on a history that is overwhelmingly prehistoric suffer from too slavish a dependence on the authority of documents, narratives, and myths and too little respect for the authority of nature itself. Science, also, is too often diverted to increase power over nature rather than understand it on its own terms. We have become too used to turning to science and technology for new expedients to overcome immediate problems as they arise.

It was as a true Victorian that Darwin sang the praises of the uniquely human intellect "which has penetrated into the movements and constitution of the solar system" (6). He went on to hope "for a still higher destiny in the distant future," but our destruction of global environments makes this an increasingly unlikely hope. The ability to comprehend and model the workings of all sorts of natural (including astronomical) processes implies a still larger, self-conscious capacity and potential for understanding natural environments and human affairs. I would argue that our niche-stealing nature severely compromises our ability to do this. Niche-stealing is especially relevant when it comes to managing habitats and behavior, not on our own self-interested terms but in the light of nature's own rules. We need to understand the structure of our environments, past and present, a lot better than we do today before we can begin to fulfill the potential Darwin had hoped for. Above all, we need to discover our biological history and get to know our evolutionary nature.

Before turning to some of the environmental and social implications for a Darwinian outlook on human origins, I must return to the task of coming up with a concluding definition of what sort of an animal I think I and you are. That is a necessary conclusion to the self-imposed task of exploring my bipedal legacy. In the biographical format of the preceding chapters I presented a succession of animal or primate forms. Each displayed an anatomical specialization that signified a property or characteristic that I could identify and integrate into my self-perception of what it is for me to be a vertebrate, a mammal, a primate, a cousin of the apes, or a human. These details have had to be teased out of diverse matrices and, for their first appearance in the body of the text, illustrate quite specific points. If my collection of bits and pieces has tended to resemble a spare parts catalogue, that is because I wanted to stress the incremental nature of evolution as a whole. Step by step, I have assembled those parts into a sort of evolutionary collage of my typically human anatomy. Here, they have been put together in that graphic convention known as the "double-page spread": images clipped out of the chapters in which they first appeared and then displayed as scattered visual details of my multiple self-portrait (figure 10.1). Likewise, previous chapters have examined the fossil record piecemeal, effectively postponing a general overview and a broad outline of the fossil evidence for human evolution. It may be just as well to remember that a printed book may not be the most appropriate and well-adapted medium for such a task, but it is necessary at this point to pinpoint data and conclusions, adding observations that help sum up my composite self-portrait. A chronological list and time chart of generally recognized fossils follows with the briefest of summaries or reminders of their apparent significance.

My account owes a lot to the most recent of fossil discoveries. These have forced fundamental changes in the way human evolution is seen: most notably, the tree of human descent was, until recently, envisaged as a single trunk with some small twigs to accommodate fossils that were portrayed as aberrant offshoots. Today, at least 18 species of fossil humans and bipedal hominins have been described, a high proportion of them from the eastern half of Africa (7). It is no longer possible to believe that all these bipeds are on our own line of descent. More important, our own line may more truthfully represent an early aberrant offshoot, while most of the others maintained more of their ape heritage in spite of taking to the bipedal way of life.

To backtrack to ape beginnings in Africa, a more mobile lineage of apes found their way out to Asia, and after several million years of separate development outside Africa, a single species found its way back to its mother continent. There, it gave rise to an astonishing variety of apes and hominins, all of which have become extinct except for humans (and, for the present, gorillas and chimps). Their common ancestor was likely to have been a widely distributed and relatively unspecialized ape. It can be predicted to have inhabited the moist woodlands that ranged from the Atlantic to the Indian Ocean, because similar woodlands were the preferred habitat of its most immediate precursors: the Miocene tree apes, or Dryopithecines, of western Eurasia. Of all its descendants, chimps are probably closest to that ancestor in general appearance, but it is in their adaptation to exploiting giant trees in lowland rainforests (rather than the lighter vegetation of less humid woodlands) that chimps probably differ most from their earliest African ancestors. So far, the only fossils from this period are the 9 million-year-old fragments of the "Samburu Ape," *Samburupithecus koptalami*, and the much younger 6–7 million-year-old "Toumai," *Sahelanthropus tchadensis*, a short-faced hominid with blunt, somewhat human-like teeth. In its cranial anatomy this fossil from north of Lake Chad provides a rough approximation for the common ancestor of chimps and humans.

The earliest, very fragmentary fossils of the "Millennium Ancestor" (*Orrorin tugenensis*) dated about 6 mya, come from the Eastern Rift Valley near Lake Baringo, Kenya. Found as recently as August 2000, this species seems to have been upright, and its teeth are more like those of humans than those of apes. There is still a lot to be learned from these newly discovered specimens, but a part of their significance is that they put a date on the earliest appearance of "uprightness." They also confirm the importance of East Africa as a pivotal region for human origins. What little we know of this hominin's anatomy and environment is broadly consistent with my proposal that eastern forests were the setting for hominin emergence. At the very least, the Millennium Ancestor's provenance in equatorial East Africa narrows the choice of where hominins differentiated from their ape ancestor to the eastern side of the continent, probably within the tropics. Its habitat, insofar as it has been described, was a mosaic of wooded and more open country; *Orrorin* almost certainly inhabited the wooded part. Because of its very early date, this species may be relatively close to the point at which human and chimpanzee ancestors diverged, the point from which all other hominins derived. However, its thick tooth enamel could hint at *Orrorin* being more directly ancestral to humans. A specifically human ancestral position could also imply that the eastern Rift Valley, where it was found, was within the human lineage's region of origin. In the absence of anything more than a few fragments from a single fossil locality, it is impossible to guess how widespread these, the earliest type of erect-backed hominins, were. Such an early date has pushed the timetable of human evolution back, making all the other "early" fossils look a lot younger and increasing the likelihood of fragmentation into two, three, or even more lineages well before these species appear in the fossil record.

The name of the next oldest fossil, *Ardipithecus ramidus*—the "rootstock ground ape"—reflected the hope of its describers that they had found an ancestor to humans. Before the discovery of the 6 million-year-old millennium ancestor, it was plausible to model the "rootstock ground

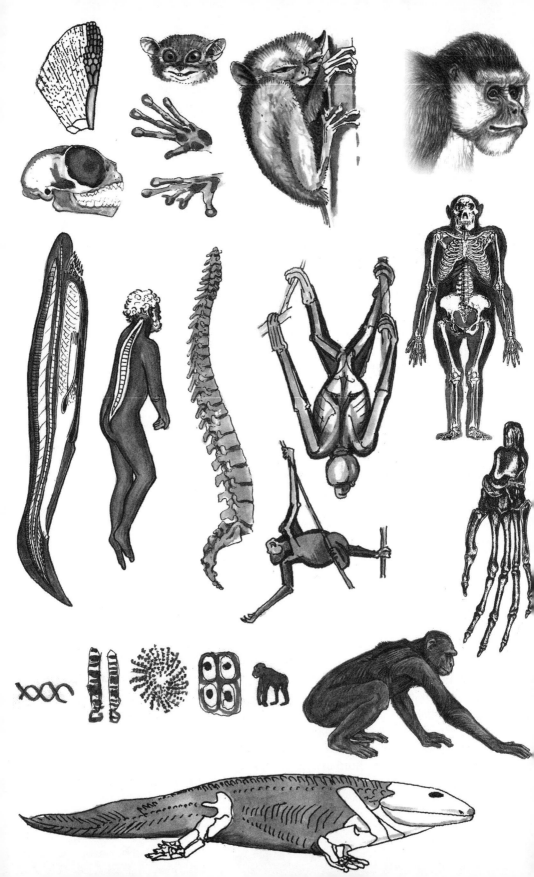

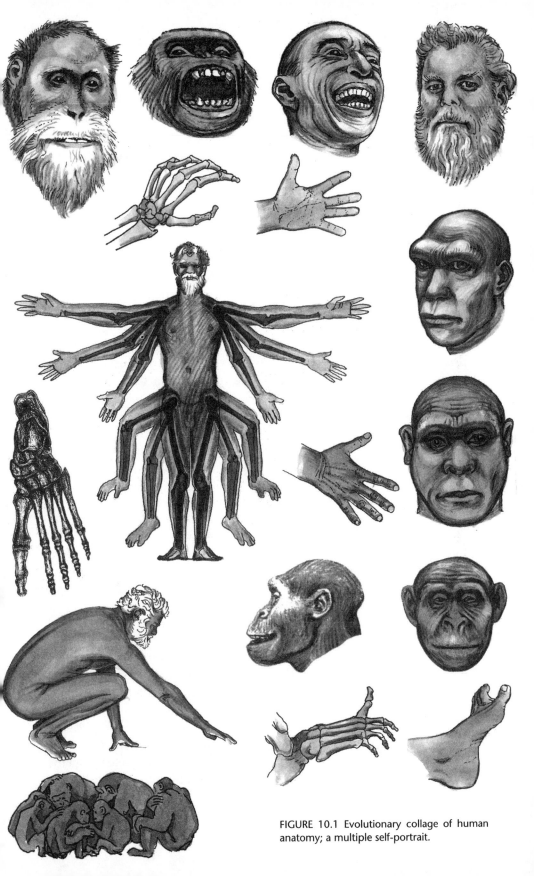

FIGURE 10.1 Evolutionary collage of human anatomy; a multiple self-portrait.

ape" as the ancestor of all hominins. However, its age, at 5.8 to 4.4 million years, is later than Orrorin, and its discovery on the furthest northeastern slopes of the Ethiopian mountain massif suggests that this was the stranded relict of an early phase in the evolution of bipedality. Quite possibly, it resembled many other Ethiopian species in being the only known member of a lineage unique to this north eastern region, a local, forest-dwelling endemic. Unlike the Millennium Ancestor, the teeth of this species are thinly enameled, more like those of apes, and it is possible that the species was not fully bipedal. Nonetheless, its existence hints at an intermediate condition of ground dwelling that may help us bridge the gap between four-legged apes and two-legged hominins.

The volcanic Eastern Rift Valley in Kenya, with its valley walls and hills eroding away into shallow inland lakes, is a rich source of fossils. The next significant hominin is the 4.1 million-year-old "Kanapoi hominid" from the shores of Lake Turkana. Because its discovery dates back to a time when the hominin tree was still thought to have a single trunk, this species was named *"Australopithecus" anamensis* (because all prehuman fossils were lumped together as australopithecines). At 4.1 million years, the "Kanapoi hominin" is too close in time and too different in its anatomy to be any kind of descendant of the "rootstock ground ape."

The clearest affinities of the "Kanapoi hominin" are with the species commonly called *"Australopithecus afarensis."* Much confusion surrounds the name of the latter species, because it (or something extremely similar) was originally described in an obscure German paper and from some fragmentary fossils that had been collected in the 1930s and named *"Praeanthropus."* Under its later name, *"Australopithecus afarensis"* became world famous when the greater part of a skeleton was discovered in Afar in Ethiopia and nicknamed "Lucy." "Lucies" first appear in the fossil record at about 3.6 mya or earlier. On present evidence, the "Lucy" and Kanapoi lineages seem to have given rise to the "Nutcracker" or Robust australopithecines. These are generally acknowledged as belonging to a distinctive lineage separate from that of humans. If the "Kanapoi hominid" and "Lucies" were indeed the earlier representatives of this very successful line of ape-brained hominins, we can envisage the entire lineage as the pioneers of a bipedal way of life. On present evidence, plausible human bipeds appear later in the record and might even have begun as inferior walkers and runners.

A mandible fragment from Chad has been called *"Australopithecus" bahrelghazali*, a form that may be closely related to the Lucies but needs more material and more secure dating before it can be interpreted properly.

The "Nutcrackers" evolved later than "Lucy" and diverged into several species that are called *Australopithecus* by some authorities and *Paranthropus* by others. Some of these are probably local regional forms, but some of the earlier types, notably *A. garhi and A. aethiopicus*, are broadly intermediate between Lucy and the extremely robust species. Nutcracker fossils have been found between 2.5 and 1 mya, overlapping various other species of humans and near-humans. Judging from the frequency with which they turn up as fossils, they must have been quite common, and a few must have been quite widespread. These were big-headed, massive-jawed bipeds with huge molar teeth; they are thought to have been adapted to process a diet of roots and other subterranean plants and animals. Robust skulls illustrate a recurrent trend in hominin evolution, namely a tendency for adaptation to specialized diets to be reflected in anatomy. This trend is normal for most animals, but not for our own Modern lineage, in which the invention of tools and technological solutions to subsistence took precedence over anatomical adaptation. The nutcrackers probably used wood, bone, and horn tools to excavate but still needed big teeth and powerful muscles to chew their foods.

The original "South African Man-ape" or *Australopithecus africanus*, was the first species of hominin to be discovered in Africa (in 1925). It survived, as a variable or gradually changing species, from 3.6 to 2.3 million years. It was exclusively south African and (with one or two exceptions) is generally regarded as being separate from "Lucy" and the "Nutcrackers." Its teeth and forehead are less apelike than those of Lucy and somewhat closer to those of early humans. The "South African Man-ape" has appropriate characteristics to be ancestral to a larger brained, tool-making East African species. In 1964, when Louis Leakey discovered the remains of this very humanlike creature in Olduvai Gorge, East Africa, it was a sensation. The fossils were more than 2 million years old and were found in the same layer as crude stone tools. Because tools were then thought to be an exclusive marker for humanity, Leakey called his fossil *Homo habilis*. This diminutive "human" was nicknamed "the Handy-man" because of his supposed skill in chipping pebbles. The "Handy-man" has remained a model for the earliest human until very recently, but that status is now being questioned (8). As it turns out, this was not the only larger brained species to develop from an earlier small-brained look-alike. There is now another candidate for the early ancestry of humankind. Aged at 3.5 million years, the "Kenya Flat-face" (*Kenyanthropus platyops*) is similar in age to both the "South African Man-ape" and "Lucy." Yet it is unequivocally different from both, and its molar teeth are particularly humanlike. How adept it was at walking is not known, but its habitat on the shores of Lake Turkana (formerly Lake Rudolf) suggests mainly terrestrial activity.

Apart from its small brain, the Kenya Flat-face is remarkably similar to a much later but big-brained species, "Rudolf man," or *Homo rudolfensis* (after the older name for Lake Turkana). Both species have long faces that are peculiarly flat with proportionally large front teeth. "Rudolf man" lived between 2.4 and 1.85 mya—a span very similar to that of the "Handy-Man" which is known from 2.3 to 1.7 mya. This species also inhabited the Lake Turkana shore.

The discovery that brain enlargement has taken place independently in two distinct hominin lineages has removed yet another criterion for the evolution of modern humans. Now, it seems, at least two types of humanlike hominins began to get big-brained, but which one led on to wholly modern humans? The two competitors for human ancestry, Rudolf man and the Handy-man, in spite of seeming to derive from different, more primitive ancestors and with different characteristics, could on present evidence develop into *Homo*, but the bias is for *habilis*.

Ecological differences between the two candidates for human ancestry are significant. Flat-faces and Rudolfs are known only from the shores of lakes on the floor of the Rift Valley. For many years now, there have been suggestions that humans went through a semiaquatic phase of evolution. Were this the case, Rudolf man is the only lineage with any potential for such a history; but it is a mark of our ignorance that we still have no more than a couple of skulls and no knowledge whatsoever of the lifestyle of this species.

By contrast, we do know that the South African Man-ape lived in a temperate climate. The fossils of its descendant (an alternative candidate for human status), the "Handy-man," are typically found in upland eastern Africa during cooler periods.

Given that either lineage, but not both, could be human ancestors, we have a rather dubious choice between lakeshores and uplands as our ancestral environments. This in turn implies that opposite climates might have favored the separate emergence of the two apparent humans.

In one model where humans emerged from a "Handy-man" with southern origins, our evolution should, theoretically, have been favored by globally cooler temperatures. During an ice age, large areas of upland Africa would have become much more temperate and consequently attractive for temperate-adapted species.

If humans emerged from tropic-originating "Rudolfs," our evolution was more likely to have been favored by humid climates because surface waters would have been much more widespread. Overall temperatures would have been substantially higher so that a formerly shade-dependant primate might have offset a more exposed existence by living close to lakes and rivers.

Two parallel trends toward bigger brains? Which one led to us? Our uncertainty underlines how the whole field of human origins is in a state of upheaval (figure 10.2).

It is at this point that a significant gap occurs—but it is a gap in the physique and anatomy of the fossils, not a gap in time. Both the previous hominins overlap with another species that is quite clearly ancestral to modern humans. As was stressed earlier, this species seems to arrive in the fossil record, fully formed, and much more like Moderns than anything that went before.

The first hominin to leave Africa, its Asiatic representatives, were Java and Pekin Man, types of *Homo erectus*. In Africa, where the earliest fossils are found, at about 1.8 mya, this form of human is now known as *Homo ergaster*, literally "The Work-man," but both African and Asian members of this cluster are known colloquially as "Erects." Apart from the size of their brains (and some other, much less conspicuous but very important differences), Erects were almost identical in general body build to modern people. There is still a school of thought that believes regional Erects gave rise to modern human "races"; however, both fossil and genetic evidence is now overwhelmingly against such a belief. We can choose to call distinctively African, Javan, and Chinese Erects "species" or "subspecies," but only one of them, the African line, gave rise to Moderns. During the nearly 2 million years that Erect humans were around, they developed hand axes and other stone tools, they probably devised some form of watercraft, and they learned to control fire. It was probably their use of fire that assisted their rapid spread from one side of Asia to the other. There is now a broad consensus that although humans have inhabited both Africa and Eurasia for nearly 2 million years, the evolution of wholly modern humans remains primarily an African and much more recent story.

A modified form of the "Work-man," known as "Heidelberg man," or *Homo heidelbergensis* (after a German fossil of this big-jawed human), emerged about 1 million years ago and became widespread in Eurasia.

In Europe, Heidelbergs developed, step by step, into those massive, big-brained ice-age humans well known as "Neanderthals." Neanderthals differentiated from their more tropical cousins during a protracted ice age and became adapted to cold northern habitats in western Asia and Europe. In their anatomy, their habits, and their habitat, Neanderthals are a vivid illustration of the influence of climate on human evolution. In this

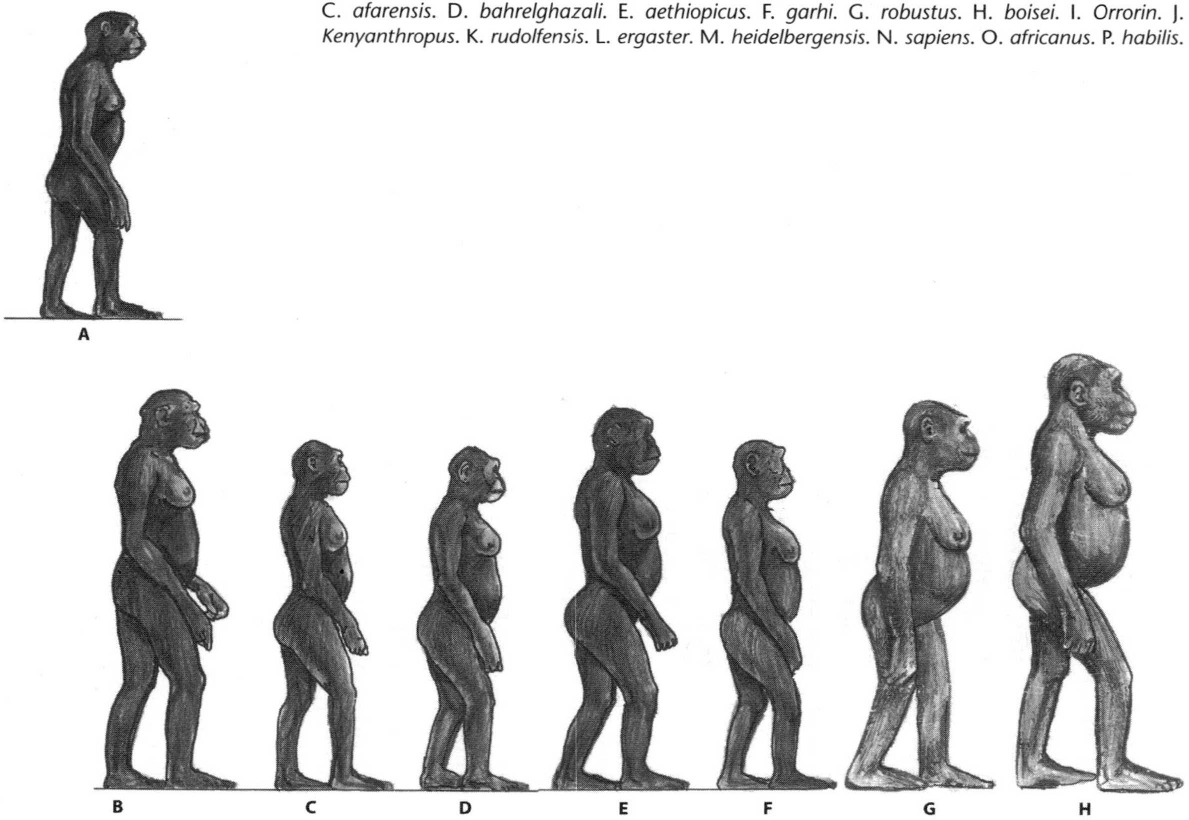

FIGURE 10.2 The certainty of ancestors, the uncertainty of ancestry. A. *ramidus*. B. *anamensis*. C. *afarensis*. D. *bahrelghazali*. E. *aethiopicus*. F. *garhi*. G. *robustus*. H. *boisei*. I. *Orrorin*. J. *Kenyanthropus*. K. *rudolfensis*. L. *ergaster*. M. *heidelbergensis*. N. *sapiens*. O. *africanus*. P. *habilis*.

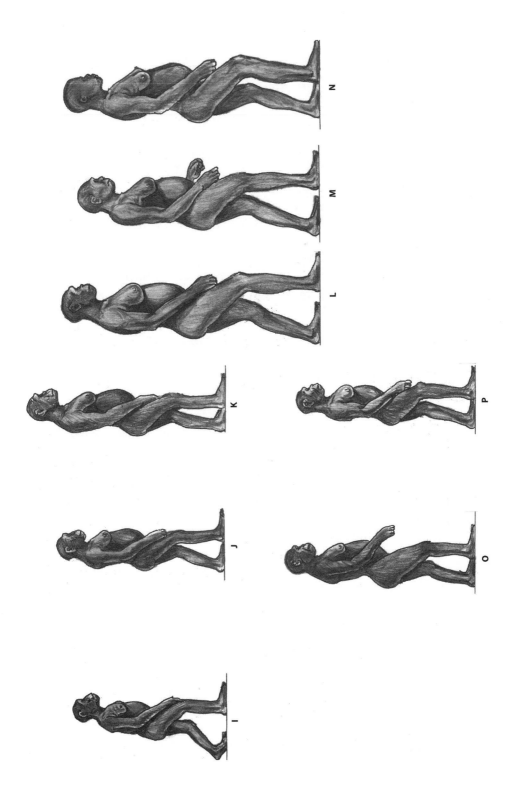

case a slow descent into a long, bitter ice age (which lasted about 20,000 years) demanded quite extraordinary levels of both anatomical and cultural adaptedness from a species that was derived directly from a tropical primate.

In Africa, anatomically modern humans evolved from another population of Heidelberg-like humans, but the exact area of Africa, the cause, and the mechanism remain a mystery. What *is* suggested, from contemporary genetic studies, is that our ancestors went through something of a genetic bottleneck at this point. How Moderns dispersed out of Africa and arrived at their present diversity is a story that is only just beginning to unfold.

This brief summary of ideas surrounding an abbreviated list of fossils offers a personal guide to a necessarily controversial field. The new discoveries—"Toumai," the "Millennium Ancestor," the "Rootstock ground ape," and the "Kenya Flat-face"—render our own evolutionary progress through an ever-bushier thicket substantially more complex. An entirely new perspective has intruded. For the greater part of their history, we must now see humans and near-humans as relatively abundant, diverse, and regionally localized animals. Our ancestors were as African as the numerous species of antelopes or elephant shrews are today. Furthermore, we are not just Africans, we are eastern Africans.

It is a novel concept to find that we are African in the same sense that kangaroos are Australian or toucans are South American. That discovery gives a special importance to trying to understand what it was about Africa and its habitats that gave rise to bipedal apes in the first place and then favored their radiation into such an astonishing variety. It is a variety that will, no doubt, be augmented by the discovery of still more species of fossil humans in the years ahead.

In the preceeding chapters, I have tried to relate the fossil record to some of the more enduring features of African geography and ecology in a quest to reveal more about the setting for hominin evolution. The discovery that we are the single survivor of at least 18 species is startlingly new and renders redundant much of what has been published before. To know that we are just one of many types of humans or near-humans has many repercussions for the way we look at ourselves and at our relationship with the natural world (9). It also raises entirely new questions as to *why* we occupy such a lonely position after such a diverse history. It raises important issues for the future of humanity. Astronauts tramping the moon for knowledge may have been a distraction from the much more urgent task of discovering how we fit into the order of things back here on Earth.

As we learn to acknowledge that we have emerged from the very specific ecology of Africa, we must recognize our self-interest in learning much more about Africa and its primates. A distinction that has been very important for the evolution of ourselves and all its other primates (which are essentially tropical animals) is that Africa, of all the continents, has by far the largest land area within the tropics. Because it has drifted less from its equatorial position than other continents, there has been more scope for the evolution of a complex primate community.

The tropics of Africa are therefore densely crowded with primate species. When I lived in Uganda, I could find as many as 16 species of primate within about 10 km of a local landmark (Mongiro hot springs, Semliki, the only locality in the world with such primate richness) (10), and there are more than 80 species in the continent as a whole (11). Early humans, as one more primate, therefore had to evolve their peculiarities in a climate of intense competition, not only from other primates but also from a rich community of large tropical mammals. As in a great city, a crowded environment is a complex, demanding, and challenging place to be. Africa owes much of its biological complexity to this fact, to its relative stability in relation to the equator, and to its ancient pattern of basins.

Basins and the rifting that has been associated with such a rumpled land surface are fundamental to understanding Africa's geology and geography. Basin ecosystems are also profoundly relevant to speciation in animals and plants.

Modern habitat types are still a fair guide to the main categories of past vegetation, but we know that there have been huge fluctuations in climate and that the boundaries of these zones have changed dramatically. I have stressed that the primary dynamic is of arid climates and arid-adapted communities moving north-south, while equatorial humid climates expand and contract west-east. For both extremes and for many intermediate environments, there are residual areas or pockets, appropriately visualized as refugia (or places where species accumulate or retreat). For various reasons, these are localities that escape the full impact of climatic change and conserve locally adapted communities right through successions of climatic vicissitudes. The proliferation of fossil hominin species has forced us to consider the possibility that coexistent humans, like many other species, can belong to older or more recent communities and may owe their survival to repetitive fluctuations in climate. Thus, when temperate communities spread up from the south to central Africa, or when tropical forest communities extended far and wide during humid periods, all the hominins involved in each upheaval were integral members of their community. Like populations of other animals, those of hominins always originated in specific localities; some remained strictly local types, but others spread widely once conditions were right for them.

Thus, the nutcrackers never got out of sub-Saharan Africa, and the Neanderthals never left Europe and western Asia, their region of origin. By contrast, Erect and modern humans spread right across continents. In the first case, it would seem that species constrained by a limited range of foods and specialized feeding techniques remained restricted to zones where their specializations yielded a good living.

Species with more adaptable feeding and foraging skills could spread over a wider range of habitats and territory. It is clear that our own ancestors belonged to the latter category. We know that later human fossils were much less tied to specific habitats and regions, but there is still much to be learned about the ecological background of all hominins, from the Millennium Ancestor, to Erects in Asia, to Neanderthals in Ice Age Europe.

Among the many lessons that the very new proliferation of fossils offer us is the need for ever greater attention to be focused on the details of ecology and geography if we are ever to understand the long drawn-out origins of bipedalism and its immensely complex consequences. This book joins a long tradition of trying to characterize the human condition, but its ecological and African perspectives (and recurrent comparisons with other African animals) has put special emphasis on the ecological role of humans as members of a once species-rich group of primates. I have pursued my primates' highly specific habitats and places, not consigned them to generalized expanses.

Like the explanations for bipedalism that were lampooned in Tuttle's titles, there have been numerous efforts to come up with single-character one-liners to sum up the human animal (12). These are equally open to satire: Killer-Ape, the Great Cooperator, Aquatic Ape, Naked Ape, Sexual Ape, Moral Animal, Ludic Creature, Man/Woman the Hunter, Hominid Scavenger, Chimp Studying to Become a Jackal, the Third Chimpanzee (figure 10.3). The best I have been able to come up with is "Niche-Thief" for the whole hominin lineage and "Self-made Man" as a process-title to help explain the emergence of Modern humans in all our variety (13). No single attribute can satisfactorily define such a complex animal nor provide a key to what made human beings uniquely different from other animals.

Nonetheless, the thieving of niches implies an ecological role (comparable to a frigate bird that specializes in robbing other sea birds of their catches), a process (of more or less continuous expansion in the exploita-

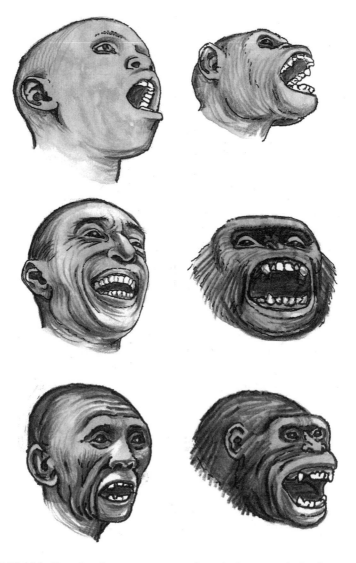

FIGURE 10.3 Changing the way we see ourselves: singing apes, singing humans.

tion of resources), and a means (technology and the aptitudes to go with it) of becoming a successful tool-using specialist.

In common with most students of human origins, I have drawn on informed analysis of both living and fossil primates as an aid to "imagining" our evolutionary past. I have set out to recast, in the light of some of the latest discoveries, Darwin's demonstration that humans are the modified descendants of preexistent forms. In illustrating incremental evolution, I try to echo his insistence that "we can partly recall in imagination the former condition of our early progenitors; and can approximately place them in their proper place in the zoological series." When Darwin wrote those words, in 1871, the word *imagination* had yet to be proscribed from the vocabulary of scientists. He wrote the 620-page *Descent of Man and Selection in Relation to Sex* in the total absence of ancestral fossils or relevant archaeological sites, and without any of the wonderful insights of modern genetics. He also lacked all the other tools that modern biological anthropologists take for granted. Yet his book was far from lacking facts, and he drew innumerable inferences from the biology of many species that, on the face of it, seemed irrelevant to human evolution. And irrelevant a lot of them may well have been (particularly some of the many species illustrating his special interest in sexual selection). Nonetheless, he confidently looked to members of the whole animal series, their affinities, classification, geographic distribution, and succession to demonstrate our community of descent with other animals.

An important aspect of his vision all too easily gets lost when the main emphasis is on "self"—on being human. Among the noblest of human qualities, in Darwin's view, was "sympathy which feels for the most debased" and benevolence that extends to the humblest living creatures. Implicit in this sympathy and benevolence is the recognition that the life that surrounds us represents a host of related organisms—our only cousins in a vast and lonely universe. It is a vision that seems to recede ever further from view. I have approached "community of descent" from the viewpoint of a field mammalogist from Africa and in the very different style of an artist attempting to examine my progenitors as if I could not just recall, but in some sense relive, their condition. In effect, this has had to get reduced to a vision-based anatomical reconstruction (which short-cuts the three-dimensional computer modeling that is such a boon to modern archaeologists and prehistorians). In this task, the fossil record has been essential, but I think it has been my lifelong involvement with other African mammals that has emboldened me to paint some sort of picture in which my ancestors can be very approximately placed in "a zoological series." In using my imagination, I have tried to remember that "the value of the products of our imagination depends of course on the number, accuracy, and clearness of our impressions; on our judgment and taste in selecting or rejecting the involuntary combinations; and to a certain extent on our powers of voluntarily combining them" (6).

My self-portrait is in many respects a conventional one. I am first and foremost a vertebrate with a backbone, nervous system, and gut to prove it. My acuity of vision is a reptilian inheritance. My mother confirmed my status as a mammal from womb to weaning. I share my hands with the primates (and probably the earliest arboreal mammals) and my skull form and taillessness with apes. If my central thesis is correct, I owe my erectness to the still hypothetical "eastern ground ape" or something resembling the "Millennium Ancestor." My ability to walk and run belongs to later *Australopithecus/Homo* ancestors. My capacity to talk, think, reflect, and draw evolved along the way, but I am convinced that these are all properties that owe their existence to many staged increments; they are not single mutant genii (14). They have not been central to my story partly because such abstract entities are not readily expressed in a self-portrait that is conventionally mute, and partly because the succession of changes in anatomical evolution must be symbolic for the infinitely more subtle process that developed my mental and neural powers. In any event, I suspect that human mental properties followed their own developmental path, one that may not be readily tied in with the gross anatomical story, except at the very crudest level. Matter and mind are clearly interrelated, but my zoological background has best equipped me to explore the former to the near exclusion of the latter.

This is not to say that there are no philosophical or political conclusions to be drawn from a predominantly physical, geographic, and ecological study of human evolution. One of the concepts that I challenged in my earlier volume (13) is the oft-repeated statement that humans have devised or evolved a special niche that is variously characterised as intelligent, technical, rational, and reciprocal (15, 16). By contrast, I take the view that the evolution of bipedalism and the emancipation of hands to become the servants of a greedy but ever-enlarging intelligence has led on to the elaboration of an ever more comprehensive set of tools and techniques. Human ingenuity has had the long-term effect of depriving other species of their livelihoods by consuming the same resources or taking up more and more of the living space of other organisms. This expansion in scope may not have begun as a deliberate assault on another species, but for more and more species the end result has been to be elbowed out of prime habitats and eventually out of existence. The evolution of technology in prehistory may have proceeded by fits and starts, but it has had a recognizable theme and direction. The adaptation of technology mimics adaptation in evolution, in that every improvement or refinement of a tool enables it to proceed faster or take in a wider arc than its predecessors. It is not that early humans broadened their niche; it is truer to represent the invention of each new technique as the acquisition of a new niche and its addition to the others already possessed. The effect of this multiplication has been to exclude and eventually replace a succession of other species, *including our closest relatives*. One incentive in the writing of my previous book was to explore when, how, and where the world became "home" (the Oikos of ecology) to humans, but I also explored the antithesis to that idea: to what extent humans had been involuntary vandals of that home.

What are the consequences of visualizing *Homo sapiens* as "having a niche" versus being a collector or "thief of niches"? In practice, the first category tends to be a static laudatory phrase that compares humans favorably *vis-a-vis* other animals, and it presupposes comparability between equally matched entities. Multiplying the number of human-occupied niches instead implies that our biological role on Earth is a dynamic but intrinsically problematic one for the ecosystems we occupy and, to an increasing degree, is invasive of habitats and regions from which we were once wholly or partially excluded. As bulldozers and chainsaws become as cheap as lawnmowers, the scale of vandalism enlarges, ever more insistently reaching into the last ramparts of nature. Our Oikos becomes more and more like the wartime billet of an army of occupation.

When it comes to self-portrayal, the combination of niche-thief and self-maker implies two very different traits in ones makeup. The first represents a blind, de facto role in the ecosystem, in which every technical innovation gallops to the limits of its exploitative potential before it is rendered redundant by a faster or more comprehensive technique (17). The niche-thief has no concern for the consequences of his inventions, his actions, or his appetites. By contrast, I posit self-making as an evolutionary property whereby human culture (including our niche-stealing proclivities) creates the circumstances to which descendants must adapt, both physically and mentally. Because niche-thieving and the invention of new technologies quickly modify ecosystems and therefore create new circumstances, and because we have evolved rational minds, there are both biological and moral imperatives that require we take stock of where our tool-assisted appetites are taking us. All very well, but the vast majority of humans, especially those that determine policies, is wholeheartedly committed to enterprises dedicated to modern tool-assisted niche-thieving. Circumstances are changing and moving too fast for anything other than a *conscious* adaptation in which we must remake ourselves in some fashion that retains and develops the countless benefits of technology and culture, yet does not cut us off from or destroy all the physical processes that created us as animals. Many of our ancestral habitats are still recognizable in Africa, and there is much still to be learned from the ecology of African natural communities. There are also whole universes to be discovered in the behavior, signal, and communication systems of other animals. In their worlds, we wander around like deaf mutes, bereft of any clue to what we are missing.

It is very likely that my birth and upbringing in tropical Africa have given me a tendency to give African ecology and our African evolutionary history more importance than would be the case with natives of other continents (18). Nonetheless, I hope to have shown in this and my previous books that there are solid rational reasons for my claims to Africa, particularly its eastern half, being the "Center of the World" with respect to human evolution.

When it comes to the best place to study the impact of humans on other forms of life and on the environment as a whole, Africa is probably not the best example: our evolutionary history there is so deep, with so many layers laid down that they obscure what went before. If we are to seek out the data that might allow us to reconstruct the ecology of continents as they were before the arrival of humans, the best place for detailed reconstruction of prehuman ecosystems is Australia.

Of all continental masses, Australia has great potential for illustrating the precise course of our impact on natural ecosystems with a lot less clutter than Africa or Eurasia. The Australian biologist Tim Flannery has made a start with his brilliant ecological history of the Australasian lands and peoples, a book perceptively entitled *The Future-Eaters* (19). Future-eating is a projection of our appetite for appropriating resources to the point at which it compromises our long-term existence. Flannery and others have unearthed evidence that the earliest human arrivals had a massive negative impact on indigenous fauna, not only in Australia (20) but also in the Americas (21)—a conclusion that has run foul of some political postures. Notwithstanding these objections, neither veniality nor sentimentality should be a part of the effort to define prehistoric relationships with nature. The search for underlying patterns demands that the relationship should not be trivialized, and we should at least acknowledge that our failings are endemic. It is not just ourselves but our fathers, too, that are on trial.

Flannery has identified three well-documented outcomes from the arrival of humans in what have come to be called "Naïve Lands": large mammal extinctions, a dwarfing of the surviving species, and a huge increase in the frequency and impact of fire, with corresponding changes in vegetation. The exploitation of naïve mammals and birds led to large but unsustainable populations of humans that eventually took over the animals' ranges. As the main source of their sustenance declined and finally went extinct, the hard-pressed surviving people turned more and more to the plant world. By contrast, large mammals in the Old World seem to

have been better able to withstand the onslaught of humans, partly because they had coevolved together. In support of this proposition, Flannery points to the fact that almost all domestic animals derive from the
Old World, whereas a high proportion of plant domesticates are from the
New World.

The history of domestication is probably better exemplified by the histories of Central American, Pacific, or Indonesian economies than by
those of aboriginals from the Australian mainland. Even so, early explor-
ers in Australia found aboriginal foragers harvesting "wild" grass seeds in
ways that were strongly reminiscent of cereal cropping in the Middle
East. As for tropical plant husbandry, the idea must have arisen in areas
where growing conditions were very favorable, where human densities
were relatively high and territories were probably small—*not* a state of affairs typical of Australia. Because women were more tied to the home
base by their children and were better placed to tend plants, horticulture
was likely begun by women while men stuck with trapping, fishing, or
hunting (13). It is also possible that the scope of horticulture was inhibited and its spread delayed because the foraging neighbors of horticul-
tural societies would have been slow to take up or imitate practices characterized by sedentariness, monotony, and a much narrower range of
subsistence activities. The sustained work of digging, weeding, guarding,
and processing was unlikely to have been adopted by nomadic foragers
except under duress. In many parts of the world other than Australia, the
adoption of agriculture may well have been influenced by declining
yields from hunting; but the main mechanism favoring agriculture would
have been the demographic success of its practitioners. In parallel with
herbivores and carnivores, plant-consuming niche-thieves would always
have outnumbered their animal-consuming equivalents.

A major proposition of both this and my previous book has been that
thieving the niches of other organisms has been integral to humanity's
ecological success from the very first invention of tools. This is not to say
that turning all niches over to our own use is inevitable. On the contrary,
niche-thieving is an action and a process; future-eating becomes the consequence of that activity only when taken to its blindest extremity—and
that is not an inevitable outcome. Flannery suggests that this recognition
could help furnish guidelines for the future management of all the "new
lands," with some lessons to be learned too in the "old lands" (22).

Our relationship with the environment is not the only topic with possible lessons to be learned from the study of prehistory, but it is certainly
the one that is most unambiguously relevant and urgent. Much less clear-
cut is our heritage as neotenous animals with a strongly playful compo-

nent to our makeup. There does seem to be a relationship between neoteny, or a childish delight in play, and the way in which we approach tools and technology (23). I have already suggested functional links between inventiveness or innovation and play. If that relationship has a strong evolutionary impetus behind it, the tendency to exercise technical skills "for their own sake" becomes a significant part of the human makeup.

For example, weapons and vehicles are devices with very practical uses, but everyone will be familiar with the way in which both these artifacts have become toys or the objects of cult sports for children and adults alike. In these instances, the players are mostly males; however, there are equally striking historical cases of predominantly female cults in which artifacts or skills that were connected with cultivation or the domestic uses of fire played a central role in religious and social ceremonies. Significant for the issue at hand is the way in which almost any activity can be robbed of its functionality when it is converted into a ritualistic game. There are numerous examples of how this tendency to transform the practicalities of life into "sport" and ritual spills over into political and economic action. Autocrats of all kinds have given free rein to their taste for play, enlisting people, other animals, habitats, or resources to serve trivial pursuits such as gladiatorial tournaments, shooting, military tattoos, and races (17). It is a well-worn refrain in history that potentates and patriarchs have repeatedly sent the people they control to die in wars that are as much interoligarchic games as they are squabbles over resources. The democratization of recreation can also serve to augment such misuse, as when rare habitats are gobbled up by golf courses, or a fashion in dress or food endangers species of animals or plants.

As I emphasized earlier, play includes wordplay; we all know and have shared in the delight of children playing with the rythms and imagery that speech engenders (24). We know too that the internal "virtual" world that we call imagination is shared and developed through speech. Less generally accepted is the fact that the beginnings of speech, structured conceptual thought, and effective verbal communication must have developed from faculties that are possessed by many other animal species that still exist around us. Such capabilities have yet to be intensively studied from this point of view. We must learn how a capacity to share virtual worlds evolved, because individuals and cults that become unable to distinguish between the virtual and the real frequently become a problem—sometimes an extremely dangerous one—to those around them.

Of the communities that attach inordinate importance to "The Word" or "The Book," few acknowledge biological origins for language. For them to study the biological matrix that gave rise to language and thought is sacrilege. Instead of being subjects of objective study, language and structured thought become intellectual toys, so divorced from their history that "The Word" becomes the ultimate and only reality. The absurdity of this belief was explicit in the words of Louis Agassiz, famous for his King Canute–like reaction to publication of *The Origin of Species*—("we must stop this!")—and infamous for his cop-out belief that species went extinct when their creator ceased to think of them! Agassiz wrote, "What are thoughts but specific acts of mind? Why should it be unscientific to infer that the facts of nature are the result of a similar process since there is no evidence of any other cause?" (25). Sadly, such myopic delusion remains an aggressively rampant force in the contemporary world.

If human beings are to survive in the world that gave birth to them, they can ill afford to drift, en masse, into a series of other virtual worlds. If virtual worlds effectively blind entire peoples to the ecological realities of their environment, then the need to understand humanities' perverse relationship with nature will have to include an effort to understand the perversities of our mental and conceptual processes.

A more cryptic part of our makeup is our tendency to put topics into mental compartments. The routing of neural circuits and the wiring of the brain may make compartmentalization an inevitable and structural part of the thinking process, but it has profound consequences for the way human societies go about ordering their existence. The allocation of every aspect of human activity and affairs to specific categories, disciplines, professions, and ministries is partly a procedural necessity for political and economic efficiency and partly an order that is equally necessary as an efficient mental procedure. Computerization serves to accentuate and amplify the compartmentalization of knowledge and action in modern societies. This effect may improve ease of access and efficiency, but the ability to make connections between separate and self-contained "subject areas" may well become rarer. There are signs that just as the Internet and other means of communication open up vast new realms of knowledge, the boundaries between spheres of influence are made increasingly difficult to cross within the many formal institutions of contemporary societies. These difficulties could be reinforced as much by the way we think as by border patrols set up by various species of thought-police (mental equivalents of actual nationalistic, commercial, or religious gatekeepers). A long-term outcome for the increasing compartmentalization of knowledge (and the rapid development of specialized skills that goes with it) could be the erection of more and more obstacles for any sort of overview on human affairs, especially at the global scale. Such myopia is readily exploited by those with narrow or regional vested interests.

So how is this discussion remotely relevant to a book mainly concerned with bipedalism and its role in our evolution? First, my emphasis on specific geographic settings, concrete physicalities, and my own struggle to conceive of a personal history that runs into millions of years is not easily abandoned. A long historical perspective implies projections into the future as well as the past. Theoretically my/our descendants should be able to survive for a few more centuries or millennia at least. In acknowledging that possibility, I am an optimist; but my generic self-portrait has left me in no doubt that the special biological role of my species is that of a niche thief. Future survival demands much greater self-consciousness and more than repentance for our endemic greed for the livelihoods of others. Previous chapters may have demoted some supposed ancestors, but in my determination that I come from a bush of speciation (and not a Chain of Being), I have confidently introduced my readers to our host of cousins: cousin-humans, cousin-hominins, cousin-primates, and many more (26). In a world where biotic niches have been steadily proliferating, most should still be with us. Chimps and orangutans *are* still with us, but I have shown how our own rise to dominance has implicated the disappearance of so many more of our biological cousins. Our role in their extinction may have been direct or oblique, but our lonely status as the only surviving hominin should give us pause for thought. It is not impossible that the human-induced processes that have served to remove so many of our cousin competitors may eventually contribute to our own removal. In making life impossible for so many species, we may eventually make life impossible for ourselves.

Among our endemic strengths and weaknesses, there are at least three biological traits that could bring ruin in their wake. The first characteristic that threatens the future of our descendants is massive impoverishment of resources through our use of ever more ingenious tools to steal niches—by my reckoning our most fundamental biological characteristic and the most primitive of all our skills. By continuously diverting more and more resources away from other species towards ourselves, we are well on the way to appropriating and degrading the entire world. There has, as yet, been no decisive break from our prehistoric relationship with the rest of nature; the only distinction in the contemporary world is in the scale and wastefulness of its manifold expressions.

A second liability is our neotenous love of play, which tends to transform every activity—from politics to war—into games that have less and less relationship to the ecological realities of the playground, let alone to the environments of the larger, physical world. A delight in our primary adaptation, wordplay easily transmutes into dangerous obsession with its multifarious and hugely stimulating cultural expressions. This can lead on to voyages, by whole communities, into their own "virtual realities" divorced from any effort to engage with the all too real problems of the biological world (of which more later).

A third endemic danger is the human mind's predilection for compartmentalization, which inhibits holistic approaches to the multistranded, supercomplex problems posed by our uneasy relationships with nature and with each other: we are congenitally prone to elaborate and get distracted by minor or even irrelevant details. Examples of negative consequences for each of these shortcomings are not hard to find, and each will bear some examination in the pages that follow.

On the much more immediate timescale of writing this book, I have been reminded, again and again, of the need for more multidisciplinary exploration into the origins of humans and their relationship with nature. I could not help but be aware of a general failure to set up any such enterprises. An inability to address the simplest features of human evolution is glaring evidence for chronic short-sightedness in current cultural leadership. The blinkers of market-oriented values may have something to do with it, but it is also possible that compartmentalization and the distractions of a neotenous brain are at play. A lack of consensus on the reasons and mechanisms for uprightness, our single most obvious characteristic, mocks our pretentions to overarching knowledge. Further evidence for our failures, if it were needed, is that research into our origins is a very minor field of study still confined to a mere handful of universities in a mere handful of nations.

A more personal reason for pleading environmental responsibility begins with my deep unease over the niche-stealing proclivities that I think are such an inescapable part of my birthright. I'm also aware that being playful and having a tidy pigeon-holing mind are traits as innocently familiar as the exploitation of resources so much admired in pioneers and entrepreneurs. Yet, in Flannery's *New Lands* we can already see that the danger of ecological collapse lies concealed beneath that innocence (19). If science and scientific world views have grown out of much simpler technological aptitudes, that historical progression is now dwarfed by the scale of science and technology's alliance with the subsistence problems of an exploding human population. Here, too, the assumptions that underlie science's purposes and the priorities set for scientists are directed toward extracting new sources of materials, foods, fuel, and other resources. There has been no clear break between proto-science as a by-product of subsistence and science in the service of subsistence. Indeed, the expectation that science is the handmaiden of technological progress is seldom challenged. Any possibility of addressing the problems that these shortcomings pose to our survival depends on a vast improvement in our efforts to understand our ecology and evolution as details—extraordinarily interesting and relevant details—of the evolutionary process.

Process: what distinguishes our own generation from all preceding ones is that we now have the intellectual and technological tools to view ourselves and our activities, our past and our future, in the context of natural processes. These processes include the astronomic, physical, climatic, chemical, and cultural, each the focus for entire clutches of disciplines. It is true that in the case of climatic change and with pollution, there has been some recognition, from *outside* the leadership of industrial nations, of the impact of present practices on the future; but virtually all serious calls to action have been subverted by irresponsible politicians and industrialists. However, it has been in addressing our biological nature that we have failed most pathetically (although the International Human Genome Project triumphantly mitigates that judgment). Even so, in the vocabulary of self-portraiture, our antievolutionary culture has put many obstructions in the way of painting any sort of truthful picture.

Why do we go to such lengths to disguise or deny the fact that we are animals? Is it because we are so profoundly embarrassed, even humiliated, to be such an odd sort of ape animal. Is there some strange subliminal angst that hates to admit that we might be the performing poodles of the primate world, dancing in an atavistic circus to tunes set by some all-too-human, all-too-venial ringmaster? So long as we remain ignorant about how, where, why, and when our ancestors reared up on their hind legs, the sense of our own absurdity will remain, fed by a choice of more or less credible, but mostly incompatible, theories.

Acknowledging our existence as a mere detail in the evolutionary process may seem humiliating, but it could be a first step in a species-specific program of "Coming Home." Such a process can only be a part of the larger enterprise of getting to know the universe we find ourselves in and learning to find our past and present place in it. Eventually, if we care for our survival into the future, we must learn to regulate our relationship with nature through a real knowledge of how nature works.

The acquisition of knowledge is an enterprise that requires action—in this instance, action toward specific objectives. One priority could be the setting up of a global process-oriented database with its primary rationale being the study and achievement of sustainability in all environments, worldwide (27). That means storing knowledge about system dynamics, not just species lists. To survive as a species within a diversity of other species requires that life processes be sustained from one generation to another in both human and nonhuman communities. To achieve even a semblance of that aim would require a common process-based analysis that covered the entire range of global habitats. It would also have to study all such entities in a time frame that greatly exceeded the lifetime of any individual. Some such combination of process analysis and inventory would be the only reference able to provide a sufficiently objective foundation for the many political and strategic judgments that will have to be made by future generations. Models of natural process performance will require massive datasets and a broad comparative base to enhance and back up the authority of future environmental policies. Unlike today's policies, which revolve around the production of commodities, these new policies will be designed to enhance sustainability. Apart from helping to ensure the future survival of species, communities, and habitats, such a bank of knowledge could eventually provide objective criteria essential for many other areas of environmental and development policy, including land use planning. From the perspective of this book and human self-portraiture, such a store could provide crucial information relevant to the retrieval of our evolutionary history.

The database could have four branches or divisions, each of which is already the province of and finds expression in several established academic disciplines. It is precisely because the data that have already been accumulated have been turned to so many disparate purposes and originated in so many other contexts that they need to be reexamined, redefined, and developed from the very focused perspective of achieving sustainability. By organizing data along a limited number of "process parameters," much established knowledge could be recast into a format that would be more accessible and more relevant to maintaining and managing environments. Such management should not only ensure the long-term survival of habitats but should enhance the survival of humans as they have been from the beginning—inhabitants of a diverse natural world, not a virtual world. In that world, our intrinsic intellectual and physical poverty could be addressed, not out of desperation or out of a bad conscience, but as part of the many challenges posed by our biological nature.

We could begin by setting up process-oriented databases organized into four main divisions. The first would concern physiology, or the workings of organisms. This would include genetics, biochemistry, microbiology, reproduction, ontogeny, and so forth. The second would focus on evolution (the transformation of organisms through time), to include speciation, biodiversity, genetics, biogeography, and paleontology. Third would be ecology (the interactions of organisms in a habitat), to include community dynamics, population biology, succession, and sustainability. For want of a better title, the fourth division could be called geophysiology, or the interaction of Earth systems, including topics such as soil sciences, geology, and climatology. Learning to understand and maintain the processes that underlie the existence of life on Earth is inseparable from understanding ourselves—inseparable from the effort to paint more truthful portraits of ourselves and of the organic world from which we emerged.

Very practical benefits could accrue from the creation of central process databases. One would be the stimulus they could provide for research itself; another would be the elaboration of new and more relevant formats in which to store, present, and retrieve new knowledge. Knowledge of process parameters—that is, the structural essentials, the functional boundaries, and the operational tolerances of processes—would have obvious utility in defining environmental policies and what can and cannot be done.

Each database division can be examined from the viewpoint of the sustainability of process and its contribution toward the maintainance of biodiversity and its role in enhancing self-knowledge.

A. Physiology. An information system that would center on understanding the conditions required for the operation of all types of physiological processes. Of course, these have already been studied widely, but seldom as functions of global sustainability.

Very precise preconditions must be met by all organisms before they can reproduce, grow, and mature. The parameters for microorganisms are generally simpler than for more complex species, but the principles involved are the same. These conditions are sometimes known for particular species, and some of the general principles involved find their way into textbooks. However, there has been no systematic effort to gather a central bank of working principles as they might apply to the physiology and molecular structure of organisms as a whole.

The category of "physiology" might serve to embrace all those entities that cannot exist outside larger living containers. At one end of the scale, this might include some parasites and symbionts; in the middle, organisms such as cyanobacteria; and at the other end, genes. As the most fundamental and irreducible of all message-transmitting systems, the genetic code is both an abstraction and a biological and chemical structure. Each gene's coded message uses the carbon cycle as its medium. Clearly, the instructions contained in the genes rely on a very exacting environment within the cell to operate. Likewise, all other sensory and information-exchanging systems are built from other chemical composites within the carbon cycle. Both unicellular and genetic structures tend to retain some of the most fundamental building blocks of life on Earth, yet their functions are directed at processes of astounding sophistication. Adaptive diversity has taken unicellular forms into the most extreme environments on Earth—in the oceans depths, around the mouths of volcanic vents, and under arctic ice. All such biota occupy precise, often rare, environments and deserve special consideration (28). In documenting the parameters and distribution of microenvironments, the physiological database would, in effect, have to chart the boundaries of many living realms, from understanding the chemistry of stromatolites to analyzing calcium layering in fish otoliths (an important source of information on climate).

Biodiversity is sometimes treated as if it were synonymous with genetic resources—understandably, because genes do not naturally exist outside their containers: bigger organisms. However, the development of genetic engineering demands that we recognize the gene as an identifiable entity. As we learn, with increasing unease, it is even a unit of existence subject to unpredictable manipulation and to patent claims.

Some form of physiological or genetic database authority could be envisaged as the central point of reference for assessing the nature, viability, and permissability of using or manipulating genetic material, including implants and mutants. It is a matter of some urgency to develop criteria for determining objective limits in the artificial manipulation of genetic material. Such a database as this might be developed to provide just such criteria. Because biodiversity and genetic diversity are not entirely separable, this division interdigitates substantially with the next one.

B. Evolution. Many institutions already exist to study species, conserve habitats, and document biodiversity. There are also many small-scale databases, especially in universities and national institutes. These are dwarfed, of course, by those that are dedicated (often in the name of fighting "poverty") to the manipulation and ultimate destruction of sustainable biotas and their conversion into unsustainable commodities. By contrast, the central raison d'etre for this division would be to understand the process that created and continues to create biodiversity. One species with a high priority for research and a role model for attention to detail would be *Homo sapiens*. This species would have the advantage of a good head start, what with the Human Genome Programme and masses of material to be culled from medical and anthropological institutions worldwide.

Human beings are now the most influential of all species in determining the evolutionary fate of other species. The first formal institutional recognition of human impact worldwide found expression in the "Man and Biosphere" program of UNESCO, launched in 1963. Subsequent initiatives have been disappointing, and there has been virtually no effort to assess the impact of humans as just one of many species that have, at one time or another, had a major impact on other animals and their environments. Dinosaurs, mammoths, elephants, and other large mammals once dominated very extensive tracts of country, and through their selection or rejection of food plants determined which plant species might flourish and how. Studying and documenting impending as well as actual extinction would also be an important part of this division's remit. Selective extermination of the sort that Flannery has described in Australia has not only degraded habitats but has also reduced the prospects for survival for an ever widening circle of species. Some of the defunct ones may have played the role of "keystone species": not only were they components in a living ecosystem, but their influence shaped the evolutionary history and potentially the future of some of the other species with which they came into contact.

In Africa, it is still possible to study the influence of megalife as an influential force in the speciation of smaller organisms; but this situation may not last long. We need to know to what extent the removal of large, long-lived mammals and trees from the land, and whales and large fishes from the sea, alters the evolutionary climate for all other species. There are enough examples of the collapse of fishing and whaling grounds and other productive ecosystems to justify much more detailed investigation of the evolution of such important species. We also need to know in much greater detail what favored the evolution of giant forms in particular lineages, such as whales, giraffes, and elephants, while others, such as shrews, never got heavier than a few grams. It is in this connection that we need to know a lot more about the role of megalife in shaping our own evolutionary history.

With geographic, ecological, or physical isolation a primary requirement for speciation, island biota would be exemplary subjects for this division. The endemic fauna and flora of actual oceanic islands, landlocked lakes, mountains, and other discrete ecological enclaves would have a special value as subjects of study and exemplars of speciation. Island biogeography would be an important sphere of activity within this division.

Adaptive specialization is at the root of biodiversity such that the more extreme examples of specialization would also attract special attention. When animals or plants are very large, very numerous, or very specialized, they have usually evolved special techniques for taking up nutrients. Accordingly, the distribution of organisms in relation to nutrients must also be a part of this division's remit—all the more so in the light of our niche-stealing tendencies, which lead us to plunder nutrient-rich communities without appreciating the evolutionary adaptations of the species that we exploit. That appreciation could be materially improved through an evolution database.

C. Ecology. The idea of a global commission for the conservation of species and their habitats was first mooted at the International Zoological Congress in Graz in 1910. However, the first practical proposal to apply systematic scientific procedures to the classification and study of natural environments was expressed in the International Biological Programme (IBP), founded in Vienna in 1963.

In 1980, "ecological processes and life-support systems" were identified by the United Nations Environmental Programme (UNEP) in a document called the World Conservation Strategy (29). Databases that map the distribution of species and ecoregions have been assembled by the World Conservation Monitoring Centre and the United Nations Environmental Programme (the original sponsor for the WCS). These and other organizations have been active in making inventories of natural resources and mapping vegetation types from various points of view, using a variety of different techniques. None of these initiatives has set out to document ecosystems in terms of process; yet, of all divisions, an ecology information system should be the most amenable to a process-oriented approach. Successions and dynamic complexity are fundamental properties of all ecosystems, requiring deep time frames for their study and interpretation.

From the perspective of process and sustainability, the WCS has been a fatally flawed initiative, effectively erecting demolition notices around all the richer natural habitats. This extraordinary outcome was the result of treating the study of ecological processes as a mere adjunct to the direct management of economically productive landscapes. The temporary conservation of pockets of residual natural vegetation was defended by the authors of the WCS less as natural reserves than as "control plots" where some of the effects of agriculture could be assessed by comparing converted areas with the as-yet unconverted "controls." According to the WCS (or rather to its food and agriculture-oriented participating consultants), the potential of natural habitats for conversion to cropland was their most valuable ecological characteristic. In areas that were still marginal for agriculture, the WCS explicitly valued such lands as gene banks for potentially useful animals and plants. Today the agronomists' narrow perspectives on ecology may seem less pervasive than they were 20 years ago, and agriculture's claims to exclusive priority in the use of land face increasing challenge. Nonetheless, a process-oriented database would have to contend with latter-day proponents of just such blinkered brands of "ecology." The leadership of many international development agencies is still dominated by similar primitive attitudes, but it must be acknowledged that they are under huge political pressure to respond to the increasing land needs of an exploding human population.

It is generally acknowledged that the management and conservation of habitats and biota are dependent on a theoretical knowledge of how ecosystems and other natural systems work. It is further acknowledged, and given institutional recognition, that we must monitor some diagnostic parameters of process—notably the survival of species, levels of radiation, atmospherics, or water quality. What is less widely acknowledged is how little we know about any of these processes in the longer term. We know least, for obvious reasons, about processes that are very slow (i.e., in very cold, high and dry environments), very complex (i.e., some tropical ecotypes), very long lived and stable (i.e., some mountain or desert foci), or very large scale (i.e., whale, elephant, and mahogany communities). Entirely new perspectives on the world, especially in relation to its age and physical and biotic limits, are prerequisites for a global program on ecological processes.

A rounded knowledge of global ecology and evolutionary history must rely on models of the full spectrum of natural systems. Natural parameters (e.g., latitude, altitude, temperature, precipitation, geology, and nutrients) can provide a series of criteria by which environments could begin to be classified and understood. The human bipeds' dispersion over the globe and invasion of "naïve" lands must, of course, be seen as integral to the shape and structure of contemporary New World landscapes.

In making an inventory of the world's great ecotypes, many are already as extinct as the mammoth or the dodo: the great swamps of the Yangtse Delta, the upland basins of Mexico, and the savannas of the Atlas foothills, to name but a few. The target of global understanding would require that even these extinct landscapes and ecotypes would need to be reconstructed, if only in the form of theoretical models. Wherever possible, reserves devoted to reconstructing some semblance of former eco-types should be attempted. These vestigial reconstructions, not viable natural environments, should be the agronomist's "controls." Reclamation of natural ecosystems from the ravages of agricultural malpractice could become a potent symbol of repentance for global vandalism.

D. Geophysiology. This word was recently coined to describe the Earth's environments as systems in which land, sea, atmosphere, radiation, geochemistry, and climate all interact. Inasmuch as these systems are understood, that knowledge has tended to emerge from the practicalities of weather prediction, shipping forecasts, defense, aviation, geography, and so forth.

In addition to understanding the planet surface's behavior, the overall objective of studying the inorganic realm would be to provide physical and time-depth contexts for organic life. Knowledge of the many vicissitudes of physical environments, from climates to soil, vulcanicity, and nutrient patterns, is clearly essential for the integration of a process-oriented environmental database.

The responses of many organisms to seasonal, sometimes daily, fluctuations in temperature, insolation, or nutrients can be so finely tuned, so precise, that a detailed record is laid down in the organisms' wood, leaf, shell, bone, or teeth. Such records permit the study of many important interactions between the organic and inorganic world. There are also many as-yet unexploited possibilities for reconstructing past climates in great detail. Such discoveries can be integrated into databases to generate much more accurate chronologies and cross-referenced models of past environments.

The physical processes determining organic life have long been studied and measured by students of all life sciences. The International Geophysical Year was set up by the International Council of Scientific Unions (ICSU) in 1957, and this was the first program to assume a global perspective. An important initiative in trying to understand Earth sciences as a whole, an integrated approach, has been the recent founding of the Geophysical Society. This will bring together established knowledge as well as fostering and promoting further research on the planetary environment. Among the many national and international climate studies, Australia's Commonwealth Scientific and Industrial Research Organization (CSIRO) has initiated a "climate through time" program called CLIMANZ, which has been a pioneer in exploring past and present climates as a major factor in understanding the biota of Australia.

If this book has attempted to hold up a mirror to what it might mean to be a human being, these proposals for process-oriented databases merely extend the concept of a mirror. We need to know who we are, we need to know what we have emerged from, and we need to know what we are doing to that matrix that gave us our existence. Databases would bring focus to the job of ensuring that the world we inhabit can be sustained *in spite of* our history.

The only way we can escape using up our resources is to recognize that they are all finite. Our society must also learn to regulate its own greed and its own numbers, and in doing so, come to understand, to respect, and to live within the limits imposed by nature. Our technology must be turned away from extracting, fighting, and consuming every resource and must instead be rebuilt to mimic the processes that make this planet a home for humans as one of many extraordinary and precious beings.

It is obvious enough that self-interested opportunism is a fundamental trait of all organisms. Among humans, this biological trait may have been an underlying motor for action, exploration, and expansion; but it is an inadaquate axiom on which to base a political credo, such as permanent growth, nor is it a recipe for survival. If we have been able to work out how some physical and living systems "work," there is now a new subject that environmentalists have set up for serious study. How do we live in the world without consuming it? It is a practical question demanding both intellect and techniques to solve, but there is no escaping the fact that it brings environmentalists into direct confrontation with just about every major establishment in the world: governments, banks, multinationals, and other big business. These bodies continue the prehistoric tradition of asset stripping, but they have at their command horrifically destructive tools to steal—not just a niche or two, but entire ecosystems.

In the context of this book, the fate of Africa's savannas, as the primary habitats of humankind, are of obvious concern. Like tropical forests, the savannas are targets for some of the most destructive so-called "development" that is taking place anywhere on Earth. The extinction of some of Africa's remaining ungulates and the spectacular carnivores that prey on them, as well as the severing of our last link with our ancestral habitats, could ultimately be the responsibility of international markets that promoted the conversion of viable and very ancient ecosystems into chemical-dependent beef lots. They can also disappear under the impact of more localized holdings of domestic stock. These immense herds, as measures of the wealth and prestige of their owners, are even less amenable to restraint than the livestock industry.

For the most part, the cowboys and loggers are not picturesque outlaws in remote ranches and log cabins: they are "the Great and the

Good"—presidents, ministers, bankers, economists, and captains of industry. In their bespectacled and innocent-seeming faces, we must see ourselves as we have always been: primitive people in possession of increasingly unprimitive tools. Only a revolution in the way we see ourselves and our relationship with our environment will change us. If we ponder where our unbridled appetites may take us, there can be no escaping the need for change and action.

If, on behalf of our descendants, we consign *their* world to biological poverty, their legacy will include many explicit signs of impoverishment. Their thoughts and emotions will be starved of many experiences. They may be able to read about *our* present (*their* past), but they will be unable to check out their own ideas about natural processes because *their* cauterized, vestigial nature will be full of voids and distortions. They will reproach us for depriving them of information about *their* (and our) history. They will know that true imagination must be based on observable information, yet too many components of *their* natural world will be missing. Breaking *their* links with *their* own animal past would serve to condemn us as more primitive and more short-sighted ancestors than the most appetite-driven ape—because we should have known better. There will be a decline in range, scope, and potential for *all* our descendants because the faculties we variously call intellect, imagination, spirit, creativity, and soul need to be fed with the sort of diversity of experience that nurtured our own emergence as thinking, emoting, and creative beings. It will be *our* fault.

Are we able to combine the exercise of contemporary minds with probing the world with much older senses? Can we hope to find the physical, ecological, and prehistoric roots of our still very partial intelligence if we do not respect our older selves? Can we not respect our countless evolutionary cousins a little more? They are living analogs of our former selves. Without such respect, we will continue to do them great injustice and have no hope of mitigating, let alone arresting, our piratical progress as all-consuming "Future-eaters."

My saga of origins, with all its congregations of cousins, dead ends, bottlenecks, and uncertain ancestors remains a tentative and ongoing tale. However imperfect and incomplete my own attempt at interpreting that story, it is grounded in the solid authority of bones, flesh, and genes. Its main virtue may be less in its snapshots of details than in its pioneering intentions. I hope that there will be many more such self-portraits to be painted by future authors. They will be equipped with more facts, more fossils, newer gene maps, more sophisticated tools, and fresher, better-prepared minds. I hope that they will go on reminding their audiences that the study of what it is to be human has no boundaries and is as much about writing the future as it is about excavating the past. It is also about rewriting, again and again, and in the light of an ever-expanding science, the old tribal stories of Genesis—all of them earnest and often wise efforts to understand our origins. Nevertheless, so long as the origins of our unique stance with all its technological and mental consequences remain in the realm of myth and magic, our performance as an ill-balanced, two-legged ape will remain a tragicomedy. Sooner or later, we must make reparations for stealing the world, not only from our descendants but from our cousins, from apes to antelopes. If we ever get there, it will be action that will involve considerable conflict with our own nature—and this may be the most paradoxical and certainly the most difficult of our many legacies of becoming the last remaining hominin biped.

REFERENCES

1. Dawkins, R. 1995. *River Out of Eden*. London: Weidenfeld and Nicolson.
2. Stringer, C., and R. McKie. 1996. *African Exodus*. London: Jonathan Cape/ Pimlico.
3. Fagan, B. M. 1990. *The Journey from Eden*. London: Thames and Hudson.
4. Dawkins, R. 1998. *Unweaving the Rainbow*. London: Allen Lane.
5. Stangroom, J. 2000. Darwin Queen. *The Philosophers' Magazine* Summer Edition.
6. Darwin, C. 1871. *The Descent of Man and Selection in Relation to Sex*. London: John Murray.
7. Bayley, E. 2000. Only Human? We used to share the Earth with many other types of human. How did we come to survive them? *Focus* 94: 44–52.
8. Wood, B. A. 1993. Early *Homo*: how many species? In *Species, Species Concepts, and Primate Evolution*, ed. W. H. Kimbel and L. B. Martin, 485–522. New York: Plenum.
9. Griffin, D. R. 1992. *Animal Minds*. Chicago: Chicago University Press.
10. Kingdon, J. 1971. *East African Mammals. An Atlas of Evolution in Africa*, Vol. 1: Primates. London: Academic Press.
11. Kingdon, J. 1997. *The Kingdon Field Guide to African Mammals*. London: Academic Press.
12. Tuttle, R. H., D. M. Webb, and N. I. Tuttle. 1991. Laetoli footprint trails and the evolution of hominid bipedalism. In *Origine(s) de la Bipédie chez les Hominidés*, ed. Y. Coppens and B. Senut, 187–198. Paris: CNRS.
13. Kingdon, J. 1993. *Self-made Man. Human Evolution from Eden to Extinction?* New York: John Wiley & Sons.
14. Cartmill, M. 1998. The gift of the gab. *Natural History* November 1998: 56–64.
15. Birdsell, J. B. 1972. *Human Evolution: An Introduction to the New Physical Anthropology*. Chicago: Rand McNally.
16. Pilbeam, D. 1970. *The Evolution of Man*. London: Thames and Hudson.

17. Diamond, J. 1998. Guns, Germs, and Steel. New York: Random House.
18. Kingdon, J. 1983. Kilimanjaro, Animals in a Landscape. London: BBC Publications.
19. Flannery, T. F. 1994. *The Future Eaters. An Ecological History of the Australasian Lands and People*. Chatswood NSW: Reed.
20. Roberts, R. G., and T. F. Flannery. 2001. New ages for the last Australian megafauna: Continent-wide extinctions about 46,000 years ago. Science 292[5523]: 1888–1893.
21. Alroy, J. 2001. A multispecies overkill simulation of the end-Pleistocene megafaunal mass extinction. Science 292[5523]: 1893–1897.
22. Flannery, T. F. 2001. *The Eternal Frontier: An Ecological History of North America and its Peoples*. New York: Atlantic Monthly Press.
23. Gibson, K., and T. Ingold, eds. 1992. *Tools, Language and Intelligence*. Cambridge, MA: Cambridge University Press.
24. Andrew, R. 1962. Evolution of intelligence and vocal mimicking. *Science* 137: 585–589.
25. Agassiz, L. 1874. Posthumous essay. *Atlantic Monthly* December 1874.
26. Griffin, D. R. 1976. *The Question of Animal Awareness: Evolutionary Continuity of Mental Experience*. New York: Rockefeller University Press.
27. Kingdon, J. 1997. The Process Principle. Do we need a process-oriented global environmental data-base and associated policy advice unit? *Occasional Paper: Green College Centre for Environmental Policy and Understanding*, Radcliffe Observatory, Oxford.
28. Wilson, E. O. 1983. Sociobiology and the Darwinian approach to mind and culture. In *Evolution from Molecules to Men*, ed. D. S. Bendall, 545–553. Cambridge, UK: Cambridge University Press.
29. UNEP 1980. The World Conservation Strategy. Nairobi: United Nations Environmental Programme.

APPENDIX

Plants Known to Be Especially Favored
by Humans and Other Primates

About 2000 species of food plants have been recorded as being eaten by three primates in sub-Saharan Africa, namely humans, chimpanzees, and baboons (1). Some 50 genera are not only common and widespread, but their choice by these primates, particularly humans, recurs repeatedly in the records of many researchers (1–9).

A list of these genera, with the plant parts chosen, indicates that fruits are by far the most important and consistent plant food, but no more than half of these fruits grow in tall or tallish trees. Likewise, slightly less than half of the genera can be gathered without climbing. Nearly half of the genera yield leaves, shoots, and stems. About a quarter of the genera have edible underground parts, such as roots, rhizomes, bulbs, or corms. Flowers are an occasional "extra."

The eastern coastal and montane forests are known to have at least 130 species of plants yielding foods fit for humans. Again, fruits are far and away the main attraction. In these forests, somewhat more than half the fruit species grow in tallish trees, but a remarkably high ratio (about 35%) can be harvested on or close to the ground.

MONOCOTS

Yams (*Dioscorea* spp.) mainly tubers, also leaves, stems, fruits
Dragon's blood agaves (*Dracaena* spp.) shoots, flowers, fruits

Various grasses, mostly grains, roots, tillers. Notably:
    Shama millets (*Echinochloa* spp.) grains
    Finger millets (*Eleusine* spp.) grains
    Pearl millets (*Pennisetum* spp.) grains
Guinea corns, durra (*Sorghum* spp.) grains
Various lilies and irises (Hyacinthaceae and Iridaceae) bulbs, corms
Banana-herb (*Ensete ventricosum*) fruits, stems, flowers
Various palms (Arecaceae) fruits, saps, stems, flowers. Notably:
    Palmyra palm (*Borassus ethiopum*)
    Wild date palms (*Phoenix* spp.)
    Doum, gingerbread palms (*Hyphaene* spp.)
Bulrushes (*Typha* spp.) rhizomes, stem-piths
Grains of paradise (*Aframomum* spp.) fruits, piths

DICOTS

Cattails (Amaranthaceae) leaves, shoots
Live-longs or false marulas (*Lannea* spp.) fruits, roots, resins
Wild currants or karrees (*Rhus* spp.) fruits
Marula plum (*Sclerocarya birrea*) fruits
Custard apples (*Annona* spp.) fruits
Rubber-vines (*Saba* spp.) fruits, shoots, piths
Hegligs (*Balanites* spp.) fruits
Baobab (*Adansonia digitata*) fruits, leaves, shoots
Saucer berries (*Cordia* spp.) fruits
Myrrhs (*Commiphora* spp.) fruits, roots, gums
Shepherds' trees (*Boscia* spp.) fruits, roots, flowers, leaves
Cats' whiskers (*Cleome* spp.) leaves, shoots, stems, flowers
Bush cherries (*Maerua* spp.) fruits, roots, flowers
Mobola plums or greyapples (*Parinari* spp.) fruits
Leadwoods (*Combretum* spp.) gums, fruits, shoots
Morning glories (*Ipomoea* spp.) tubers, leaves, shoots
Various gourds and cucurbits: fruits, leaves, flowers, roots. Notably:
    Wild cucumbers (*Cucumis* spp.)
    Wild squashes (*Coccinia* spp.)
    Wild calabashes (*Lagenaria* spp.)
Ebonies (*Diospyros* spp.) fruits
Guarris or black ebonies (*Euclea* spp.) fruits
Sweetberries (*Bridelia* spp.) fruits
Sugar plums (*Uapaca* spp.) fruits

Governor's plums (*Flacourtia* spp.) fruits
Camel's foot (*Bauhinia* spp.) fruits, flowers, leaves, shoots
Indian date (*Tamarindus indica*) fruits
Acacias (*Acacia* spp.) gums, fruits, leaves, shoots, flowers
Cowpeas (*Vigna* spp.) fruits, leaves, tubers
Monkey apples (*Strychnos* spp.) fruits
Jutes (*Hibiscus* spp.) flowers, leaves, shoots, fruits
Figs (*Ficus* spp.) fruits
Wild cloves or waterwoods (*Syzygium* spp.) fruits
Monkey plums (*Ximenia* spp.) fruits
Knotweeds (*Polygonum/Persicaria* spp.) leaves
Jujubes (*Ziziphus* spp.) fruits
Turkey berries (*Canthium* spp.) fruits
Bush medlars (*Vangueria* spp.) fruits, leaves
Toothbrush tree (*Salvadora persica*) fruits, leaves
Milk plums (*Bequartiodendron* spp.) fruits
Cross berries or wild raisins (*Grewia* spp.) fruits, shoots, flowers
Water lilies (*Nymphaea* spp.) rhizomes, fruits, leaves, flowers

REFERENCES

1. Peters, C. R., E. M. O'Brien, and R. B. Drummond. 1992. *Edible Wild Plants of Sub-saharan Africa*. Kew, UK: Royal Botanic Gardens.
2. Lovett, J. C. 1993. Eastern Arc moist forest flora. In *Biogeography and Ecology of the Rainforests of Eastern Africa*, ed. J. C. Lovett and S. K. Wasser, 35–55.
3. Maguire, B. 1980. The potential vegetable dietary of Plio-Pleistocene hominids at Makapansgat. *Palaeont. Afr.* 23: 69.
4. Nishida, T. and S. Uehara. 1983. Natural diet of chimpanzees (*Pan troglodytes schweinfurthii*): Long-term record from the Mahale Mountains, Tanzania. *African Studies Monographs* 3: 109–130.
5. Mabberley, D. J. 1987. *The Plant Book*. Cambridge, UK: Cambridge University Press.
6. O'Brien, E. M., and C. R. Peters. Ecobotanical contexts for African hominids. In *Cultural Beginnings: Approaches to Understanding Early Hominid Life-ways in the African Savanna*, ed. J. D. Clarke, 1–15. Bonn: Rudolf Hablet GMBH.
7. O'Brien, E. M., and C. R. Peters. 1998. Wild fruit trees and shrubs of southern Africa: Geographic distribution of species richness. *Economic Botany* 10: 245–256.
8. Cunningham, A. B. 2001. *Applied Ethnobotany*. London: Earthscan Publications.
9. Cunningham, A. B. 1998. Collection of wild foods in Thembe-Thonga society: A guide to Iron Age gathering activities? *Annals of the Natal Museum* 29: 433–466.

Index

Note: Page numbers followed by letters *f* and *t* refer to figures and tables, respectively.

Acanthostega, 26
Achulean culture, 273, 278–79, 318
Achulean hand axe, 278–79
activity compartments, and anatomical change, 64
Adam, 334
Adamson, Joy, 326
Adapids, 56, 58*f*
adaptations: biological, to technology, 322–29; dental, to specialized diets, 168–69, 184–87, 198, 201, 243–44, 341; enhanced, 190
Adrar Mgom, 44
adulthood, delayed, in hominins, 236–39, 292–97, 305–12
Aegyptopithecus zeuxis (Egyptian monkey), 62*f*–63*f*, 62–65, 66*f*–67*f*, 69–70, 80
Africa: ape emigration from and immigration back to, 82–83, 85*f*, 92–99, 93*f*–94*f*; "Atlas of Evolution in Africa," 11–13; biogeographic regions of, 181–82; biotic divides of, 95; contemporary understanding of, importance of, 346–48; ecological barriers in, 101, 106–7; emergence of mammals in, 44; genetic backflow of prehistoric humans to, 315, 317–18, 320–23; human origins in, 280–82, 281*f*, 286–87, 304–5; isolation of, continental drift and, 45, 45*f*, 47, 70–71, 71*f*; prehis-

toric human (Modern) migration from, 313–24; prehistoric human (Modern) return to, 315, 317–18, 320–23; primate emergence in, 47, 63, 70–71, 71*f*; rivers and river basins of, 158–64, 159*f*, 161*f*, 166*f*, 347 (*see also* rivers and river basins); surface patterns of, 11, 12*f*. See also *specific regions*
African climate: and ape range in Africa, 92–101, 100*f*; and bipedalism, 11, 12*f*, 18–21, 20*f*, 137–38, 138*f*, 140, 156; and ground apes, 118*f*, 118–20, 145–47; and hominin-chimpanzee divergence, 102; and hominin migration and evolution, 165–66, 194–98, 206–8, 208*f*, 241–42, 247–48, 261, 343–48; and *Homo* species, 247, 342–43
African east coast region: bipedalism beginning in, 5–6, 19, 109, 346; as Center of Endemism, 117; distinctiveness of, as habitat, 116–25, 121*t*–122*t*; as evolutionary center, 11–13, 107–9, 346–47, 353; extent of forests in, 158; as genetic melting pot, 305; geographic isolation of, 18–21, 20*f*, 107–9, 116, 118, 118*f*, 164; ground ape emergence in, 115–25; human origins in, 305, 337; isolation of, 116, 118, 118*f*, 164; prehistoric human evolution in, 304–5; resources and food supply of,